Open Data and Models for Energy and Environment

Open Data and Models for Energy and Environment

Editors

Benedetto Nastasi
Massimiliano Manfren
Michel Noussan

MDPI • Basel • Beijing • Wuhan • Barcelona • Belgrade • Manchester • Tokyo • Cluj • Tianjin

Editors

Benedetto Nastasi
Department of Planning, Design
Technology of Architecture
Sapienza University of Rome
Rome
Italy

Massimiliano Manfren
Faculty of Engineering and
Physical Sciences
University of Southampton
Southampton
United Kingdom

Michel Noussan
Future Energy Research Program
Fondazione Eni Enrico Mattei
(FEEM)
Milan
Italy

Editorial Office
MDPI
St. Alban-Anlage 66
4052 Basel, Switzerland

This is a reprint of articles from the Special Issue published online in the open access journal *Energies* (ISSN 1996-1073) (available at: www.mdpi.com/journal/energies/special_issues/open_data_energy_analytics).

For citation purposes, cite each article independently as indicated on the article page online and as indicated below:

LastName, A.A.; LastName, B.B.; LastName, C.C. Article Title. *Journal Name* **Year**, *Volume Number*, Page Range.

ISBN 978-3-0365-1756-8 (Hbk)
ISBN 978-3-0365-1755-1 (PDF)

Contents

About the Editors

Benedetto Nastasi

Benedetto Nastasi (PhD) is Senior Energy Planner and Assistant Professor at Sapienza University of Rome Previous affiliations include TU Delft University of Technology, TU/e Eindhoven University of Technology, The Netherlands, and International Solar Energy Society and Guglielmo Marconi University, Italy. His work is related to Power-to-What solutions for energy systems design with a specific focus on the built environment. He has developed expertise on hydrogen technologies, energy efficiency, hybrid systems, energy efficiency in buildings, distributed generation, as well as micro and smart grids. He holds a PhD with honors in Energy Systems Planning and Design at Sapienza University of Rome.

Massimiliano Manfren

Massimiliano Manfren (PhD) is Lecturer in the Sustainable Energy Research Group (SERG), within the Faculty of Engineering and Physical Sciences of the University of Southampton (UK). His previous affiliations include Politecnico di Milano (IT) and University of Bologna (IT). His research focuses on analytics and predictive models for energy system design and operational optimization at multiple scales, from individual users to communities. His research aims to establish a convergence between scientific disciplinary knowledge in energy demand modelling at multiple levels; energy-efficient technologies; and advances in machine learning and operation research techniques, through an integrated use of simulation, optimization, statistics, and data mining on case studies. He holds a PhD in "Programming, Maintenance, and Rehabilitation of Buildings and Urban Systems" from Politecnico di Milano.

Michel Noussan

Michel Noussan (PhD) is Senior Research Fellow at Fondazione Eni Enrico Mattei (FEEM) Future Energy Research Program and Affiliate Professor of Sustainable Transport at Sciences Po's Paris School of International Affairs (PSIA). His current research activities are focused on the analysis and comparison of different mobility solutions in the framework of decarbonization and digitalization trends of the transport sector. He has developed expertise on energy systems analysis, combined heat and power, district heating, energy efficiency and local energy planning. He was a researcher and university lecturer at Politecnico di Torino in the domain of energy systems analysis, and he has a track record of several publications in international journals and conferences. He holds a PhD in Energy Engineering from Politecnico di Torino.

energies MDPI

Editorial

Open Data and Models for Energy and Environment

Benedetto Nastasi [1,*], Massimiliano Manfren [2] and Michel Noussan [3]

1 Department of Planning, Design and Technology of Architecture, Sapienza University of Rome, Via Flaminia 72, 00196 Rome, Italy
2 Faculty of Engineering and Physical Sciences, University of Southampton, Boldrewood Innovation Campus, Burgess Rd, Southampton SO16 7QF, UK; m.manfren@soton.ac.uk
3 Fondazione Eni Enrico Mattei, Corso Magenta 63, 20123 Milano, Italy; michel.noussan@feem.it
* Correspondence: benedetto.nastasi@outlook.com

check for updates

Citation: Nastasi, B.; Manfren, M.; Noussan, M. Open Data and Models for Energy and Environment. *Energies* **2021**, *14*, 4413. https://doi.org/10.3390/en14154413

Received: 29 May 2021
Accepted: 15 July 2021
Published: 22 July 2021

1. Overview of the Articles in This Special Issue

An increasing number of data sources and models to handle them call for transparency and openness in assessing their goodness and practical use for people. The simplest and most robust tools to collect, process, and analyse data to offer solid data-based evidence for future projections in building and district and regional system planning are of interest. For this purpose, and following the success of the first Special Issue "Open Data and Energy Analytics", the Special Issue "Open Data and Models for Energy and Environment" has been launched, intended for energy engineers and planners. Among a very high number of submissions, 10 articles were selected for acceptance and published.

The first paper by Noussan and Neirotti [1] provides a quantification of the potential influence of different charging strategies on the average emission factor of the electricity supplied to electric vehicles. The next paper by Prina et al. [2] is related to the application of the EPLANOPT model to the Italian energy system, showing the difficulties to meet the Paris Agreement target of limiting the temperature increase to 1.5 °C.

The third paper in this special issue, by Neshat et al. [3], presents an optimization framework of a multi-mode wave energy converter to be tested in a small island in the west of Sicily, Italy, in the Mediterranean Sea. Cardone and Gargiulo [4], in the fourth paper of this special issue, describe a semiempirical model of a scroll compressor to predict the power consumption and the mass flow rate by considering leakages and mechanical losses. The next paper, by Amini et al. [5], performs a parametric study on wave energy converter layouts, investigating the distance influence and the effect of rotation regarding significant wave direction in each arrangement compared to the predefined layout. The sixth paper of this special issue, by Chiosa et al. [6], proposes an innovative anomaly detection and diagnosis methodology to automatically detect anomalous energy consumption in buildings, in addition to performing a diagnosis on the sub-loads that are responsible for anomalous patterns. In the next paper, Henrich et al. [7] analyse the impact of energy models in decision making processes for energy transitions in ten municipalities in the Netherlands. In the eight paper, Manfren et al. [8] review the role of energy modelling and analytics for energy transitions in the construction sector. Skeie and Gustavsen [9] investigate the use of geospatial data to improve the level of definition of weather variables used in data-driven building thermal performance characterization. Finally, in the tenth paper, Agostinelli et al. [10] illustrate the use of cyber-physical systems, Internet of things, and machine learning to achieve optimized energy management for a residential district in Rome.

Author Contributions: Conceptualization, B.N.; writing—original draft preparation, B.N., M.M. and M.N.; writing—review and editing, B.N., M.M. and M.N. All authors have read and agreed to the published version of the manuscript.

Funding: This research received no external funding.

Conflicts of Interest: The authors declare no conflict of interests.

References

1. Noussan, M.; Neirotti, F. Cross-Country Comparison of Hourly Electricity Mixes for EV Charging Profiles. *Energies* **2020**, *13*, 2527. [CrossRef]
2. Prina, M.; Manzolini, G.; Moser, D.; Vaccaro, R.; Sparber, W. Multi-Objective Optimization Model EPLANopt for Energy Transition Analysis and Comparison with Climate-Change Scenarios. *Energies* **2020**, *13*, 3255. [CrossRef]
3. Neshat, M.; Sergiienko, N.; Amini, E.; Majidi Nezhad, M.; Astiaso Garcia, D.; Alexander, B.; Wagner, M. A New Bi-Level Optimisation Framework for Optimising a Multi-Mode Wave Energy Converter Design: A Case Study for the Marettimo Island, Mediterranean Sea. *Energies* **2020**, *13*, 5498. [CrossRef]
4. Cardone, M.; Gargiulo, B. Numerical Simulation and Experimental Validation of an Oil Free Scroll Compressor. *Energies* **2020**, *13*, 5863. [CrossRef]
5. Amini, E.; Golbaz, D.; Amini, F.; Majidi Nezhad, M.; Neshat, M.; Astiaso Garcia, D. A Parametric Study of Wave Energy Converter Layouts in Real Wave Models. *Energies* **2020**, *13*, 6095. [CrossRef]
6. Chiosa, R.; Piscitelli, M.; Capozzoli, A. A Data Analytics-Based Energy Information System (EIS) Tool to Perform Meter-Level Anomaly Detection and Diagnosis in Buildings. *Energies* **2021**, *14*, 237. [CrossRef]
7. Henrich, B.; Hoppe, T.; Diran, D.; Lukszo, Z. The Use of Energy Models in Local Heating Transition Decision Making: Insights from Ten Municipalities in The Netherlands. *Energies* **2021**, *14*, 423. [CrossRef]
8. Manfren, M.; Sibilla, M.; Tronchin, L. Energy Modelling and Analytics in the Built Environment—A Review of Their Role for Energy Transitions in the Construction Sector. *Energies* **2021**, *14*, 679. [CrossRef]
9. Skeie, K.; Gustavsen, A. Utilising Open Geospatial Data to Refine Weather Variables for Building Energy Performance Evaluation—Incident Solar Radiation and Wind-Driven Infiltration Modelling. *Energies* **2021**, *14*, 802. [CrossRef]
10. Agostinelli, S.; Cumo, F.; Guidi, G.; Tomazzoli, C. Cyber-Physical Systems Improving Building Energy Management: Digital Twin and Artificial Intelligence. *Energies* **2021**, *14*, 2338. [CrossRef]

energies

MDPI

Article

Cross-Country Comparison of Hourly Electricity Mixes for EV Charging Profiles

Michel Noussan [1,*] and **Francesco Neirotti** [2]

1 Future Energy Program, Fondazione Eni Enrico Mattei, Corso Magenta 63, 20123 Milano, Italy
2 Department of Energy, Politecnico di Torino, Corso Duca degli Abruzzi 24, 10129 Torino, Italy; francesco.neirotti@polito.it
* Correspondence: michel.noussan@feem.it

Received: 9 April 2020; Accepted: 14 May 2020; Published: 16 May 2020

Abstract: Electric vehicles, when coupled to electricity generation from renewable energy sources, can become a viable solution to decarbonize the transport sector. However, given the high variability of electricity mixes on a daily and seasonal basis, high-resolution profiles are needed for a precise analysis of the impacts of electric vehicles in terms of greenhouse gases emissions. This paper presents a comparison of different charging profiles evaluated on 10 European countries over four years, to highlight the effects of national electricity mixes and of the type of charging location on the specific emissions of EVs charging. This study, based on three archetypal charging profiles, provide a quantification of the potential influence of different charging strategies on the average emission factor of the electricity supplied to electric vehicles. The results show that the variability related to charging profiles is generally limited, with an average variation range of 6% for any given country and year, while in several countries the variability from one year to another is much larger, with an average range of 18% for any given country and charging profile.

Keywords: electric vehicles; electricity mix; charging profile; emissions; energy

1. Introduction

Electrification of final sectors is among the main studied solution to mitigate the climate-change related issues together with the pollutant emissions in urban environments. In fact, the electricity sector is experiencing a steep growth rate in final energy consumption, and its renewable share is increasing [1]. At the same time, renewable energy sources (RESs) stress the stability and balance of the grid, due to their unpredictable and variable production.

The transport sector is among those that could see a significant increase of electricity penetration, increasing its RES use [2] and helping the grid balancing [3]. Moreover, electrification of transport leads to significant benefits related to the decrease of some pollutants in cities. These factors are giving an important boost to the electric vehible (EV) market [4], to radically convert the environmental impact of the transport sector [5] and their interaction with new technologies [6,7].

Electric vehicles (EVs) are not new technology, since they first appeared during the nineteenth century, but they were overcome by internal combustion engines thanks to their better reliability and available range. Moreover, in 1908 Henry Ford spread the gasoline engine vehicles around the globe with his affordable and revolutionary Ford Model T [8], putting an end to the EV market. Today, thanks to the new conditions discussed above, there are around 5 million electric cars running on the roads, marking a +40% increase with respect to the 3 million of 2018, with China as leading market [9], followed by the United States and the northern EU regions [10], confirming the current evolution trend towards the electric mobility. The global COVID-19 crisis of early 2020, together with the current low oil prices, is having a short-term impact on this positive trend, but the medium- and long-term effects are not yet clear.

Due to the increasing amount of electrical energy derived from fluctuating resources, the timing of electricity demand is getting high importance to fully exploit the electrification benefit potential, since production and consumption profiles show large variations on a daily, weekly and seasonal basis. This oscillating behaviours lead to significant variations of the actual energy mix during even in small time windows, with consequent variability of the main energy indicators used to assess the impact of final energy sectors, such as the Primary Energy Factor (PEF) or the CO_2 emission factor (EF) [11].

Accurate temporal analyses have been carried on in different works [12,13]. In [12], the authors present the discrepancy between the annual energy mix, which is often used as reference, and the real energy mix of the actual consumption of different heat pumps based on hourly calculations. In [13], the temporal changes in electricity conversion factors supporting a life cycle assessment (LCA) are evaluated, underling the importance of dynamic conversion factors to correctly estimated the entire life cycle impact of a process. Fluctuating RESs are also producing undesired effects related to grid stability. Transport electrification has been pinpointed as a possible solution to reduce the stress and improve the grid reliability [14], especially at the distribution level. A smart integration and management of the charging strategies could bring important benefits for the network operators. At the same tame, unplanned and uncontrolled management of future EVs charging may produce severe impacts over the electric network in terms of efficiency and reliability [15].

In order to properly forecast the impact on the grid, both in terms of RES electricity usage and grid stability issues, detailed analyses based on different charging profiles are needed. Since EV adoption is still at an early stage, there is a small amount of real profiles freely available [16], and they are generally limited to specific conditions (e.g., private fleets, limited geographical coverage, single user type, etc.). Thus, large part of modelling and simulation studies has been performed with artificial charging profiles created making different assumptions or using stochastic and mathematical techniques. In [17], a two-step modelling framework extracts information from a small amount of real data, in order to create reliable charging profiles to asses the potential impact on the grid. Similarly, [18,19] use stochastic simulations to generate realistic profiles based on real data collection. In particular, in [19] the authors underline the difference between the week and week-end consumption patterns, and they stress the importance of a stochastic approach to effectively manage the EV charging periods. This last finding is supported also in [20], where a data-driven machine-learning algorithm is used to replicate historical loads pattern and effectively manage the power system to avoid disservice. Schauble et al. [21] deeply investigated three large datasets in order to underline the large information hidden in charging patterns data, with the aim of spotlighting the most important ones to be used to create reliable profiles. The resulting load profiles are in line with other scientific works. Nevertheless, validation and generalization of the resulting profiles are reported as difficult tasks to be performed due to the wide differences between the analysed data and the observed fleets. Instead, in [22], a user behaviour simulation is used to test potential price incentives. The results highlight that variable pricing can be used to modify the user behaviour and so the charging profiles allowing for a more secure profile forecasting.

While much attention has been paid to charging profiles variation related to grid stability, a comprehensive assessment of the impact on greenhouse gas (GHG) emissions on multiple countries is still lacking. In this paper, we evaluate this impact by considering detailed temporal analyses for both electricity generation and EVs' charging profiles. The EVs' electrical consumption, based on three main charging archetypes, will be compared with several national energy production mixes on an hourly basis, to evaluate the differences between the average annual energy mix and the actual EV consumption patterns. In order to forecast the potential benefit of transport electrification, different countries with specific production portfolios have been considered (Austria, France, Germany, Switzerland, Italy, Netherlands, United Kingdom, Ireland, Denmark, Poland).

2. Methods

The high-resolution analysis of the GHG emissions related to the electricity supplied to EVs requires multiple steps. Different charging profiles and electricity generation mixes are considered and compared, to calculate the weighted average of the hourly emission factors based on different generation technologies. This section presents a description of the methods and the hypotheses that have been used.

2.1. EVs' Charging Profiles

The charging profiles considered in this study have been chosen with the aim of comparing archetypal profiles related to EV charging in three different locations: at home, at work and in public charging stations (usually associated with different services, such as restaurants, shopping centers, cinemas, etc.). The profiles are the result of an analysis based on several data from real charging profiles in Germany, based on a complex model that considered different car segments, households characteristics and charging behaviours, to build up representative profiles for home, work and public charging [23]. The profiles are based on an available power of 3.7 kW for home and work chargers, and 22 kW for public charging points.

The profiles have been normalized on the total demand of a weekday for each type of charging, to analyze comparable results considering the same amount of energy supplied to a vehicle. Figure 1 represents the three different profiles compared for a weekday and a weekend, measuring the contribution of each hour of the day on the total charging over a weekday. An alternative approach may have been the analysis of a large number of real-world EV charging profiles, which may be collected by charging station companies or EV fleet owners. Unfortunately, the availability of statistically significant datasets for research purposes remains limited, and they are often limited in terms of charging locations, temporal spanning and country spanning, which may result in a limited variability. For this reason, this work has been performed on standard profiles defined from literature studies, but future developments of this work may assess the effect of considering real data on the results.

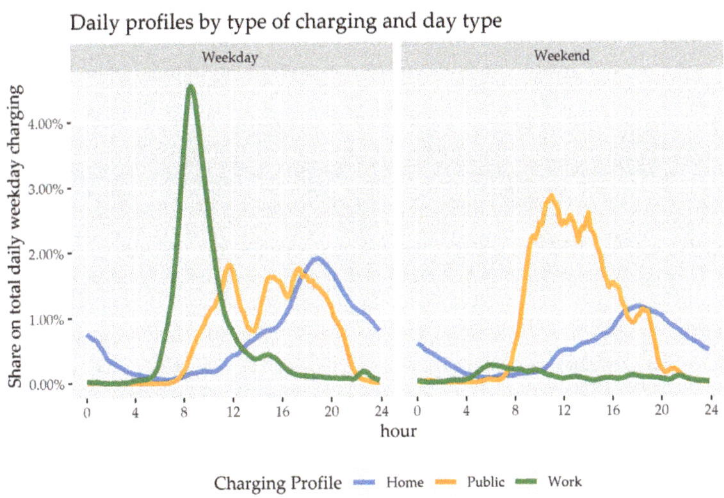

Figure 1. Different electric vehicle (EV) charging profiles.

Another potential limitation lies in the fact that the considered profiles are representative for Germany and may differ from a country to another. Again, this approximation is in part related to the unavailability of a large set of data over different countries, especially considering the limited penetration of EVs. Nevertheless, we believe that the assumptions on which these charging behaviours have been build can still be an acceptable representation for the European countries that are being considered in this study.

All the results are presented in relative terms, considering the electricity mixes that lead to specific emission factors for the electricity supplied to EVs, differentiated per each country and type of profile. The resulting emission factors may be used to quantify the impact of the electricity supply to different EVs, with a greater accuracy in comparison with the common practice of considering average annual performance indicators.

2.2. Electricity Generation Mixes

Together with EV charging profiles, the quantification of actual emission factors requires also high-resolution electricity generation mixes, which have been calculated for several countries. The methodology is described below.

2.2.1. Data Sources

The power generation data per each country has been retrieved from the European Network Transmission Operators for Electricity (ENTSOE) Transparency Platform [24], which makes production data freely available for most European countries. The ENTSOE platform groups together 43 transmission system operators (TSOs) from 36 countries (complete list available at [25]). The main objective of the ENTSOE project is to ensure an optimal functioning of the electricity market as well as support the implementation of the ambitious EU targets on renewable energy and climate change policies [26,27]. Finally, it represents the largest open EU dataset on electricity, including generation data, actual load data and production forecasts for some sources. For some countries, the geographical detail is even higher, with some information available at the level of the different bidding zones, which vary from a country to another.

The analysis performed in this research work has been limited to the generation profiles and electricity mixes at the country level, without taking into account the additional information related to import/export data. To increase the accuracy of the estimation, a deeper investigation of traded electricity may lead to slightly different results. However, to obtain meaningful results it would be necessary to consider all the countries that have international power trading, and in some cases not all the necessary data are available.

2.2.2. Data Processing

We took 10 European countries into account in this work, namely: Austria, Denmark, France, Germany, Ireland, Italy, Netherlands, Poland, Switzerland and the United Kingdom. The generation data were available with 60, 30 and 15 min timestamps, depending on the country. The wide variety of the energy production portfolio of these regions will highlight the effect of EV profiles in very different electricity production mixes. The power generation by source was collected for the years 2016 to 2019 in order to highlight the differences that may occur among the years, thus considering the variability of the generation mix within each country. The data presented a large number of energy sources (hard coal, brown coal, wind on/off-shore, etc.) but in order to get meaningful electricity mixes to be compared among the different countries they were aggregated into 11 main sources: biomass, coal, geothermal, hydro, natural gas, nuclear, oil, solar, waste, wind and other.

The dataset presented a certain amount of "NA" data, both representing missing values as well as the lack of a specific generation technology in any specific country (e.g., Nuclear power in Italy or Denmark is always represented by NA values). Since no detailed information was reported for all the considered countries, all the NA were replaced with zeroes, being aware of the potential under-estimation that can arise from this simplification.

The data were elaborated to build an hourly dataset by calculating the hourly average power for each source in each country, to perform a comparable analysis across all the available countries.

Moreover, the hourly share per energy source $Share_{i,h}$ was calculated as the ratio between the energy produced by a specific source i and the total energy produced in the considered hourly time interval $E_{tot,h}$:

$$Share_{i,h} = \frac{\sum_i E_{i,h}}{E_{tot,h}} \quad (1)$$

The hourly shares are used to calculate the emission factors described in the following sections.

A general comparison of the electricity mixes in different countries is reported in Figure 2, where the average values for the year 2019 are reported for each country. The RES category includes solar, hydro, wind, biomass and geothermal, while the fossil one includes coal, gas, oil and other. Electricity generation from waste is allocated in equal parts to fossil and RESs, in accordance to statistical rules that are applied in some countries, to account for the biological share of municipal solid waste.

The chart shows the considerable variability of electricity mixes across countries, which in turn could lead to significant differences in the emissions related to EV charging. While only 2019 is represented in the chart, additional differences arise from one year to the other, as will be better discussed in the following sections. It has also to be noted that the categories *fossil* and *RES* include different sources that have a broad range of specific emission factors as well as other characteristics (e.g., availability, predictability, dispatchability etc.).

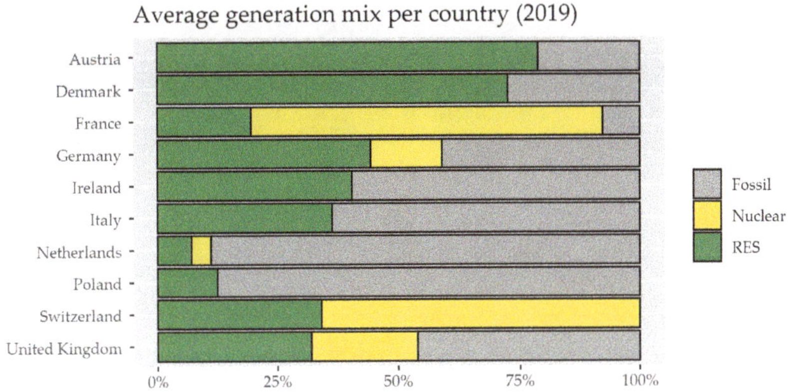

Figure 2. Average electricity mixes in 2019 on an annual basis in different European countries.

2.2.3. Data Quality Issues

ENTSOE dataset is the largest and most comprehensive source of data about the European electricity market, but data quality has been criticized due to lack of completeness and consistency [28,29].

The EU report [29] is focusing on four main issues: data completeness, accuracy, timeliness and user-friendliness. A detailed analysis is carried on based on online surveys with users, interviews with experts, statistical analyses of the data in 2015/2016 and other documents in the literature. The main outcomes are that the generation and load data present inconsistency and missing values with poor documentation about the reasons ad the origin of these gaps. As an example, in the Aggregated Generation per Type data the "Other" category is abused. Natural Gas Combined Cycle (NGCC) plants are aggregated in "Other" instead of "Natural Gas", probably due to the efficiency difference with the usual gas plants as also highlighted in [30]. Moreover, this choice was not explicitly documented in the dataset, but has been discovered by the authors of [30] only after a specific inquiry. This choice, of course, can lead to major errors when computing the emission factor (EF) or the primary energy factor (PEF) of different countries in which this kind of plants have a significant share of electricity generation (e.g., Italy). For this reason, in our work the parameters related to this category have been mostly related to NGCC plants.

Nevertheless, a data quality improvement has been highlighted over the years, especially in 2018 and 2019, and different countries present coherent data with official national data [29]. Finally, it is worth to mention that ENTSOE developed and maintains the platform and the database, but it is not directly responsible for the quality of the data. The data providers are the National transmission system operators (TSOs), sometimes supported by other organizations which, after a proper intermediate service for data cleaning and organization, submit the electricity data. Sometimes the TSO does not have direct access to very small plants data or do not take into account the back-up plants (the ones that are turned on just in specific cases) leading to the underestimation of the generation from specific sources [30]. This is a problem especially for distributed solar plants, whose generation may not be correctly represented in ENTSOE data, since each country has different methods and categories to evaluate and quantify the energy generation from final users.

2.3. Emission Factors

Hourly emission factors for electricity were calculated on the basis of the actual electricity mix of power generation in each country, as described in the previous section, for the years 2016–2019 considered in this analysis. The analysis is limited to domestic generation in the country, since the calculation of the share of imported electricity would have considerably complicated the analysis, with limited additional benefits. Moreover, it would have been necessary to consider other countries for which not all the required data were available.

The hourly emission factors ef_h were calculated as a weighted average of the electricity generation by different sources, in accordance with the following equation:

$$ef_h = \frac{\sum_i ef_i * E_{h,i}}{E_h} \tag{2}$$

where ef_i are the electricity emission factors for the different energy sources, and $E_{h,i}$ is the hourly electricity production from each source and E_h is the total electricity generated in this hour.

Emission factors have been calculated by taking into account direct emissions as well as LCA emissions, based on specific average coefficients available in the literature, reported in Table 1. The LCA data have been chosen from international standards, by considering the values provided by IPCC [31], which is the leading reference for emission factors. A different approach has been chosen for the direct emissions, since data have been obtained by real calculations for the Italian power plants in operation in 2019 [32]. We believe that this approach reflects better the operational conditions of power plants in European countries, although the availability of differentiated coefficients for each country would further improve our results. The emissions factors include the contribution of CO_2, N_2O and CH_4, and they are expressed in g_{CO2eq}/kWh.

An approximation is introduced by choosing average emissions factors by energy source, which may be variable both across countries and across and within years. Considering any energy source, but especially fossil fuels, the electricity emission factor is related to the conversion efficiency of the plant, which shows variations related to the technology, the size of the plants, the outdoor temperature and other design and operational parameters. Moreover, in combined heat and power (CHP) plants the emissions may be allocated properly considering both heat and electricity [33], and multiple methods are available based on different parameters.

Thus, a more precise approach would require to calculate the actual efficiency of each power plant in each given hour, based on a number of data on fuel consumption and heat production that are unfortunately not available over such a wide range of countries and with the required high temporal detail. Still, we believe that using annual average data from real plants represents a better estimation than considering nominal values from power plants design parameters. A future research work focused on a single country may provide additional information on the potential variability related to annual operations and CHP allocation methods.

Table 1. Emission factors for electricity generation from different sources [31,32].

g_{CO2eq}/kWh_{el}	Direct Emissions	LCA Emissions
Biomass	0	230
Coal	870	910
Gas	368	490
Geothermal	0	38
Hydro	0	24
Nuclear	0	12
Oil	545	650
Other	368	490
Solar	0	45
Waste	555	620
Wind	0	11

3. Results and Discussion

3.1. Calculation of Hourly Emission Factors

The first step of the analysis led to the calculation of hourly emission factors, which show significant ranges of variation, as reported in Figure 3 for direct emissions. With the exception of Switzerland, which was totally relying on RES (mostly hydro) and nuclear energy, the other countries showed a very large range of emission factors that was the result of seasonal and daily variations in their electricity mixes. These results underline the importance of performing a more detailed assessment considering actual EV charging profiles rather than considering simple average annual emission factors.

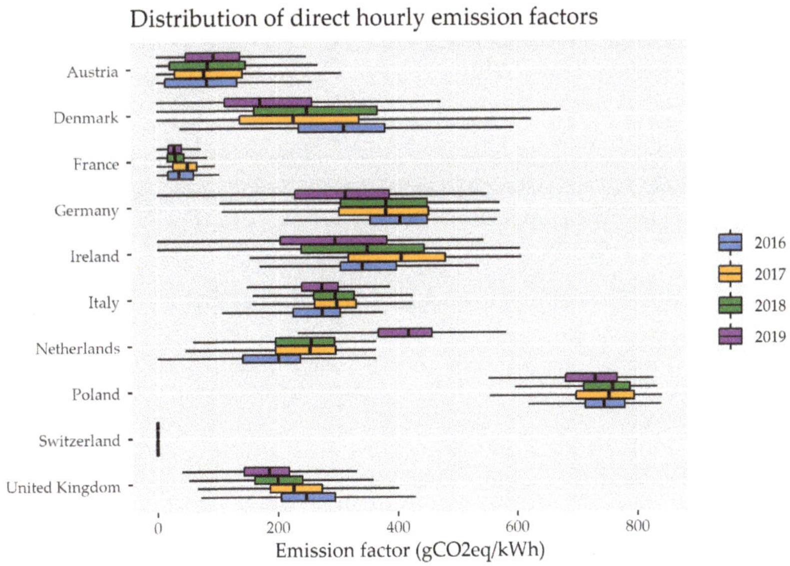

Figure 3. Comparison of direct hourly emission factors in selected European countries.

This variability also emerged while considering the daily patterns of the hourly emission factors of the electricity generation, as reported in Figure 4. Considering some random working days (e.g., the second Wednesday for selected months in 2019) it is clear that the generation mix in each day showed a different behaviour for each country, which was related to multiple factors. The calculation of an annual emission factor starting from hourly resolution allowed us to take into account this variability, but at the same time provided a single indicator that quantified the combined effect of this variability when considering an entire year.

It is important to remember that some results may be affected of the quality of the available data that has been discussed in the previous sections, especially for 2016.

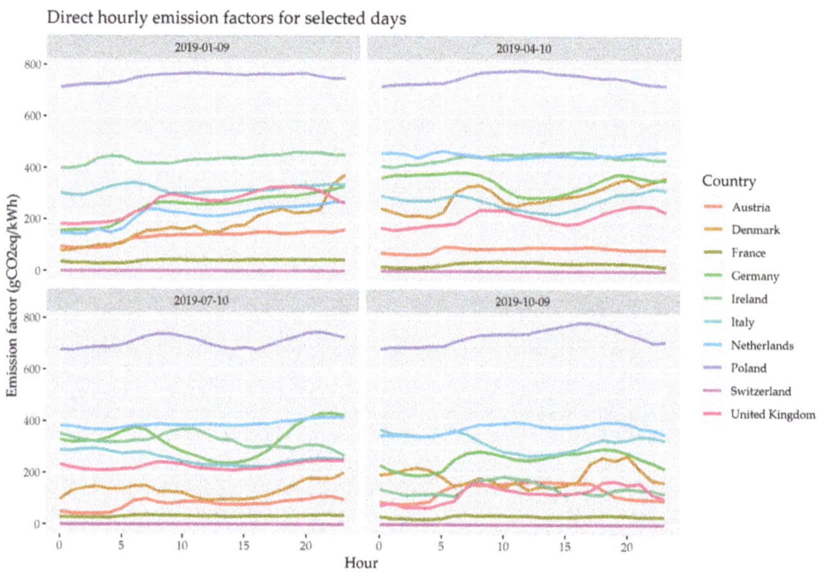

Figure 4. Comparison of direct hourly emission factors for selected days (the second Wednesday of different months in 2019.)

3.2. *Variability of Emission Factors*

As described above, annual emission factors for the electricity supplied to EVs have been calculated for each country considering three different EV charging profiles, and evaluating them in four different years. The combined results, highlighting the variability of direct and LCA emission factors for each country, are showed in Figure 5. A summary of the main information related to the range of variation for each country is also reported in Table 2.

While in some countries the variations remain under 10% (in Switzerland, Poland and Italy), in other cases the maximum variation from the mean can reach values higher than 20%, due to the combined effect of different charging profiles on the demand side as well as different electricity generation mixes on the supply side. Considering all the combinations, the increase of emissions when considering the LCA perspective is on average around 56 g/kWh (with variations in the range 15–97 g/kWh), which represents a 30% increase of the direct emissions. Still, with the exception of Switzerland, which totally relies on nuclear and hydropower, the largest share of GHG emissions in the other countries is related to direct emissions from power plants.

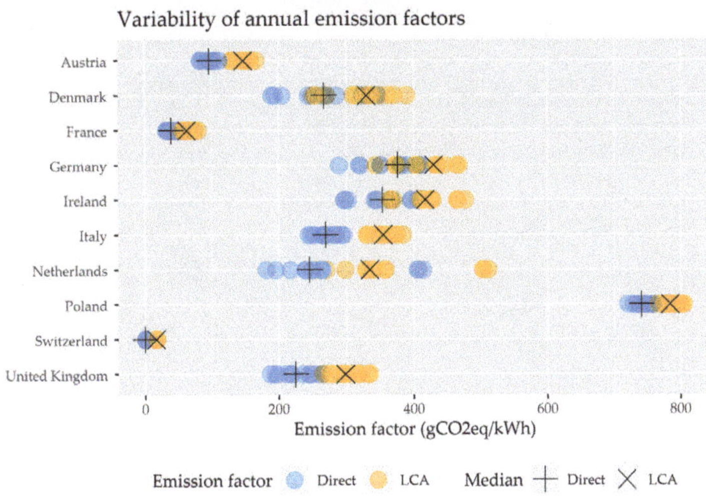

Figure 5. Variability of annual emission factors of EV charging in each country (with variable year and charging profile).

Table 2. Variation of annual emission factors over charging profile and years for each country

g_{CO2eq}/kWh_{el}	**Direct Emission Factor**				**LCA Emission Factor**			
	Mean	**Min**	**Max**	**Range vs. Mean**	**Mean**	**Min**	**Max**	**Range vs. Mean**
Austria	96.9	83.1	111.4	−14%/15%	147.5	129.1	166.2	−13%/13%
Denmark	264.0	189.7	346.4	−28%/31%	332.4	251.8	389.0	−22%/21%
France	41.0	31.6	54.0	−23%/32%	65.2	55.8	79.8	−14%/22%
Germany	363.1	288.7	410.6	−20%/13%	417.1	342.1	466.3	−18%/12%
Ireland	351.6	296.5	405.5	−16%/15%	418.0	363.6	475.8	−13%/14%
Italy	269.7	244.7	295.1	−9%/9%	354.5	329.0	383.7	−7%/8%
Netherlands	277.1	181.6	413.6	−34%/49%	365.3	253.7	510.5	−31%/40%
Poland	742.2	720.2	759.2	−3%/2%	786.6	766.8	804.7	−3%/2%
Switzerland	0.0	0.0	0.0	-	16.7	15.5	17.4	−7%/4%
United Kingdom	227.2	186.8	263.1	-18%/16%	301.7	265.4	333.2	−12%/10%

While the variation of the emission factors for electricity from a country to another has been widely discussed in the literature, these results point out that significant variations also existed when considering the variations within a single country, due to a combined effect of different electricity mixes over the years and for different profiles. The problems related to the quality of data discussed above may also have an impact on the highest variability ranges, and the availability of better data in the next years may help in isolating this potential effect.

Moreover, Figure 6 illustrates an additional seasonal variability related to the monthly emission factors, that have been calculated for each month, year and country. This chart illustrates the fact that the very same charging profiles had different impacts depending on the period of the year that was considered. While in some cases there might be some variations related to the car usage over the year (e.g., vacation periods, different leisure activities, etc.), in general it can be assumed that private car use remains relatively constant over the year. For this reason, the main results of our work were considered on an annual basis rather than on a monthly basis.

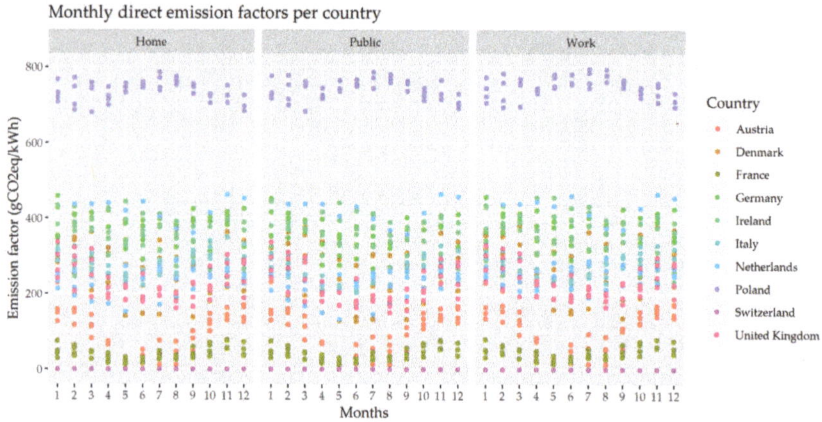

Figure 6. Monthly emission factors of EV charging in each country and each month.

3.3. Variations in a Single Year

To further evaluate the relative weight of annual variations and charging profiles in the variability of the annual emission factors, an analysis focusing on a single year can allow to highlight the effect of the latter aspect. For this reason, Figure 7 reports a comparison of emission factors for each profile considering the data of 2019, that were the most recent available. Moreover, given the issues of data quality discussed above, 2019 appeared to be less affected by these potential inaccuracies. However, it is important to note that the analysis on the other years resulted in very similar results.

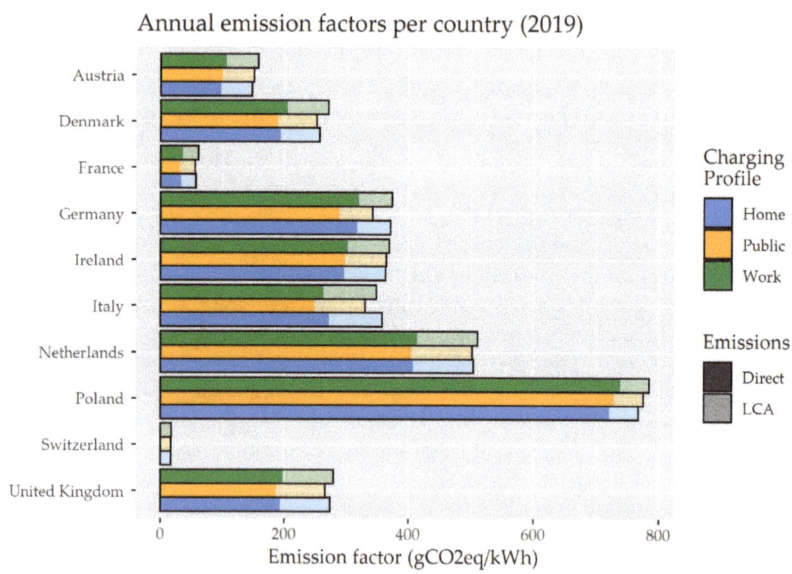

Figure 7. Annual emission factors of EV charging considering different profiles, year 2019.

The chart shows clearly that the emissions associated with the Work profile are higher in all the countries but Italy, for both direct emissions and LCA emissions. This charging profile is strongly concentrated in the beginning of the morning, when people arrive to their work location and plug-in their electric cars. In most countries the morning hours are already characterized by a peak of electricity consumption, which may require the operation of dispatchable fossil plants to guarantee the energy supply. For this reason, a better approach would be to develop strategies to distribute the peak demand over a higher number of hours, considering the fact that people will remain to work for large part of the day. However, it is important to underline that future conditions may vary, in particular with the possible implementation of electricity storage and other flexibility options. The charging profile with minimum impact varies from a country to another, with a slight majority of countries showing lower impacts for Public charging.

Moreover, when considering the range of variation of the different profiles against the mean value, we observe that it remains rather limited for any given country and year. With the exception of the Netherlands in 2016, which may also be affected by some low-quality data, all the other combinations of country and year show ranges of variation (calculated as the ratio between the difference of maximum and minimum values and the average value) lower than 10% (with an average value of 6% and a minimum value of 1%). On the other hand, when considering the variations of the emission factor of each charging profile over the years for any given country, the ranges of variations are significantly higher. Excluding the values for the Netherlands, due to the potential issues with data quality for 2016, the average range of variation is 18%, much higher than the 6% variation associated to the profiles.

Thus it is clear that, within the data considered in this study, the variation related to the year has a higher impact that the variation associated to the charging profiles. Given the fact that the three charging profiles that we have chosen are already representing very different behaviours, we expect that a sensitivity analysis would still confirm this limited variability, when evaluating aggregated profiles.

Another interesting result is that in most cases, the three different charging profiles result in higher annual emission factors in comparison with the mean emission factor for each country, calculated as a simple average of all the hourly values over the year. This result suggests that these charging profiles end up in slightly higher emissions than those that can be calculated by using simplified values, which would then underestimate these impacts. At the same time, the differences remain generally limited, suggesting that also quite different profiles may not lead to dramatic changes in the emissions associated to EVs charging.

On the top of that, it is useful to remind that the aggregated EV charging profile in any country is generally a mix of the different profiles that have been considered here, since users are generally relying on multiple EV charging solutions. As a result, the combined effect on the average EV charging profile in a country may be even more similar to the average value of the electricity mix for this country. Still, the information of the effectiveness of different profiles may remain an important insight for policy choices to support specific charging strategies.

These results suggest that under the current conditions the potential benefits of smart charging strategies based on emission savings may remain limited, although in future power systems characterized by higher shares of RES their role may prove to be very effective. Moreover, smart charging solutions are often designed to fulfill other benefits, such as avoiding excessive loads on the grid or provide flexibility services. Those aspects may thus represent a higher priority for the definition of proper charging profiles, given the limited variability of GHG emissions.

A final note is the fact that all the calculations in this work are performed with the current conditions of the electricity networks and generation mixes. The future will likely be characterized by an increase of the role of variable RESs, but at the same time the energy demand by EVs, in the case of large penetrations, may have a direct impact on the electricity dispatching and on the power mixes. In many countries these demand may remain marginal in comparison with other sectors, and it may also be compensated by energy efficiency measures in other applications, but further work may be required to assess the potential importance of this aspect.

4. Conclusions

This research work presents an analysis of the annual emission factor of different EV charging profiles by considering an hourly time step over one year. The analysis is performed on 10 different

European countries and four years of operation, by comparing three alternative charging profiles that are representative of EV charging at home, at the work location and in public charging facilities.

The results show that in addition to the well-know variability between countries, which is related to the different mixes of electricity generation, there is an additional variability related to both the profiles and the years of operation. In particular, for some countries this latter effect appears to be more strong than for the charging profiles.

However, while there is generally an agreement on the importance of supporting smart charging strategies for EVs, the results of this study show that when assessing the effect of different charging profiles in a given country and a given year, the variability remains limited, with an average range of variation around 6%. Thus, the difference of different charging strategies appears lower than expected. Still, other optimized profiles may provide better results, but they need to be specifically considered in each different case.

On the other hand, for any given country and charging profile, the variation of the year of analysis resulted in an average range of variation of 18%, highlighting the significant variability in the electricity mix of these European countries over the years. This fact underlines the importance of considering multiple years when estimating the emissions related to the use of EVs in any given country, accounting for the potential uncertainty related to varying electricity mixes over the years.

The results of this study provide a preliminary assessment of the difference in GHG emissions associated to multiple EV charging profiles. The results of this work may be strengthened by considering a large amount of real EV profiles rather than the archetypal profiles considered in this work. This could provide additional insights on the variability across multiple changing strategies. Unfortunately, statistically significant datasets from real cases and from multiple countries are seldom available for research purposes. Additionally, an improvement of the time resolution going beyond the hourly data that have been used here could provide additional insights, especially when applied to fast-charging systems.

Author Contributions: Conceptualization, M.N. and F.N.; methodology, F.N. and M.N.; data curation, F.N. and M.N.; visualisation M.N.; writing—original draft preparation, F.N. and M.N.; writing—review and editing, F.N. and M.N. All authors have read and agreed to the published version of the manuscript.

Funding: This research received no external funding.

Conflicts of Interest: The authors declare no conflict of interest.

Abbreviations

The following abbreviations are used in this manuscript:

CHP	combined heat and power
EF	emissions factor
GHG	greenhouse gases
LCA	life cycle assessment
NGCC	natural gas combined cycle
PEF	primary energy factor
RES	renewable energy sources
TSO	transmission system operator

References

1. International Energy Agency. *World Energy Outlook 2018–Part B "Special Focus on Electricity"*; International Energy Agency: Paris, France, 2018.
2. Bellocchi, S.; Klöckner, K.; Manno, M.; Noussan, M.; Vellini, M. On the role of electric vehicles towards low-carbon energy systems: Italy and Germany in comparison. *Appl. Energy* **2019**, *255*, 113848. [CrossRef]
3. Hansen, K.; Mathiesen, B.V.; Skov, I.R. Full energy system transition towards 100% renewable energy in Germany in 2050. *Renew. Sustain. Energy Rev.* **2019**, *102*, 1–13. [CrossRef]

4. Guo, J.; Zhang, X.; Gu, F.; Zhang, H.; Fan, Y. Does air pollution stimulate electric vehicle sales? Empirical evidence from twenty major cities in China. *J. Clean. Prod.* **2020**, *249*, 119372. [CrossRef]
5. Clauß, J.; Stinner, S.; Solli, C.; Lindberg, K.B.; Madsen, H.; Georges, L. A generic methodology to evaluate hourly average CO 2eq. intensities of the electricity mix to deploy the energy flexibility potential of Norwegian buildings. In Proceedings of the 10th International Conference on System Simulation in Buildings, Liege, Belgium, 10–12 December 2018.
6. Noussan, M.; Tagliapietra, S. The effect of digitalization in the energy consumption of passenger transport: An analysis of future scenarios for Europe. *J. Clean. Prod.* **2020**, *258*, 120926. [CrossRef]
7. Alonso Raposo, M.; Ciuffo, B.; Alves Dies, P.; Ardente, F.; Aurambout, J.-P.; Baldini, G.; Baranzelli, C.; Blagoeva, D.; Bobba, S.; Braun, R.; et al. The Future of Road Transport—Implications of Automated, Connected, Low-Carbon and Shared Mobility. 2019. Available online: http://publications.jrc.ec.europa.eu/repository/bitstream/JRC116644/fr-future-road-transport-online.pdf (accessed on 15 April 2020). [CrossRef]
8. Alhelou, H.H.; Hayek, G. 9.4 Electric Vehicles Grid Integration. In *Handbook of Research on Smart Power System Operation and Control*; IGI Global: Hershey, PA, USA, 2019.
9. Bunsen, T.; Cazzola, P.; D'Amore, L.; Gorner, M.; Scheffer, S.; Schuitmaker, R.; Signollet, H.; Tattini, J.; Paoli, J.T.L. *Global EV Outlook 2019 to Electric Mobility*; OECD iea.org; IEA: Paris, France, 2019; p. 232.
10. Salvucci, R.; Petrović, S.; Karlsson, K.; Wråke, M.; Uteng, T.P.; Balyk, O. Energy Scenario Analysis for the Nordic Transport Sector: A Critical Review. *Energies* **2019**, *12*, 2232. [CrossRef]
11. Noussan, M.; Roberto, R.; Nastasi, B. Performance indicators of electricity generation at country level—The case of Italy. *Energies* **2018**, *11*, 650. [CrossRef]
12. Neirotti, F.; Noussan, M.; Simonetti, M. Towards the electrification of buildings heating—Real heat pumps electricity mixes based on high resolution operational profiles. *Energy* **2020**, *195*, 116974. [CrossRef]
13. Vuarnoz, D.; Jusselme, T. Temporal variations in the primary energy use and greenhouse gas emissions of electricity provided by the Swiss grid. *Energy* **2018**, *161*, 573–582. [CrossRef]
14. Mozafar, M.R.; Amini, M.H.; Moradi, M.H. Innovative appraisement of smart grid operation considering large-scale integration of electric vehicles enabling V2G and G2V systems. *Electr. Power Syst. Res.* **2018**, *154*, 245–256. [CrossRef]
15. Deilami, S.; Muyeen, S.M. An Insight into Practical Solutions for Electric Vehicle Charging in Smart Grid. *Energies* **2020**, *13*, 1545. [CrossRef]
16. Lee, Z.J.; Li, T.; Low, S.H. ACN-Data: Analysis and Applications of an Open EV Charging Dataset. In Proceedings of the Tenth International Conference on Future Energy Systems (e-Energy '19), Phoenix, AZ, USA, 25–28 June 2019.
17. Xydas, E.; Marmaras, C.; Cipcigan, L.M.; Jenkins, N.; Carroll, S.; Barker, M. A data-driven approach for characterising the charging demand of electric vehicles: A UK case study. *Appl. Energy* **2016**, *162*, 763–771. [CrossRef]
18. Brady, J.; O'Mahony, M. Modelling charging profiles of electric vehicles based on real-world electric vehicle charging data. *Sustain. Cities Soc.* **2016**, *26*, 203–216. [CrossRef]
19. Flammini, M.G.; Prettico, G.; Julea, A.; Fulli, G.; Mazza, A.; Chicco, G. Statistical characterisation of the real transaction data gathered from electric vehicle charging stations. *Electr. Power Syst. Res.* **2019**, *166*, 136–150. [CrossRef]
20. Pan, Z.; Wang, J.; Liao, W.; Chen, H.; Yuan, D.; Zhu, W.; Fang, X.; Zhu, Z. Data-Driven EV Load Profiles Generation Using a Variational Auto-Encoder. *Energies* **2019**, *12*, 849. [CrossRef]
21. Schäuble, J.; Kaschub, T.; Ensslen, A.; Jochem, P.; Fichtner, W. Generating electric vehicle load profiles from empirical data of three EV fleets in Southwest Germany. *J. Clean. Prod.* **2017**, *150*, 253–266. [CrossRef]
22. Canizes, B.; Soares, J.; Costa, A.; Pinto, T.; Lezama, F.; Novais, P.; Vale, Z. Electric Vehicles' User Charging Behaviour Simulator for a Smart City. *Energies* **2019**, *12*, 1470. [CrossRef]
23. Heinz, D. *Erstellung und Auswertung repräsentativer Mobilitäts- und Ladeprofile für Elektrofahrzeuge in Deutschland*; Working Paper Series in Production and Energy; KIT: Karlsruhe, Germany, 2018; p. 30. [CrossRef]
24. European Network of Transmission System Operators for Electricity. Entsoe Website. Available online: https://www.entsoe.eu/ (accessed on 15 April 2020).

25. Entsoe. ENTSOE TSO. Available online: https://www.entsoe.eu/about/inside-entsoe/members/ (accessed on 5 April 2020).
26. European Commission. Clean Energy Package for all Europeans—Website. Available online: https://ec.europa.eu/energy/en/topics/energy-strategy-and-energy-union/clean-energy-all-europeans (accessed on 15 April 2020).
27. Entsoe. ENTSO-E Annual Report-2018. 2019. Available online: https://annualreport2018.entsoe.eu/ (accessed on 15 May 2020).
28. Hirth, L.; Mühlenpfordt, J.; Bulkeley, M. The ENTSO-E Transparency Platform—A review of Europe's most ambitious electricity data platform. *Appl. Energy* **2018**, *225*, 1054–1067. [CrossRef]
29. European Commission. *A Review of the ENTSO-E Transparency Platform*; European Commission: Brussels, Belgium, 2017; pp. 1–58.
30. Gotzens, F.; Heinrichs, H.; Hörsch, J.; Hofmann, F. Performing energy modelling exercises in a transparent way—The issue of data quality in power plant databases. *Energy Strategy Rev.* **2019**, *23*, 1–12. [CrossRef]
31. Schlömer , S.; Bruckner, T.; Fulton, L.; Hertwich, E.; McKinnon, A.; Perczyk, D.; Roy, J.; Schaeffer, R.; Sims, R.; Smith, P.; et al. Annex III: Technology-specific cost and performance parameters. In *Climate Change 2014: Mitigation of Climate Change. Contribution of Working Group III to the Fifth Assessment Report of the Intergovernmental Panel on Climate Change*; Cambridge University Press: Cambridge, UK; New York, NY, USA, 2014.
32. ISPRA. *Fattori di Emissione di Produzione e Consumo di Elettricità—2019*; ISPRA: Rome, Italy, 2019. (In Italian)
33. Noussan, M. Allocation factors in Combined Heat and Power systems—Comparison of different methods in real applications. *Energy Convers. Manag.* **2018**, *173*, 516–526. [CrossRef]

energies

MDPI

Article

Multi-Objective Optimization Model EPLANopt for Energy Transition Analysis and Comparison with Climate-Change Scenarios

Matteo Giacomo Prina [1,*], Giampaolo Manzolini [2], David Moser [1], Roberto Vaccaro [1] and Wolfram Sparber [1]

1 EURAC Research, Institute for Renewable Energy, Viale Druso 1, I-39100 Bolzano, Italy; david.moser@eurac.edu (D.M.); roberto.vaccaro@eurac.edu (R.V.); wolfarm.sparber@eurac.edu (W.S.)

2 Dipartimento di energia, Politecnico di Milano, Via Lambruschini, 4, 20156 Milano (MI), Italy; giampaolo.manzolini@polimi.it

* Correspondence: matteogiacomo.prina@eurac.edu

Received: 25 May 2020; Accepted: 22 June 2020; Published: 23 June 2020

Abstract: The modeling of energy systems with high penetration of renewables is becoming more relevant due to environmental and security issues. Researchers need to support policy makers in the development of energy policies through results from simulating tools able to guide them. The EPLANopt model couples a multi-objective evolutionary algorithm to EnergyPLAN simulation software to study the future best energy mix. In this study, EPLANopt is applied at country level to the Italian case study to assess the best configurations of the energy system in 2030. A scenario, the result of the optimization, is selected and compared to the Italian integrated energy and climate action plan scenario. It allows a further reduction of CO_2 emissions equal to 10% at the same annual costs of the Italian integrated energy and climate action plan scenario. Both these results are then compared to climate change scenarios through the carbon budget indicator. This comparison shows the difficulties to meet the Paris Agreement target of limiting the temperature increase to 1.5 °C. The results also show that this target can only be met through an increase in the total annual costs in the order of 25% with respect to the integrated energy and climate action plan scenario. However, the study also shows how the shift in expenditure from fossil fuels, external expenses, to investment on the national territory represents an opportunity to enhance the national economy.

Keywords: energy scenarios; photovoltaics; wind; EPLANopt; multi-objective optimization; climate-change

1. Introduction

Energy system modeling [1] is a relevant discipline in supporting policy-makers in the definition of the energy strategy. Different European countries have already published energy strategies to meet the European climate and energy targets for 2030 [2]. Among them, there is also Italy which is selected as the case study in this paper [3].

With the aim of using an energy system model to develop different scenarios for the considered case study, the main characteristics of energy system models have been analyzed in order to choose the proper model. In particular, two features of energy system models are identified as being relevant in the scenario development process. These are the following: high temporal resolution and sector-coupling.

The hourly time-step is largely considered as high temporal resolution in energy system modeling [4]. It is particularly important when modeling energy systems with high penetration of variable renewable energy sources (VRES). Poncelet et al. [4] showed the importance of time resolution in energy system modeling. They demonstrated how the resolution in time should be prioritized compared to the resolution in techno-economic detail and how the use of a low number of time-slices (usually 12 time-slices) produces a generation mix error that cannot be considered negligible.

The second important characteristic is sector coupling. Several papers showed the advantages of sector coupling modeling compared to the single specific sector modeling approach. In this regard, it is important to mention the contribution of Aalborg University in the definition of the smart energy system concept that showed the advantages of studying the interactions and synergies between different energy sectors to maximize efficiency and reduce costs [5,6]. H. Lund in [6] and D. Connolly et al. [7] conceptualized the smart energy system definition highlighting the opportunities and synergies among different energy sectors. In [8], B.V. Mathiesen et al. analyzed the smart energy system concept focusing on the integration of the transport sector.

In [9], B. Nastasi et al. highlighted the importance of hydrogen as an energy vector to link the electricity and heating sectors. M. G. Prina et al. [10] showed the advantages of sector coupling at district heating level. R. Bramstoft et al. [11], through the studying of the decarbonization pathways of Sweden at 2050, showed the advantages of the integrated modeling of transportation, electricity, gas, fuel refinery and heat systems. S. Ben Amer et al. [12] used the Balmorel model on the Greater Copenhagen case study integrating the electricity and heating sectors. V. Heinisch et al. [13] showed the advantages of coupling the electricity, heating and transport sectors focusing on urban areas. M. Pavičević et al. [14] studied the potential of sector-coupling at European level and found out how the transport sector coupled to the power sector guarantees the highest flexibility potential in terms of power curtailment, load shedding and congestion. H. Lund et al. [15] underlined the importance of moving beyond the electricity-only approach and towards an integrated cross-sector approach.

A model which reflects these characteristics is the software EnergyPLAN [5,16,17] developed by Aalborg university [17], Denmark. EnergyPLAN software is a bottom-up single-node simulation model which allows the evaluation of different future alternatives of the energy system through the testing of different energy mixes. It implements an hourly time-step to properly describe energy systems with high penetration of renewables. Several studies coupled an optimization algorithm to the EnergyPLAN software; therefore, using the EnergyPLAN software for the simulation and dispatch over the year and an optimization algorithm for expansion capacity evaluation. This approach is largely diffuse due to the characteristics of EnergyPLAN software which requires a very short computational time. This is due to its heuristic modeling based on internal predefined priorities.

I. Batas Bjelić et al. [18] presented a single-objective (SO) optimization model coupled to EnergyPLAN and selecting the flexibility options only in the electricity sector. M. S. Mahbub et al. in [19] and in [20] presented a multi-objective (MO) optimization tool considering as flexibility options within the decision variables, only heat pumps. The EPLANopt model [21,22] developed by Eurac research through the coupling of the simulation software EnergyPLAN and a multi-objective optimization algorithm is applied at regional level in [23] selecting as flexibility options, heat pumps, batteries and power to gas. The EPLANopt tool is open-source and the full code is available at [24]. The EPLANopt model is characterized by an hourly time-step, single-node approach and by sector-coupling, i.e., the main sectors of the energy system (electricity, heating and transport sectors) are all implemented in the model. With a similar approach the EPLANopt model is applied in this study at the Italian energy system.

In particular, the aim of this paper is the following: (i) the creation of an energy system model which reflects the Italian integrated energy and climate action plan at 2030 (Piano Nazionale Integrato Clima e Energia—PNIEC [3]). This is useful to validate the model on the achievement of the energy targets and at the same time to estimate the total annual costs of the energy system. This latter is important information for the comparison with different scenarios. (ii) Through the expansion capacity optimization model, EPLANopt, this work aims to inspect future scenarios under total annual costs and CO_2 emissions minimization. The comparison of these scenarios with the Italian integrated energy and climate action plan allows the understanding of other alternatives to the energy system able to further decrease CO_2 emissions. (iii) The last scope is to compare the scenarios, result of the optimization process and the Italian integrated energy and climate action plan scenario with the climate change scenarios. Through this comparison the aim is to understand if these scenarios are in line with the Paris Agreement target [25] of limiting the temperature increase to 1.5 °C. This comparison is possible thanks to the use of the carbon budget indicator.

One challenge of energy system modeling is the connection between bottom-up energy system models and the impacts of energy transition on the environmental sphere. Considering climate change scenarios in energy system modeling is usually performed through the integration of top-down and bottom-up approaches with different degrees of linking between the two methodologies. For example, M. Rocco et al. [26] realized a soft-link between the open-source energy optimization model (OSEMOSYS) and a linear input-output model (IO) to evaluate the environmental impact of future energy scenarios. S. D. Tuladhar [27] realized a hard-link between a bottom-up and top-down model by means of an iterative process with the aim of climate change analysis. Other examples of hybrid models, results of the integration between bottom-up and top-down, are integrated assessment models (IAMs) [28] which present close loops between climate, impacts, economy and energy modules. This paper aims at implementing a simplified and fast method to connect the results of a bottom-up energy system model with climate change scenarios. Future steps will be dedicated to the integration of EPLANopt with a top-down model to assess the impacts on the economy and the environment.

The paper has the following structure: a materials and methods section presents the EPLANopt model, its main characteristics and the carbon budget indicator; a section on the Italian case study presents the assumptions and sources of the input data used for the case study; a results section shows the outcomes of the model and conclusive remarks are given in the last section. The work presented in this paper received funding from the FESR 1042 "Integrids" project and by institutional funding. The funding body did not influence the case study, assumptions or the choice of the model.

2. Materials and Methods

The materials and methods section is structured as follows: (i) the EPLANopt model, why it is chosen and its main characteristics; (ii) EnergyPLAN software main general features; (iii) explanation of the multi-objective optimization analysis used in EPLANopt; (iv) specific characteristics of the EnergyPLAN software to this particular case study and (v) the carbon budget methodology and how the results from the bottom-up energy system model EPLANopt are connected to climate change scenarios.

EPLANopt [21,22] is a bottom-up short-term energy system model which is selected because it allows the implementation of multi-objective optimization without losing resolution in time and in sector-coupling [29], which, as already mentioned, are relevant characteristics for describing energy systems with high penetration of renewables. Bottom-up models accurately describe the energy system internal relationships and allow the user to evaluate the future alternatives of the energy system and the potential synergies between energy sectors. These models do not usually describe the interactions between the energy sectors and the economics of a nation, region or municipality. These models differ from the top-down approach [30] which instead are characterized by less details in the energy sector but describe the relations with other interconnected sectors such as employment, social growth, public welfare, et cetera. Short-term models inspect the alternatives of the energy system in a future target year. These differ from long-term models [31] which study and assess the entire transition between the current state of the energy system up to a future target year.

The EPLANopt model is the result of a coupling between the EnergyPLAN software [5,16,17] and an expansion capacity optimization algorithm. The EnergyPLAN software is a deterministic simulation model, it is suited to describe future scenarios with high penetration of VRES, it simulates a one-year period with an hourly time-step and it integrates the three primary sectors of the energy system. The model was applied at different scales: at European level [32], at national level [33–40], at regional level [41], to towns and municipalities [10,42] and to small islands [43–45].

The multi-objective expansion capacity optimization algorithm is based on a multi-objective evolutionary algorithm (MOEA) [46–48] which allows the assessment of the Pareto front of optimal solutions. The multi-objective approach allows the modeler to find optimal solutions in terms of simultaneous minimization or maximization of different indicators. Therefore, the multi-objective optimization approach allows the simultaneous assessment not only of an economic objective, as usually adopted in single-objective expansion capacity optimization problems, but also considering an environmental one. The considered objectives for this particular case are the minimization of the total annual costs and the minimization of annual CO_2 emissions. Equation (1) shows the objective functions of the multi-objective minimization problem. The main constraints describe how the value of the decision variables should remain in a fixed range defined by the decision variables' lower $DV_i^{(L)}$ and upper $DV_i^{(U)}$ bounds. Other constraints such as balance between demand and generation at each time-step or storage behavior with initial content equal to final content are defined within the EnergyPLAN software.

$$\text{Optimization function } min\begin{pmatrix} Annual\_Costs \text{ [M€]} \\ Annual\_CO_2\_Emissions \text{ [Mt]} \end{pmatrix} \tag{1}$$

$$\text{Subject to } DV_i^{(L)} \leq DV_i \leq DV_i^{(U)}$$

The operational simulation of the year is performed through EnergyPLAN software while the expansion capacity optimization is achieved through the MOEA. The total annual costs are considered by EnergyPLAN as the sum between annualized investment costs, fixed operation costs and variable costs. These latter are divided into fuel costs and electricity exchange costs.

The MOEA creates an initial population of random individuals. The population is the set of different solutions which will be tested by the optimization algorithm to find the optimum. Each individual is made up of a list of values for the decision variables. For each decision variable, the value is found in the range defined by a minimum, $DV_i^{(L)}$, and a maximum bound, $DV_i^{(U)}$. The MOEA then inspects and evaluates each solution (thus each individual) by running the simulation model, in this case EnergyPLAN. EnergyPLAN is run on the selected future target year, replacing the values of the decision variables which characterized the solution in the EnergyPLAN input file of the baseline. EnergyPLAN returns the values of different indicators. Two of them, total CO_2 emissions and total annual costs, are chosen as objective functions. Each individual or solution is compared to the others based on these two indicators. Through the use of operators typical of genetic algorithms (such as selection, crossover and mutation) the optimization algorithm moves forward creating a new population of individuals. These steps are repeated until the convergence is reached and the final Pareto front is found.

EnergyPLAN is adopted in this study with the following specific characteristics: (i) The version is 12.1; (ii) the technical simulation option is selected; (iii) dump charge is chosen for electric mobility; and (iv) power-to-gas is modeled through two main variables, the hydrogen produced and the capacity of the electrolyzer. The electrolyzer will start producing hydrogen in the time-step in which there is over-generation of electricity from VRES and injecting it into the gas grid. This is implemented in EnergyPLAN through the electro-fuels sheet; CO_2 hydrogenation section, by setting to zero the parameters of the carbon recycling and the electrolyzer efficiency equal to 0.7. The variables SynGridGas [TWh/year] under output section, which corresponds to the produced hydrogen, and the MaxCap variable under the flexibility section, which represents the capacity of the electrolyzer are chosen within the optimization. The decision variables are the technologies on which the expansion capacity optimization analysis is performed. These two variables, produced hydrogen and capacity of the electrolyzer, are chosen as decision variables together with a list of other technologies such as variable renewable energy sources, electric storage, et cetera. The complete list is introduced in Section 3. The optimization algorithm varies their values in order to find the best energy mix for the considered case study, hence, finding the right combination of excess electricity production, size of the electrolyzer, produced hydrogen, et cetera.

As mentioned in the introduction, the aim of the paper is also to connect and compare the scenarios obtained through the optimization methodology to climate-change scenarios. In order to achieve this task, there is the need to introduce the remaining carbon budget concept. It is defined by J. Rogelj et al. [49] as the "Finite total amount of CO_2 that can be emitted into the atmosphere by human activities while still holding global warming to a desired temperature limit". In [49], J. Rogelj et al. estimate the remaining carbon budget on a global level equal to 480 Gt CO_2. This amount represents the overall quantity of CO_2 emissions that can be emitted for a 50% probability of limiting global warming to 1.5 °C (within 2100 above pre-industrial levels, 1850–1990). Assuming that each person on earth has the same carbon budget, the overall carbon budget for the considered case study, Italy, results in 3.8 Gt.

3. Italian Case Study

This section is dedicated to the description of the input data and assumptions regarding the selected case study. It is structured as follows: (Section 3.1) The first section concentrates on the baseline, the required data for its characterization and the assumptions. The baseline is a scenario which reflects the current state of the energy system for a reference year. (Section 3.2) The second section focuses on the Italian integrated energy and climate action plan and its definition in the EnergyPLAN software. (Section 3.3) The third section describes the decision variables and their ranges. The decision variables are the technologies on which the expansion capacity optimization analysis is performed. Their ranges define the domain of the optimization problem. For each decision variable the range is defined by a minimum bound, the value in the reference year and a maximum bound, the maximum potential for the source. Moreover, this section explains the assumptions introduced on energy efficiency of buildings, electric mobility, energy efficiency in the industry sector, et cetera. (Section 3.4) The fourth section defines the optimization problems and their assumptions.

3.1. Baseline

The first step consists of the creation of the Baseline which is the scenario reproducing the Italian energy system for a specific reference year. In this study the reference year is 2015 and the application case study is the Italian energy system. The created scenario is called Baseline 2015. It is characterized by an input file of EnergyPLAN which collects all the information of the Italian energy system for the year 2015: energy demand, installed power and capacity of different sources, efficiencies, emission factors, fuel costs, investment and operation and maintenance (O&M) costs.

The Baseline 2015 is created by the Heat Roadmap Europe 4 (HRE4) project [50] which provides the 2015 EnergyPLAN input file for 14 EU member countries (Italy included) [51]. This 2015 HRE4 baseline is modified using more precise data taken from Italian authorities: GSE [50], RSE [52] and Terna [53]. Table 1 summaries the main open sources used to collect the input data divided into the electricity, heating and mobility sectors. The data are mainly generation and consumption data, thus including capacities, efficiencies and time-series.

Table 1. Sources used for the development of the 2015 baseline for Italy.

Sector	Data	Source	References
Electricity	Capacity of renewables	GSE	[54]
	Hourly profile for renewables	Terna, GSE	[53,54]
	Capacity for other technologies	Terna	[55]
	Electricity demand	Terna, HRE	[50,53]
Heating	Generation and consumption data	HRE	[50]
Mobility	Consumption data	HRE	[50]
	Electric vehicles demand and charge profile	RSE	[52]

The electricity sector produced by HRE4 is the most modified and adapted in this work. M. Noussan et al. [56], starting from open data, presented a data analysis of the electricity generation at the Italian level, by considering some performance indicators based on primary energy consumption, share of renewable energy sources and CO_2 emissions. The same open data sources are used to update the EnergyPLAN input file for the Italian case study in 2015.

Renewable energy installed power is taken from the statistic report of GSE (Italian authority on renewable energy and energy efficiency) [54]. Hourly distributions profiles of renewable energy sources are taken from Terna transparency reports [53]. Terna is the Italian transmission system operator. However, these distributions are influenced by the commissioning and entry into service of new plants during the year. This affects the hourly distribution because, at the end of the year, the generation is characterized by a different installed capacity from the one at the beginning of the year. In order to get rid of this increasing factor, a linear increase of the installed capacity of each renewable energy source is assumed between the beginning and end of the year. The artificial value of capacity in each time-step t is given by Equation (2). $C_{m,t}$ is the capacity of technology m at time-step t. C_m^{start} is the capacity of technology m at the beginning of the considered year while C_m^{end} is the capacity of technology m at the end of the considered year. 8784 are the hours in a year (EnergyPLAN considers a leap year).

$$C_{m,t} = C_{m,t-1} + t \cdot \frac{C_m^{start} - C_m^{end}}{8784} \quad (2)$$

The generation of each renewable energy source in each time-step t is rescaled to take into consideration this increase in capacity over the year. Equation (3) shows the new hourly power output of technology m. $P'_{m,t}$ is the new rescaled power output of technology m. $P_{m,t}^{Terna}$ is the old power output of technology m taken from the Terna data (values that should be purged from the increasing capacity over the year).

$$P'_{m,t} = P_{m,t}^{Terna} \cdot \frac{C_m^{end}}{C_{m,t}} \quad (3)$$

The new generation value is equal to the equivalent power output that would have been produced in the time-step t by the total capacity installed at the end of the year. Another modification of the existing Heat Roadmap Europe data is required: the equivalent hours of the hourly distribution must coincide with the value from GSE. Equivalent hours express the full load hours of a certain source. Equation (4) shows how the equivalent hours are calculated, where P_m is the total power output of technology m over the year and C_m is the capacity of technology m.

$$h_{eq,m} = \frac{P_m}{C_m} \quad (4)$$

Table 2 shows the equivalent hours of the main Italian renewable energy sources and the final five year average value. This average value is chosen for the modification of the hourly distribution data with the aim to reconcile the year by year variability of these renewable energy sources.

Table 2. Equivalent hours for renewable energy sources in Italy: Photovoltaic (PV), wind power, river hydro and geothermal. Values from 2011 to 2015.

Equivalent Hours	PV	Wind Power	Hydro (River)	Geothermal
Equivalent hours 2011	1325	1563	4060	7324
Equivalent hours 2012	1312	1855	4379	7243
Equivalent hours 2013	1241	1793	4392	7321
Equivalent hours 2014	1211	1767	4454	7206
Equivalent hours 2015	1225	1683	4374	7534
Average equivalent hours	1263	1732	4332	7325

The method to correct the hourly distribution data is taken from the EnergyPLAN documentation. This method is shown in Equation (5) where FAC^{RES} is a correction factor. This value is unknown and has to be found in order to obtain a value of equivalent hours for the new distribution equal to the average equivalent hours of Table 2.

$$P_{m,t} = P'_{m,t} \cdot \frac{1}{1 - FAC^{RES}.(1 - P'_{m,t})} \tag{5}$$

A last modification that needs to be done is caused by the EnergyPLAN leap year approach. In order to generate hourly distributions of 8784 elements it is necessary to create an imaginary 29th February for the year 2015. This day is created by copying the previous day's distributions.

Once all of these modifications are implemented, it is possible to calculate the hourly distributions to be used in EnergyPLAN (Equation (6). P_m^{nom} is the nominal power or installed capacity of technology m in the year 2015. $P_{m,t}^{ePLAN}$ is the final hourly distribution with values between 0 and 1.

$$P_{m,t}^{ePLAN} = \frac{P_{m,t}}{P_m^{nom}} \tag{6}$$

Figure 1 shows an elaboration of the hourly distributions for PV, wind power, geothermal and river hydro in the years from 2011 to 2015. The representative day for each month is implemented calculating the average of the values of the same hours on all the days of that month. This type of graph is useful to highlight seasonal and intra-day variability. It shows the potential integration between PV and wind power. In fact while PV has a higher generation in summer wind power produces the most in the winter season. Geothermal is almost constant and river hydro has its peak during the end of spring and the beginning of summer. It is important to highlight that the seasonal variability of river hydro could change in the future as a result of climate change and melting of the glaciers [57]. The installed power of PV in 2015 was equal to 18,892 MW. Wind power installed power was 9162 MW and geothermal was 821 MW. Hydro power is divided into three categories: pumped hydro storage, reservoir hydro power and river hydro. River hydro is characterized by a small reservoir in the order of a few hours. The installed power of river hydro was equal to 5332 MW in 2015. As it is possible to see from Figure 1, the river hydro profile has a daily cycle with two peaks: one in the morning and the other one in the evening. This is typical of the daily electricity demand profile and suggests that these plants within the day can modulate their production in response to the demand and the prices of electricity.

Other renewable energy sources such as biomass plants and reservoir hydro power can modulate their output power. Biomass power plants' installed capacity in Italy is equal to 4057 MW. In this work, biomass power plants are modeled within VRES using a constant distribution profile. In support of this assumption, these types of plants, even if potentially they can produce and modulate their power to follow the load, are usually forced to produce at maximum constant output power in order to exploit the maximum of their abilities. Equation (7) shows how the new artificial capacity C'^{bio} is calculated. C^{bio} was the actual capacity for biomass power plants in 2015. h_{eq}^{bio} expresses the actual equivalent hours of biomass power plants and 8784 are the total hours in a leap year. The CO_2 emissions from biomass are considered zero (as in the Heat Roadmap Europe baseline and scenarios).

$$C'^{bio} = C^{bio} \cdot \frac{h_{eq}^{bio}}{8784} \tag{7}$$

Reservoir hydro power was characterized by an installed power of 9425 MW in 2015. In EnergyPLAN, reservoir hydro power has a dedicated component. However, this component is used to model pumped hydro storage; thus, like the biomass power plants, reservoir hydro power is modeled into the VRES category assuming a constant hourly profile of generation. Equation (7) is applied to reservoir hydro power to find the new artificial capacity.

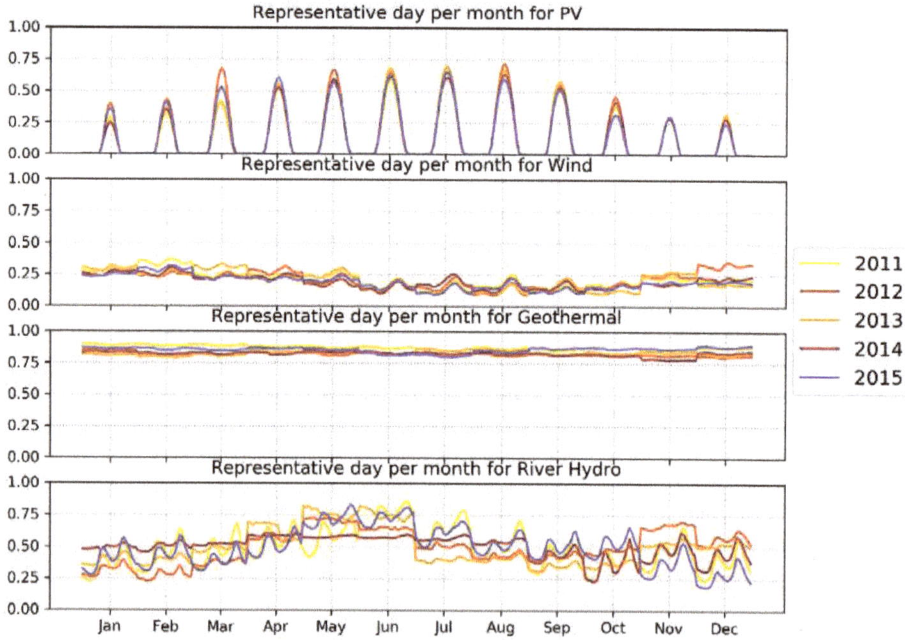

Figure 1. Representative day per month for PV, wind power, geothermal and river hydro in the years from 2011 to 2015.

Italian fossil fuel power plants are constituted mostly of combined cycle gas turbine (CCGT) systems burning natural gas. These plants have the highest efficiencies among fossil fuel power plants and are also flexible in load modulation. However, the increase in electricity generation from renewables has brought a reduction in the utilization of these systems with a drastic drop in recent years of their equivalent hours [58] (from 4000–6000 to 1000–2000 hours). Fossil fuel power plants are modeled in EnergyPLAN in the power plant (PP2) component with a value of overall capacity taken from Terna [59], a value of average efficiency taken from Heat Roadmap Europe [52] and the relative quantities of each type of burnt fuel. The overall fossil fuel power plants installed capacity of the 2015 Italian energy system is 63,863 MW, the sum of the installed capacity of CCGTs and coal power plants. The efficiency of the average fossil fuel power plant is 0.455 and the percentages of burnt fuel types are the following: 39% coal and 61% natural gas.

For electricity storage the only technology present in the energy system of 2015 was pumped hydro storage. This technology has already reached its maximum potential and for future installation of electric storage other technologies must be taken into account. In Italy, in 2015, the pumped hydro-storage installed plants allow the achievement of 700.76 GWh of available storage capacity, 6175 MW of pumps and 7815 MW of installed turbines. These data are the aggregated results of internal analysis developed by Politecnico di Milano. The average charging efficiency is set at 0.85 while the average discharging efficiency equal to 0.9. S. Mazzoni et al. [60] investigated the techno-economic impact of different storage technologies demonstrating how the use of these technologies leads to primary energy savings and high efficiency.

The overall electricity demand is the sum of various contributions: the generic electricity demand, electricity demand from the heating sector, electricity demand from the cooling sector and electricity demand from the mobility sector. Figure 2 shows the representative day per month of the overall electricity demand in the years from 2011 to 2015. It is possible to observe the two-peak daily profile, one in the morning and one in the evening. The highest values of electricity demand are in July when cooling requirements become more relevant. The total electricity demand in 2015 is equal to 316.9 TWh after transmission losses deduction, Terna data [55].

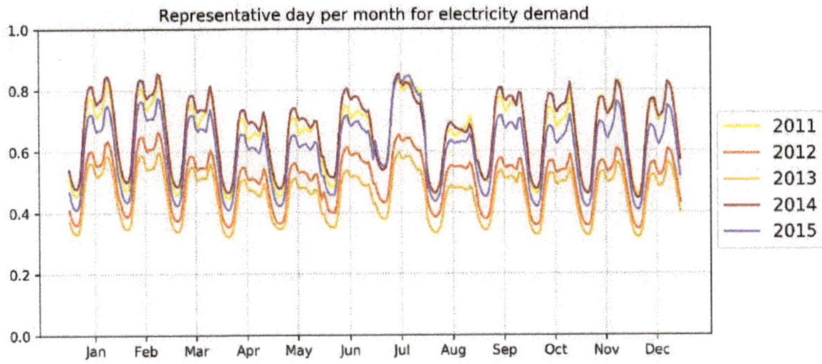

Figure 2. Representative day per month for electricity demand from 2011 to 2015.

For the heating and cooling sector, M. Noussan et al. [61] presented an insightful analysis on building heating systems for the most populated Italian region by means of open data. However, from a national perspective there is a lack of updated open data. Therefore, the Heat Roadmap Europe 2015 baseline is taken as it is without changes. Few modifications in the mobility sector are implemented instead. Benini et al. [52] analyzed the Italian mobility sector identifying an electricity demand from the mobility sector in 2015 equal to 0.87 TWh. The same report also provides the hourly distribution of electricity demand from electric mobility (shown in Figure 3). This profile is assumed to be the same for each day of the year.

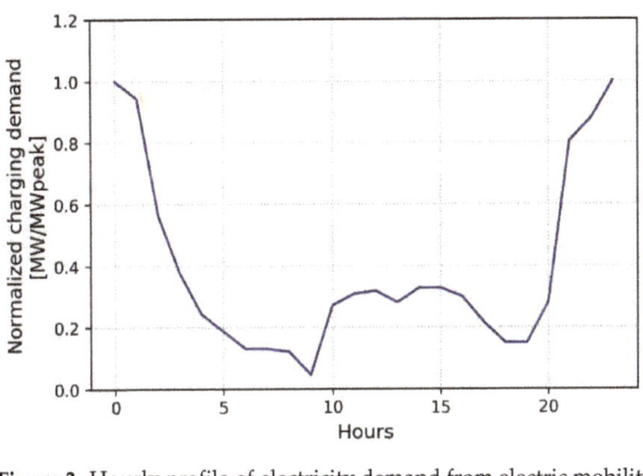

Figure 3. Hourly profile of electricity demand from electric mobility.

In 2015, Italy imported 43.7 TWh of electricity [53]. This is modeled in EnergyPLAN through the import component that allows setting the total amount of import and its hourly distribution during the year.

The Italian EnergyPLAN Baseline 2015 is validated comparing the final CO_2 output emissions with the amount estimated by different references shown in Figure 4. These values span from a minimum of 330.7 t of CO_2 emissions reported by the International Energy Agency [62] to a maximum of 339.95 estimated by the UNFCCC [63]. The value obtained after creating and running the EnergyPLAN Baseline 2015 input file is equal to 334.7 t of CO_2 emissions and thus fully included in the range given by the different analyzed sources.

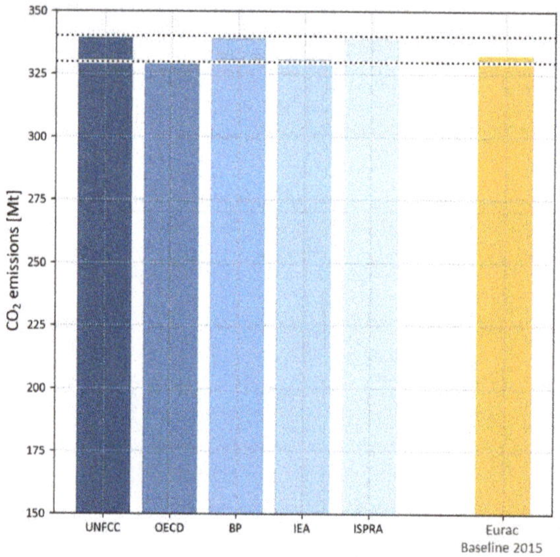

Figure 4. Comparison of annual CO_2 emissions between the value obtained through simulation in the Baseline 2015 and those given by other different sources: UNFCCC [63], OECD [64], BP [65], IEA [62] and ISPRA [66].

3.2. PNIEC 2030 Scenario

The second step is the definition of the EnergyPLAN input file for the Italian integrated energy and climate action plan [3]. All the input values are taken from the Italian integrated energy and climate action plan document. The resulting scenario is called PNIEC 2030 scenario. The major transformations regard the electricity sector in which an increase of renewable energy penetration is foreseen, mainly photovoltaics and wind power technology, capable of covering 55% of gross final electricity consumption. The same indicator for the baseline year is equal to 34%. Moreover, the PNIEC 2030 scenario implements 40 GWh of stationary batteries. It is characterized by 10% penetration of electric mobility and an increase of the consumption of advanced biomethane in the transport sector. It also foresees an increase of energy efficiency of buildings equal to 15% by 2030.

3.3. Decision Variables and Assumptions

The third step is the definition of the decision variables and their ranges. This information is provided by Table 3. The choice of the decision variables is driven by the technologies which the Italian integrated energy and climate action plan concentrates on. They are rooftop PV, utility scale PV, wind power, lithium-ion batteries, power-to-gas, advanced bio-methane in the transport sector and energy efficiency of buildings and heat pumps in the heating sector.

Table 3. List of decision variables and their lower $DV_i^{(L)}$ and upper $DV_i^{(U)}$ bounds.

Decision Variables	*CurrentValue (2015),* $DV_i^{(L)}$	*MaximumPotential* $DV_i^{(U)}$
Residential PV (GW)	15	120
Utility scale PV (GW)	4	70
Wind power (GW)	9	49
Lithium-ion batteries (GWh)	0	600
Power to gas, H_2 produced (%)	0	15
Power to gas, Electrolyzer max capacity (GW)	0	30
Advanced biomethane (TWh)	3	15
Energy efficiency of buildings (%)	0	75
Heat pumps (%)	0	100

The maximum potential is determined for each VRES through an analysis of the technical availability of installable capacity. For other variables such as lithium-ion batteries and power-to-gas electrolyzers, a number large enough to perform the optimization and small enough not to enlarge the domain of the optimization too much by increasing the computation time without an added value was selected. The assumptions on the potential are the following:

- Solar PV. For residential rooftop PV a couple of studies, Taylor et al. [67] and Vartiainen et al. [68], together with internal studies of Eurac research based on the Solar Tyrol project [69] identified a share of 2 kW per person as the maximum rooftop PV potential. Considering roughly 60 million inhabitants in Italy the final maximum potential for residential PV is assumed to be 120 GW. For what concerns utility scale PV, the maximum potential is taken from a study of the Energy Strategy Group [70] which studied the potential for the Italian territory evaluating the brownfield sites and unutilized rural areas. The overall estimated value is equal to 70 GW. An analysis of the land use for solar power by 2030 was realized by F. Mancini et al. [71]. They demonstrated how the use of 10% of the soil already consumed could be sufficient to achieve the set objectives by 2030.
- On-shore wind power. Hoefnagels et al. [72] in the framework of the RE-shaping project estimated a maximum potential of 49 GW for Italy.
- Lithium-ion batteries. The maximum potential is evaluated through a series of simulations. A value above 600 GWh brings higher costs without any benefits in terms of renewable energy integration.
- Power to gas is managed through two variables: the produced hydrogen and the capacity of the electrolyzer. The produced hydrogen maximum potential is assumed to be 15% of the overall natural gas consumption. The maximum size of the electrolyzer is taken high enough to exploit the full potential of power-to-gas and low enough to contain the domain of the optimization problem.
- The installation of heat pumps is allowed only after a deep energy refurbishment of buildings. This decision variable is the percentage of the overall buildings that switched their heating system from boilers to heat pumps. For this reason, its maximum potential is 100%.
- The energy efficiency of buildings: the potential of energy efficiency by means of passive solutions is bound to the energy efficiency cost curve and is equal to 75%. The energy efficiency cost curve and the way it is implemented in the source code of EPLANopt is explained in a previous publication [21].

Other assumptions in the heating sector are: domestic hot water (DHW) in buildings reached by district heating network is supplied by district heating itself. For the other individual buildings, only the heating demand can be reduced through energy efficiency refurbishment, while DHW share instead is not influenced by energy efficiency and kept constant. The optimization decides which share of renovated buildings should install heat pumps. In the individual sector, at the increase of the energy efficiency share, heat pumps substitute different types of boilers with the following priorities: (1) coal boilers, (2) oil boilers, (3) electric boilers, (4) natural gas boilers and (5) biomass boilers. A more detailed explanation of the modeling approach taking into account energy efficiency of buildings is provided by previous studies [21,23].

The costs related to electric mobility is considered in the model. Starting from the study of the Enel foundation [73], the cost of the electric vehicles (EV) infrastructure is estimated for different shares of penetration of battery electric vehicles (BEV). Figure 5 shows these costs which increase linearly at the increasing of the EV penetration. These costs take into account the infrastructural costs of electric mobility for urban and sub-urban areas and the costs for different type of charging stations.

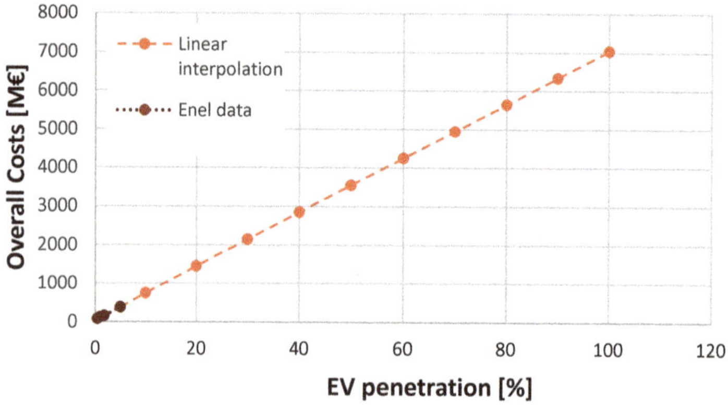

Figure 5. Infrastructural costs of electric mobility: data from the Enel foundation [73] and linear interpolation to extend these costs at higher penetration of EV.

The model also considers a decrease in energy consumption from the industry sector. Starting from the historical data provided by the Odyssee-Mure database [74] a logarithmic interpolation is implemented to inspect the business as a usual scenario for energy efficiency in the industry sector. The value is integrated in the model which therefore considers 9% of energy efficiency in the industry sector with respect to the energy consumption of 2015, the year of the baseline (See Figure 6).

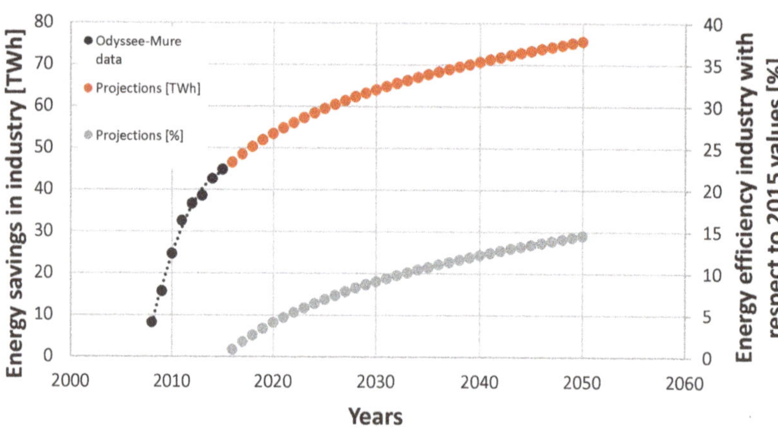

Figure 6. Energy efficiency in the industry sector: historical data and projections to 2050 through a logarithmic interpolation.

Additional assumptions are the following: (i) constant demographic situation from 2015 to 2030; (ii) export price for electricity equal to 35 €/MWh [75]; (iii) import price for electricity equal to 45 €/MWh [75]; (iv) emission factor of imported electricity equal to 270 kg/MWh [76] in 2030. (v) The electricity generation from river hydro is assumed to slightly increase from 23.1 to 24.8 TWh, while the generation from dammed hydro remain constant [3]. The generation from geothermal increases from 6.0 to 7.1 TWh [3]. The electricity production from biomass power plants decreases from 19.4 to 15.7 TWh [3]. (vi) Transport demand in terms of driven km and modal split is assumed constant, (vii) power to gas costs are those of the electrolyzer installed capacity (400 €/kW, lifetime = 15, O&M = 3% of the investment cost). Other costs of the power to gas flexibility option are not included because the injection of hydrogen goes directly into the existing gas grid.

3.4. Optimization Problems Definition

The fourth step is the definition of the optimization problems interesting to the current analysis. Two optimization problems are formulated to study the impact of different levels of penetration of electric mobility. The electric mobility share is set as an exogenous variable and the optimization problem is run to find the optimal solutions under the considered assumptions. The evaluated optimization problems are the following:

- One case considering 10% electric mobility
- One case considering 20% electric mobility

Different levels of penetration of electric mobility are considered by simply introducing some conversion factors in kWh/100 km for each fuel based vehicle [77] assuming the transport demand in terms of driven km is constant. The values used are the following: 52.78 kWh/100 km for petrol fuelled cars, 46.11 kWh/100 km for gasoil fuelled cars and 13.61 kWh/100 km for electric vehicles. The use of these conversion factors allows the evaluation of the electricity demand generated from electric mobility. This value is set in EnergyPLAN in order to evaluate the hourly operational simulation and the impacts on the overall energy system.

4. Results

The results of the two optimization problems are depicted in Figure 7. It shows the Baseline 2015, the PNIEC 2030 and the Pareto fronts of optimal solutions for the cases with 10 and 20% electric mobility penetration. One scenario is selected on the 20% electric mobility Pareto front with the same costs of the PNIEC 2030 scenario. It is named Advanced 2030. The graph allows the following remarks:

(i) The PNIEC 2030 scenario produces a relevant reduction of CO_2 emissions compared to the Baseline 2015. This reduction is in line with the CO_2 emissions reduction target in 2030. The PNIEC 2030 scenario is found by Italian authorities through an optimization process but the assumptions on costs and efficiencies of the energy system components are not public. Therefore, it is important to validate the model and the PNIEC 2030 scenario. This result allows this validation which is added to the validation of the Baseline 2015 on CO_2 emissions.
(ii) The PNIEC 2030 scenario, characterized by 10% electric mobility penetration, is almost placed on the Pareto front characterized by 10% electric mobility. Thus, it is a solution close to the optimum. As already mentioned, the PNIEC 2030 scenario is found as a result of an optimization process by Italian authorities. In this study, the difference between the PNIEC 2030 scenario and the Pareto front with 10% electric mobility can be a consequence of different costs assumptions.
(iii) With the same cost of the PNIEC 2030 scenario it is possible to reach higher CO_2 emissions reduction by selecting a solution on the Pareto front with 20% electric mobility. The Advanced 2030 scenario showed that at the same costs of the PNIEC 2030 there are solutions which further reduce the CO_2 emissions. In this case the Advanced 2030 scenario produces a further reduction of 10%.

(iv) Another consideration that needs to be done is on the impact of electric mobility. The increase of electric mobility from 10 to 20% together with the optimal energy mix found by the optimization algorithm allows a further reduction of the CO_2 emissions.

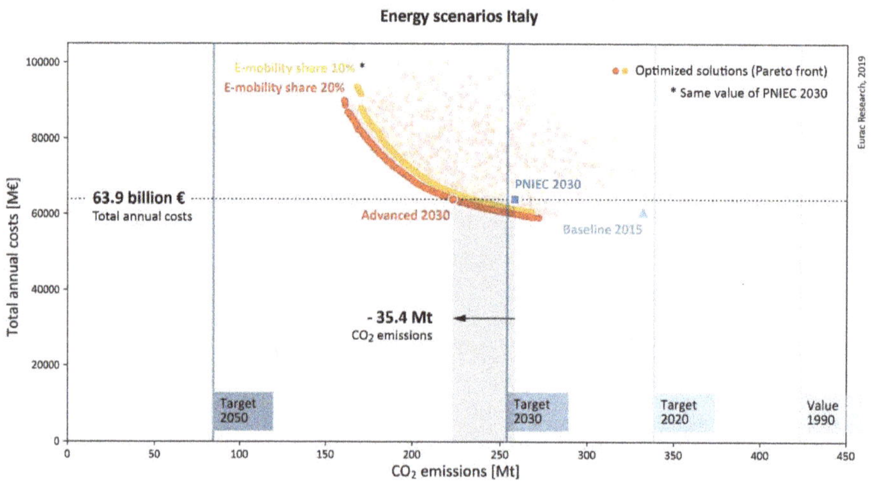

Figure 7. Total annual costs and CO_2 emissions for the evaluated scenarios: Baseline 2015, PNIEC 2030, Pareto fronts and Advanced 2030.

The analysis of the results focuses on the Advanced 2030 scenario and on the comparison with the PNIEC 2030 and Baseline 2015. In particular, the comparison of the scenarios with similar costs (PNIEC 2030 and Advanced 2030) allows the study of the best choices to decarbonize the energy system.

Table 4 shows the characteristics of the considered scenarios with the comparison of the values of the main decision variables. This shows how the Advanced 2030 scenario, when compared to the PNIEC 2030 scenario, presents a higher installed capacity of VRES and higher energy efficiency of buildings, with a lower value of capacity of stationary batteries and advanced biomethane production. Another difference is the penetration of electric mobility: 10% for the PNIEC 2030 scenario and 20% for the Advanced 2030 scenario. This results in an overall aggregated capacity of batteries in the electric vehicles which is higher in the Advanced 2030 scenario. The value of the installed capacity of batteries for electric vehicles is calculated to give an idea of the overall size. The assumptions for this calculation are the following: 39 million cars are considered and an average size of 50 kWh battery per car [78].

Table 4. Values of the main technologies in the different scenarios.

Scenarios	PV	Wind Power	Stationary Batteries	Batteries of EV	Advanced Biomethane	Energy Efficiency of Buildings
Baseline 2015	19 GW	9 GW	0 GWh	0 GWh	3 TWh	0%
PNIEC 2030	59 GW	23 GW	40 GWh	200 GWh	15 TWh	15%
Advanced 2030	86 GW	48 GW	0 GWh	400 GWh	3 TWh	30%

Figure 8 shows the evolution of the electricity demand in a week in summer and in winter for the three solutions: Baseline 2015, PNIEC 2030 and Advanced 2030. It is possible to observe the increase of electricity demand due to energy efficiency of buildings and the substitution of electric boilers with heat pumps and electric mobility. Power-to-gas electricity demand is equal to zero because the optimization does not choose it to decarbonize the energy system. The main reason is the limited amount of available over-generation from renewables that can be used by power-to-gas for the hydrogen generation. This over-generation can be noted in Figure 9 which shows the electricity generation from the different sources. The over-generation is present mostly in summer and partly exploited by the existing pumped hydro storage systems.

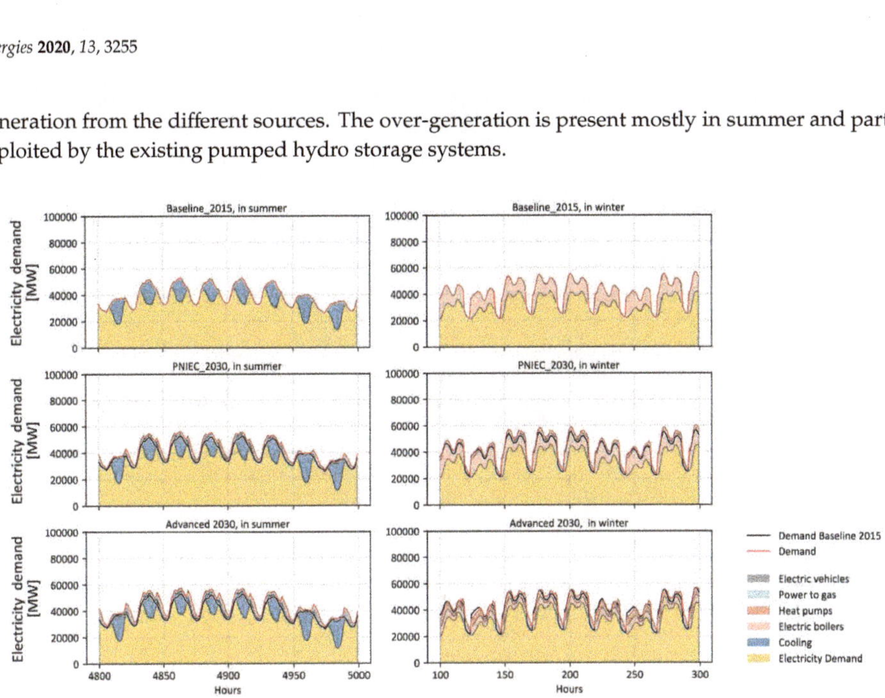

Figure 8. Hourly electricity demand: different contributions for two weeks of the year (one in summer and one in winter) for the scenarios Baseline 2015, PNIEC 2030 and Advanced 2030.

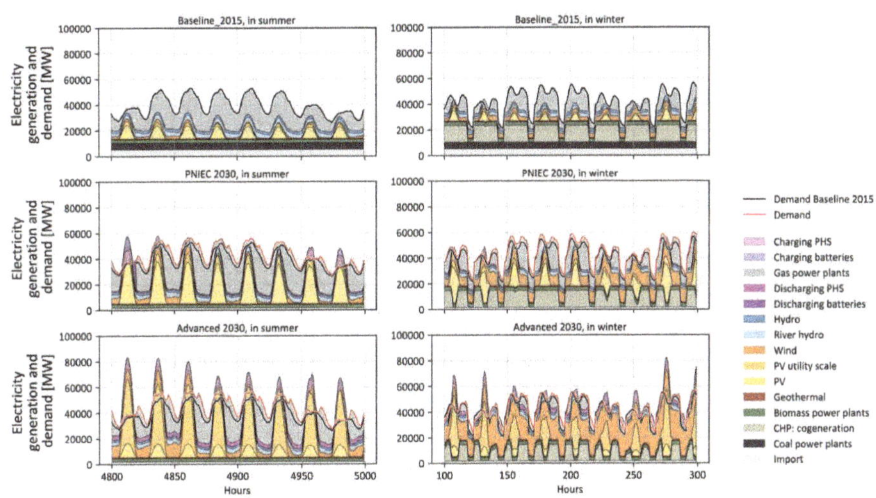

Figure 9. Hourly electricity dispatch for two weeks of the year (one in summer and one in winter) for the scenarios Baseline 2015, PNIEC 2030 and Advanced 2030.

Figure 10 shows the annual electricity generation from different sources for the tree different scenarios. The electricity generation from coal disappears in the PNIEC 2030 and Advanced 2030 scenarios due to the coal phase-out foreseen by the Italian integrated energy and climate action plan can be noted. The Advanced 2030 scenario is characterized by higher installed power of VRES and therefore a lower generation from gas power plants. The electricity demand is equal to 315.7 TWh in the Baseline 2015, 340 TWh in the PNIEC 2030 scenario and 343.6 TWh in the Advanced 2030 scenario. This is due to the higher share of electric mobility and energy efficiency of buildings with heat pumps installation in the Advanced 2030 scenario.

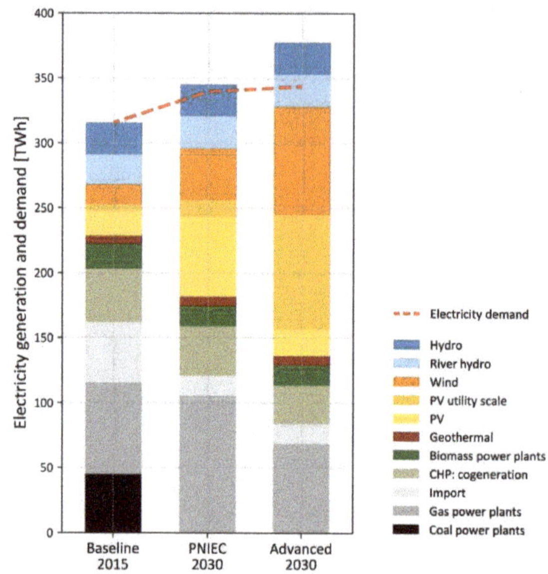

Figure 10. Annual electricity generation for the scenarios Baseline 2015, PNIEC 2030 and Advanced 2030.

Figure 11 shows the annual energy consumption for the scenarios Baseline 2015, PNIEC 2030 and Advanced 2030 with the percentages of renewables for each sector. Compared to the PNIEC 2030 scenario, the Advanced 2030 scenario is characterized by higher renewable energy sources (RES) share in the electricity sector coupled to an electrification of the heating and transport sector.

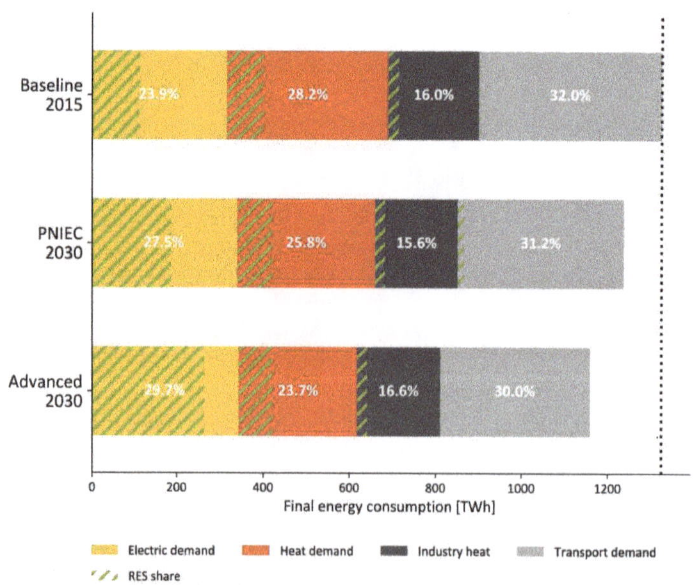

Figure 11. Annual energy consumption in electricity, heat, industry and transport sectors for the scenarios Baseline 2015, PNIEC 2030 and Advanced 2030.

Figure 12 shows the total annual costs structure in the three considered scenarios: Baseline 2015, PNIEC 2030 and Advanced 2030. The Advanced 2030 compared to the PNIEC 2030 shows a reduction of the fossil fuel costs and an increase in the costs of VRES installed power and energy efficiency of buildings.

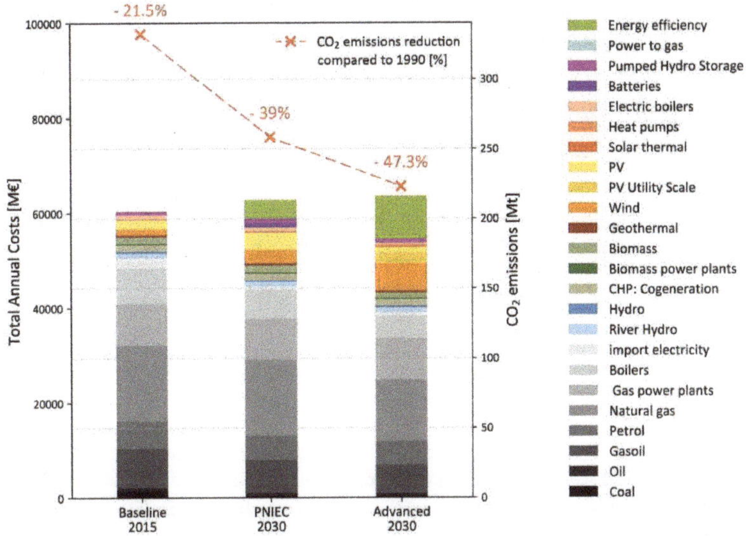

Figure 12. Total annual costs for the scenarios Baseline 2015, PNIEC 2030 and Advanced 2030.

Figure 13 allows a consideration on the nature of the costs which change in these three scenarios. Moving from the Baseline 2015 to the Advanced 2030 scenario, the costs for fossil fuels decrease and the possible domestic value creation increases. This latter represents the possible investments in the territory which could boost the local economy.

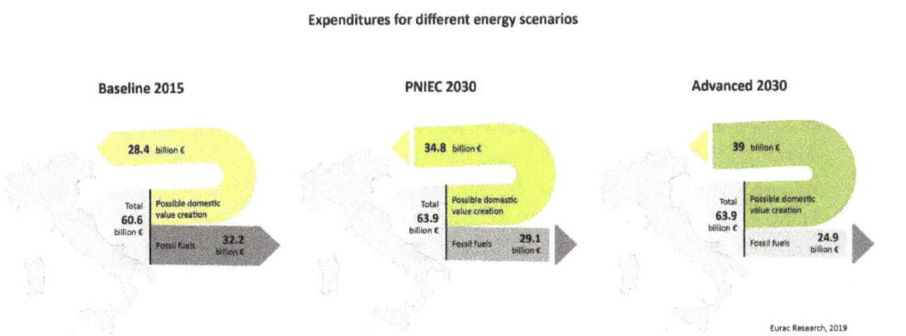

Figure 13. Subdivision of annual costs in the scenarios Baseline 2015, PNIEC 2030 and Advanced 2030.

In Section 2 the remaining carbon budget value is estimated for Italy. Figure 14 shows (i) the historical trend of CO_2 emissions of Italy from 1990 to 2018, (ii) the remaining carbon budget assuming a linear decrease from the value of 2018, (iii) the PNIEC 2030 and Advanced 2030 scenarios. The graph shows how the CO_2 emissions reduction in the PNIEC 2030 is not in line with the carbon budget limit. This is assuming a linear decrease of the CO_2 emissions. Following the trajectory of the PNIEC 2030 would require a drastic decrease of the emissions after 2030. The results show that the Advanced 2030 scenario improves the CO_2 emissions reduction compared to the PNIEC 2030, but it is still higher than the carbon budget limit, assuming a linear reduction.

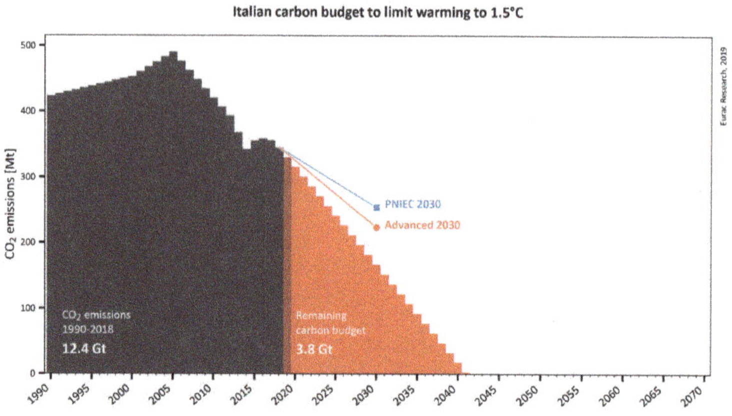

Figure 14. CO_2 emissions historical data and comparison between the PNIEC 2030, Advanced 2030 scenarios and the scenario defined by the remaining carbon budget.

This highlights the necessity of a more drastic transition. This can be achieved through the selection of an optimized scenario, among those obtained through the EPLANopt multi-objective method, with higher total annual costs and lower CO_2 emissions. Knowing the carbon budget limit, it is possible, thanks to the Pareto curve, to identify the scenario in line with the carbon budget. This scenario would require an increase of the total annual costs compared to the integrated energy and climate action plan scenario equal to about 25%.

5. Conclusions

The EPLANopt model allows the multi-objective investment optimization of the energy system. It is based on the EnergyPLAN model which considers an hourly time resolution and a sector-coupling approach. The EPLANopt energy system model is applied to the Italian case study to compare the Italian integrated energy and climate action plan (PNIEC 2030 scenario) with scenarios obtained through the optimization process.

Among the optimized scenarios, one with costs similar to the Italian integrated energy and climate action plan scenario is identified (Advanced 2030 scenario). It allows a further reduction of CO_2 emissions equal to 10% at the same annual costs. The integrated energy and climate action plan and the optimized scenario with the same costs were compared in the analysis of the results. The optimized scenario presents a higher share of renewables in the electricity sector together with a higher degree of electrification of the heating and transport sectors. It is important to underline that the difference between the Advanced 2030 scenario and the PNIEC 2030 scenario can arise from different costs assumptions.

The two scenarios were compared to climate-change scenarios through the use of the carbon budget concept. The results highlighted how both the integrated energy and climate action plan scenario and the Advanced 2030 scenario are far from the carbon budget limit. The results show that the Advanced 2030 scenario improves the CO_2 emissions reduction if compared to the integrated energy and climate action plan scenario, but it is not enough to meet the carbon budget limit. In order to meet this target, there is the need for a more drastic scenario in terms of CO_2 emissions reduction. Moving on to the Pareto front it is possible to select the first optimal scenario in line with the CO_2 emission reduction driven by the carbon budget limit, assuming a linear decrease, and with the lowest costs. This scenario requires an increase of the overall costs equal to 25% with respect to the integrated energy and climate action plan scenario.

The results also showed the economic opportunity represented by energy transition. The study of the nature of the costs showed how, moving from the Baseline 2015 to the Advanced 2030 scenario, the costs for fossil fuels decrease and the possible domestic value creation increases. This represents an opportunity of development of investments in the territory and a boost to local economy. Therefore, even if the study showed the need for an energy system characterized by higher costs compared to the Baseline 2015, it also highlighted the opportunities represented by the energy transition.

Future steps will concentrate on the integration of the bottom-up EPLANopt model with a top-down approach in order to evaluate the different policies needed to support the energy transition. This integration of a top-down into a bottom-up approach will also allow a more detailed analysis of the nature of the expenses and investments of the energy system evaluating the impacts on the territory and the local economy.

Author Contributions: Conceptualization, M.G.P., G.M., D.M., R.V. and W.S.; methodology, M.G.P., G.M., D.M., R.V. and W.S.; software, M.G.P.; validation, M.G.P.; writing—original draft preparation, M.G.P.; writing—review and editing, M.G.P., G.M., D.M. and R.V.; visualization, M.G.P.; supervision, G.M. and D.M.; project administration, D.M. and W.S.; funding acquisition, W.S. All authors have read and agreed to the published version of the manuscript.

Funding: This research was carried out through funding by the project FESR 1042 "Integrids" and institutional funding.

Acknowledgments: The research leading to these results has received funding from the EFRE/FESR Provincia autonoma di Bolzano-Alto Adige 2014–2020, under Project number FESR 1042 "Integrids". The authors wish to thank A. Segata (Eurac research) for taking care of the graphics.

Conflicts of Interest: The authors declare no conflict of interest.

References

1. Energy Planning—An Overview|ScienceDirect Topics. Available online: https://www.sciencedirect.com/topics/engineering/energy-planning (accessed on 29 October 2019).
2. Energy Strategy—European Commission. 2030. Available online: https://ec.europa.eu/energy/en/topics/energy-strategy-and-energy-union/2030-energy-strategy (accessed on 12 April 2018).
3. Ministero Dello Sviluppo Economico. PIANO NAZIONALE INTEGRATO PER L'ENERGIA E IL CLIMA. Available online: https://www.mise.gov.it/index.php/it/198-notizie-stampa/2039046-piano-nazionale-integrato-per-l-energia-e-il-clima-inviata-la-proposta-a-bruxelles (accessed on 1 April 2020).
4. Poncelet, K.; Delarue, E.; Six, D.; Duerinck, J.; D'haeseleer, W. Impact of the level of temporal and operational detail in energy-system planning models. *Appl. Energy* **2016**, *162*, 631–643. [CrossRef]
5. Aalborg University. EnergyPLAN|Advanced Energy Systems Analysis Computer Model. Available online: http://www.energyplan.eu/ (accessed on 10 February 2020).
6. Lund, H. Chapter 4—Tool: The EnergyPLAN Energy System Analysis Model. Renewable Energy Systems. 2014. Available online: https://www.sciencedirect.com/science/article/pii/B9780124104235000043?via%3Dihub (accessed on 10 February 2020).
7. Connolly, D.; Lund, H.; Mathiesen, B.V.; Østergaard, P.A.; Möller, B.; Nielsen, S.; Ridjan, I.; Hvelplund, F.; Sperling, K.; Karnøe, P.; et al. Smart Energy Systems: Holistic and Integrated Energy Systems for the era of 100% Renewable Energy. 2013. Available online: https://vbn.aau.dk/en/publications/smart-energy-systems-holistic-and-integrated-energy-systems-for-t (accessed on 10 February 2020).
8. Mathiesen, B.V.; Lund, H.; Connolly, D.; Wenzel, H.; Østergaard, P.A.; Möller, B.; Nielsen, S.; Ridjan, I.; Karnøe, P.; Sperling, K.; et al. Smart Energy Systems for coherent 100% renewable energy and transport solutions. *Appl. Energy* **2015**, *145*, 139–154. [CrossRef]
9. Nastasi, B.; Lo Basso, G. Hydrogen to link heat and electricity in the transition towards future Smart Energy Systems. *Energy* **2016**, *110*, 5–22. [CrossRef]
10. Prina, M.G.; Cozzini, M.; Garegnani, G.; Moser, D.; Filippi Oberegger, U.; Vaccaro, R.; Sparber, W. Smart energy systems applied at urban level: The case of the municipality of Bressanone-Brixen. *Int. J. Sustain. Energy Plan. Manag.* **2016**, *10*, 33–52.

11. Bramstoft, R.; Skytte, K. Decarbonizing Sweden's energy and transportation system by 2050. *Int. J. Sustain. Energy Plan. Manag.* **2017**, *14*, 3–20.
12. Ben Amer, S.; Bramstoft, R.; Balyk, O.; Nielsen, P.S. Modelling the future low-carbon energy systems-case study of greater Copenhagen, Denmark. *Int. J. Sustain. Energy Plan. Manag.* **2019**, *24*, 21–32.
13. Heinisch, V.; Göransson, L.; Odenberger, M.; Johnsson, F. Interconnection of the electricity and heating sectors to support the energy transition in cities. *Int. J. Sustain. Energy Plan. Manag.* **2019**, *24*, 57–66.
14. Pavičević, M.; Mangipinto, A.; Nijs, W.; Lombardi, F.; Kavvadias, K.; Jiménez Navarro, J.P.; Colombo, E.; Quoilin, S. The potential of sector coupling in future European energy systems: Soft linking between the Dispa-SET and JRC-EU-TIMES models. *Appl. Energy* **2020**, *267*, 115100. [CrossRef]
15. Lund, H.; Østergaard, P.A.; Connolly, D.; Ridjan, I.; Mathiesen, B.V.; Hvelplund, F.; Thellufsen, J.Z.; Sorknæs, P. Energy Storage and Smart Energy Systems. *Int. J. Sustain. Energy Plan. Manag.* **2016**, *11*, 3–14.
16. Lund, H. EnergyPLAN. 2015. Available online: www.EnergyPLAN.eu (accessed on 10 February 2020).
17. Documentation|EnergyPLAN. Available online: https://www.energyplan.eu/training/documentation/ (accessed on 24 January 2019).
18. Batas Bjelić, I.; Rajaković, N. Simulation-based optimization of sustainable national energy systems. *Energy* **2015**, *91*, 1087–1098. [CrossRef]
19. Mahbub, M.S.; Cozzini, M.; Østergaard, P.A.; Alberti, F. Combining multi-objective evolutionary algorithms and descriptive analytical modelling in energy scenario design. *Appl. Energy* **2016**, *164*, 140–151. [CrossRef]
20. Mahbub, M.S.; Viesi, D.; Crema, L. Designing optimized energy scenarios for an Italian Alpine valley: The case of Giudicarie Esteriori. *Energy* **2016**, *116*, 236–249. [CrossRef]
21. Prina, M.G.; Cozzini, M.; Garegnani, G.; Manzolini, G.; Moser, D.; Filippi Oberegger, U.; Pernetti, R.; Vaccaro, R.; Sparber, W. Multi-objective optimization algorithm coupled to EnergyPLAN software: The EPLANopt model. *Energy* **2018**, *149*, 213–221. [CrossRef]
22. Prina, M.G.; Fanali, L.; Manzolini, G.; Moser, D.; Sparber, W. Incorporating combined cycle gas turbine flexibility constraints and additional costs into the EPLANopt model: The Italian case study. *Energy* **2018**, *160*, 33–43. [CrossRef]
23. Prina, M.G.; Moser, D.; Vaccaro, R.; Sparber, W. EPLANopt optimization model based on EnergyPLAN applied at regional level: The future competition on excess electricity production from renewables. *Int. J. Sustain. Energy Plan. Manag.* **2020**, *27*, 35–50.
24. Garegnani, G.; Prina, M.G.; Vaccaro, R.; Cozzini, M.; Filippi Oberegger, U.; Moser, D. EPLANopt: EnergyPLAN Optimization Library. Available online: https://gitlab.inf.unibz.it/URS/EPLANopt (accessed on 2 February 2020).
25. The Paris Agreement|UNFCCC. Available online: https://unfccc.int/process-and-meetings/the-paris-agreement/the-paris-agreement (accessed on 20 April 2020).
26. Rocco, M.; Rady, Y.; Colombo, E. Soft-linking bottom-up energy models with top-down input-output models to assess the environmental impact of future energy scenarios. *Model. Meas. Control. C* **2018**, *79*, 103–110. [CrossRef]
27. Tuladhar, S.D.; Yuan, M.; Bernstein, P.; Montgomery, W.D.; Smith, A. A top-down bottom-up modeling approach to climate change policy analysis. *Energy Econ.* **2009**, *31*, S223–S234. [CrossRef]
28. Nikas, A.; Doukas, H.; Papandreou, A. A detailed overview and consistent classification of climate-economy models. In *Understanding Risks and Uncertainties in Energy and Climate Policy: Multidisciplinary Methods and Tools for a Low Carbon Society*; Springer International Publishing: Berlin/Heidelberg, Germany, 2018; pp. 1–54. ISBN 9783030031527.
29. Prina, M.G.; Manzolini, G.; Moser, D.; Nastasi, B.; Sparber, W. Classification and challenges of bottom-up energy system models—A review. *Renew. Sustain. Energy Rev.* **2020**, *129*, 109917. [CrossRef]
30. Herbst, A.; Toro, F.; Reitze, F.; Jochem, E. Introduction to Energy Systems Modelling. *Statistics* **2012**, *148*, 111–135. [CrossRef]
31. Prina, M.G.; Lionetti, M.; Manzolini, G.; Sparber, W.; Moser, D. Transition pathways optimization methodology through EnergyPLAN software for long-term energy planning. *Appl. Energy* **2019**, *235*, 356–368. [CrossRef]
32. Connolly, D.; Lund, H.; Mathiesen, B.V. Smart Energy Europe: The technical and economic impact of one potential 100% renewable energy scenario for the European Union. *Renew. Sustain. Energy Rev.* **2016**, *60*, 1634–1653. [CrossRef]

33. Ćosić, B.; Krajačić, G.; Duić, N. A 100% renewable energy system in the year 2050: The case of Macedonia. *Energy* **2012**, *48*, 80–87. [CrossRef]
34. Connolly, D.; Lund, H.; Mathiesen, B.V.; Leahy, M. The first step towards a 100% renewable energy-system for Ireland. *Appl. Energy* **2011**, *88*, 502–507. [CrossRef]
35. Fernandes, L.; Ferreira, P. Renewable energy scenarios in the Portuguese electricity system. *Energy* **2014**, *69*, 51–57. [CrossRef]
36. Lund, H.; Mathiesen, B.V. Energy system analysis of 100% renewable energy systems—The case of Denmark in years 2030 and 2050. *Energy* **2009**, *34*, 524–531. [CrossRef]
37. Connolly, D.; Lund, H.; Mathiesen, B.V.; Pican, E.; Leahy, M. The technical and economic implications of integrating fluctuating renewable energy using energy storage. *Renew. Energy* **2012**, *43*, 47–60. [CrossRef]
38. Connolly, D.; Lund, H.; Mathiesen, B.V.; Leahy, M. Modelling the existing Irish energy-system to identify future energy costs and the maximum wind penetration feasible. *Energy* **2010**, *35*, 2164–2173. [CrossRef]
39. Komušanac, I.; Ćosić, B.; Duić, N. Impact of high penetration of wind and solar PV generation on the country power system load: The case study of Croatia. *Appl. Energy* **2016**, *184*, 1470–1482. [CrossRef]
40. Jääskeläinen, J.; Veijalainen, N.; Syri, S.; Marttunen, M.; Zakeri, B. Energy security impacts of a severe drought on the future Finnish energy system. *J. Environ. Manag.* **2018**, *217*, 542–554. [CrossRef]
41. Bhuvanesh, A.; Jaya Christa, S.T.; Kannan, S.; Karuppasamy Pandiyan, M. Aiming towards pollution free future by high penetration of renewable energy sources in electricity generation expansion planning. *Futures* **2018**, *104*, 25–36. [CrossRef]
42. Novosel, T.; Pukšec, T.; Krajačić, G.; Duić, N. Role of District Heating in Systems with a High Share of Renewables: Case Study for the City of Osijek. *Energy Procedia* **2016**, *95*, 337–343. [CrossRef]
43. Groppi, D.; Astiaso Garcia, D.; Lo Basso, G.; De Santoli, L. Synergy between smart energy systems simulation tools for greening small Mediterranean islands. *Renew. Energy* **2019**, 515–524. [CrossRef]
44. Alves, M.; Segurado, R.; Costa, M. Increasing the penetration of renewable energy sources in isolated islands through the interconnection of their power systems. The case of Pico and Faial islands, Azores. *Energy* **2019**, *182*, 502–510. [CrossRef]
45. Marczinkowski, H.M.; Østergaard, P.A. Evaluation of electricity storage versus thermal storage as part of two different energy planning approaches for the islands SamsØ and Orkney. *Energy* **2019**, *175*, 505–514. [CrossRef]
46. Konak, A.; Coit, D.W.; Smith, A.E. Multi-objective optimization using genetic algorithms: A tutorial. *Reliab. Eng. Syst. Saf.* **2006**, *91*, 992–1007. [CrossRef]
47. Deb, K.; Agrawal, S.; Pratap, A.; Meyarivan, T. *A Fast Elitist Non-Dominated Sorting Genetic Algorithm for Multi-Objective Optimization: NSGA-II*; Springer: Berlin/Heidelberg, Germany, 2000; pp. 849–858.
48. Deb, K.; Pratap, A.; Agarwal, S.; Meyarivan, T. A fast and elitist multiobjective genetic algorithm: NSGA-II. *IEEE Trans. Evol. Comput.* **2002**, *6*, 182–197. [CrossRef]
49. Rogelj, J.; Forster, P.M.; Kriegler, E.; Smith, C.J.; Séférian, R. Estimating and tracking the remaining carbon budget for stringent climate targets. *Nature* **2019**, *571*, 335–342. [CrossRef]
50. Heat Roadmap Europe. Available online: http://www.heatroadmap.eu/ (accessed on 10 April 2018).
51. Energy Models—Heat Roadmap Europe. Available online: https://heatroadmap.eu/energy-models/ (accessed on 25 February 2019).
52. Benini, M.; Celaschi, S.; Colzi, F.; De Nigris, M.; Gianinoni, I.M.; Girardi, P.; Martinotti, V.; Micolano, E.; Pirovano, G.; Riva, G.M.; et al. E ... Muoviti! Mobilità Elettrica a Sistema. 2013. Available online: http://www.selidori.com/tech/00000-04999/724-MtMJB.pdf (accessed on 10 February 2020).
53. Terna—Transparency Report. Available online: http://www.terna.it/en-gb/sistemaelettrico/transparencyreport.aspx (accessed on 17 January 2018).
54. GSE, Statistiche. Available online: https://www.gse.it/dati-e-scenari/statistiche (accessed on 10 April 2018).
55. Terna—Statistical Data. Available online: http://www.terna.it/en-gb/sistemaelettrico/statisticheeprevisioni/datistatistici.aspx (accessed on 10 April 2018).
56. Noussan, M.; Roberto, R.; Nastasi, B. Performance Indicators of Electricity Generation at Country Level—The Case of Italy. *Energies* **2018**, *11*, 650. [CrossRef]
57. Wagner, T.; Themeßl, M.; Schüppel, A.; Gobiet, A.; Stigler, H.; Birk, S. Impacts of climate change on stream flow and hydro power generation in the Alpine region. *Environ. Earth Sci.* **2017**, *76*, 1–22. [CrossRef]

58. Prina, M.G.; Garegnani, G.; Moser, D.; Oberegger, U.F.; Vaccaro, R.; Sparber, W.; Gazzani, M.; Manzolini, G. Economic and environmental impact of photovoltaic and wind energy high penetration towards the achievement of the Italian 20-20-20 targets. In Proceedings of the 2015 10th International Conference on Ecological Vehicles and Renewable Energies, EVER 2015, Monte-Carlo, Monaco, 31 March–2 April 2015.
59. Terna Impianti di Generazione. 2015. Available online: https://www.terna.it/en/electric-system/statistical-data-forecast/evolution-electricity-market (accessed on 10 February 2020).
60. Mazzoni, S.; Ooi, S.; Nastasi, B.; Romagnoli, A. Energy storage technologies as techno-economic parameters for master-planning and optimal dispatch in smart multi energy systems. *Appl. Energy* **2019**, *254*, 113682. [CrossRef]
61. Noussan, M.; Nastasi, B. Data Analysis of Heating Systems for Buildings—A Tool for Energy Planning, Policies and Systems Simulation. *Energies* **2018**, *11*, 233. [CrossRef]
62. CO_2 Emissions from Fuel Combustion 2019—Analysis—IEA. Available online: https://www.iea.org/reports/co2-emissions-from-fuel-combustion-2019 (accessed on 16 June 2020).
63. UNFCCC, Greenhouse Gas Inventory Data—Comparison by Category. Available online: http://di.unfccc.int/comparison_by_category (accessed on 21 February 2019).
64. OECD iLibrary|Air and GHG Emissions. Available online: https://www.oecd-ilibrary.org/environment/air-and-ghg-emissions/indicator/english_93d10cf7-en (accessed on 21 February 2019).
65. Dudley, B. *BP CO_2 Emissions—BP Statistical Review of World Energy 2018*; BP Statistical Review: London, UK, 2018.
66. ISPRA. National Inventory Report 2017. In *ItalianGreenhouseGas Inventory1990–2015*; ISPRA: Rome, Italy, 2017.
67. Taylor, N.; Szabo, S.; Kona, A.; Melica, G.; Huld, T.; Jaeger-Waldau, A.; Ossenbrick, H. Deployment Pathways for Photovoltaics in the EU Towards 2020: Comparing Economic Factors with Policies at Municipal Level. In Proceedings of the 31st European Photovoltaic Solar Energy Conference and Exhibition, Hamburg, Germany, 14–18 September 2015.
68. Vartiainen, E.; Masson, G.; Breyer, C. *PV LCOE in Europe 2014–30*; European Photovoltaic Technology Platform: Munich, Germany, 2015.
69. Solar Tirol—WebGIS. Available online: http://webgis.eurac.edu/solartirol/ (accessed on 17 January 2018).
70. Renewable Energy Report 2019|Energy & Strategy Group. Available online: http://www.energystrategy.it/area-riservata/rer-2019.html (accessed on 12 February 2020).
71. Mancini, F.; Nastasi, B. Solar Energy Data Analytics: PV Deployment and Land Use. *Energies* **2020**, *13*, 417. [CrossRef]
72. Re-shaping Project. Available online: http://www.reshaping-res-policy.eu/ (accessed on 14 December 2017).
73. Azzone, G.; Piercesare, S.; Zaninelli, D. APRIAMO LA STRADA AL TRASPORTO ELETTRICO NAZIONALE. 2017. Available online: https://www.enelfoundation.org/content/dam/enel-found/topic-download/Apriamo%20la%20strada%20al%20trasporto%20elettrico%20nazionale.pdf (accessed on 10 February 2020).
74. Energy Efficiency Trends & Policies|ODYSSEE-MURE. Available online: https://www.odyssee-mure.eu/ (accessed on 21 February 2020).
75. GME—Gestore dei Mercati Energetici SpA. Available online: https://www.mercatoelettrico.org/it/ (accessed on 21 February 2020).
76. Koffi, B.; Cerutti, A.; Duerr, M.; Iancu, A.; Kona, A.; Janssens-Maenhout, G. *CoM Default Emission Factors for the Member States of the European Union—Version 2017|Knowledge for Policy*; European Commission, Joint Research Centre (JRC): Ispra, Italy, 2017.
77. Thiel, C.; Perujo, A.; Mercier, A. Cost and CO_2 aspects of future vehicle options in Europe under new energy policy scenarios. *Energy Policy* **2010**, *38*, 7142–7151. [CrossRef]
78. ACI Studi e Ricerche—Dati e Statistiche. Available online: http://www.aci.it/laci/studi-e-ricerche/dati-e-statistiche.html (accessed on 18 May 2020).

energies

MDPI

Article

A New Bi-Level Optimisation Framework for Optimising a Multi-Mode Wave Energy Converter Design: A Case Study for the Marettimo Island, Mediterranean Sea

Mehdi Neshat [1], Nataliia Y. Sergiienko [2], Erfan Amini [3], Meysam Majidi Nezhad [4,*], Davide Astiaso Garcia [5], Bradley Alexander [1] and Markus Wagner [1]

1 Optimization and Logistics Group, School of Computer Science, The University of Adelaide, 5005 Adelaide, Australia; mehdi.neshat@adelaide.edu.au (M.N.); bradley.alexander@adelaide.edu.au (B.A.); markus.wagner@adelaide.edu.au (M.W.)

2 School of Mechanical Engineering, The University of Adelaide, 5005 Adelaide, Australia; nataliia.sergiienko@adelaide.edu.au

3 Coastal and offshore structures engineering group, School of Civil Engineering, University of Tehran, 13145-1384 Tehran, Iran; erfan.amini@ut.ac.ir

4 Department of Astronautics, Electrical and Energy Engineering (DIAEE), Sapienza University of Rome, 00184 Rome, Italy

5 Department of Planning, Design and Technology of Architecture, Sapienza University of Rome, 00197 Rome, Italy; davide.astiasogarcia@uniroma1.it

* Correspondence: meysam.majidinezhad@uniroma1.it

Received: 30 August 2020; Accepted: 15 October 2020; Published: 20 October 2020

check for updates

Abstract: To advance commercialisation of ocean wave energy and for the technology to become competitive with other sources of renewable energy, the cost of wave energy harvesting should be significantly reduced. The Mediterranean Sea is a region with a relatively low wave energy potential, but due to the absence of extreme waves, can be considered at the initial stage of the prototype development as a proof of concept. In this study, we focus on the optimisation of a multi-mode wave energy converter inspired by the CETO system to be tested in the west of Sicily, Italy. We develop a computationally efficient spectral-domain model that fully captures the nonlinear dynamics of a wave energy converter (WEC). We consider two different objective functions for the purpose of optimising a WEC: (1) maximise the annual average power output (with no concern for WEC cost), and (2) minimise the levelised cost of energy (LCoE). We develop a new bi-level optimisation framework to simultaneously optimise the WEC geometry, tether angles and power take-off (PTO) parameters. In the upper-level of this bi-level process, all WEC parameters are optimised using a state-of-the-art self-adaptive differential evolution method as a global optimisation technique. At the lower-level, we apply a local downhill search method to optimise the geometry and tether angles settings in two independent steps. We evaluate and compare the performance of the new bi-level optimisation framework with seven well-known evolutionary and swarm optimisation methods using the same computational budget. The simulation results demonstrate that the bi-level method converges faster than other methods to a better configuration in terms of both absorbed power and the levelised cost of energy. The optimisation results confirm that if we focus on minimising the produced energy cost at the given location, the best-found WEC dimension is that of a small WEC with a radius of 5 m and height of 2 m.

Keywords: bi-level optimisation method; evolutionary algorithms; renewable energy; wave energy converter; geometric parameters; power take-off; levelised cost of energy

1. Introduction

Renewable energy is the fastest-growing new energy source globally. As an example, in the United States, the growth rate of this technology increased by 100% between 2000 and 2018 [1]. On a global scale, renewable energy technologies produced 26.2% of the global electricity demand in 2018, and this is expected to climb to 45% by 2040 [1]. A large number of investigations have been applied in order to optimise various characteristics of renewable energy systems such as dealing with the uncertainty in renewable energy accessibility, support decision-making in the built environment [2] and the appropriation of energy storage operations for dampening the chaotic problems [3]. Among the different renewable energy sources, ocean wave energy is the cleanest, safest, most reliable and predictable source of renewable energy [4] with a power density significantly higher than that of solar and wind [5]. However, wave energy technology is not fully developed, and their commercial penetration is still shallow. This is because the costs involved in producing energy using ocean waves are currently much higher than those for other renewables [6]. Therefore, in the last decade, a large number of investigations have been carried out to optimise wave energy converter (WEC) design and dimensions [7–12], power generation settings (PTO) [13,14], and the position of WECs in a wave farm [15–19].

The wave energy resource around the globe has been divided into six major classes depending on the wave energy potential, directional and spectral characteristics, and extreme waves [20]. However, it has been noted [20] that while wave energy developers mainly target wave climates with the highest energy content (class 5 and 6), other resource classes can provide additional benefits to the technology development. For example, the Mediterranean Sea due to its enclosed nature has low wave power availability [21–23] and belongs to the resource class 1 but the absence of extreme wave heights makes this region attractive for the initial prototype testing.

Shape optimisation is important for all types of wave energy conversion systems, including oscillating water columns [24]), and over-topping designs [25]. The majority of efforts, to date, have been restricted to analysing a few specific shapes. The main reason for this is that the computational demands of searching and evaluating all feasible designs are high. Vantorre et al. [26] evaluated and compared the performance of a set of geometries for a heaving point absorber in a Belgian coastal area. These included a hemisphere and some conical geometries. The authors proposed that the best power efficiency was related to a cylindrical extension with a 90° cone. Later work by Goggins and Finnegan [27] contemplated a vertical cylinder of various heights and radii under wave conditions off the west coast of Ireland. They found that the most substantial significant heave velocity response was that of a trimmed cylinder with a hemisphere joined to its foundation, with a whole draft to the aspect ratio of 2.5. In other recent publications, a wide range of asymmetrical buoy designs has been proposed, including a concave buoy face which is better able to absorb power than a flat or convex model [28]. Another recommendation of a surface described by bi-cubic B-spline [29] outperforms conventional WEC models. However, in these studies, the main objective was to maximise the harnessed power of the WEC, and the authors did not consider the design, installation and maintenance costs of these asymmetric converters.

Other work has taken into account the trade-offs between absorbed power and the cost of building and deploying the WECs. These analyses have considered the cost-efficiency or levelised cost of energy (LCoE) [30]. This metric is one of the most reliable indices for the evaluation of energy investments. Recently, Piscopo et al. [31] combined an LCoE minimisation with a power take-off (PTO) control optimisation based on point-absorber dimensions in five Mediterranean Sea sites. This refined earlier work, optimising LCoE through optimisation of both WEC geometry and PTO settings [32,33].

In this work, we consider a single fully submerged, three-tether, cylindrical wave energy converter. This WEC is under development by Carnegie Clean Energy Limited, Australia. Two initial attempts [12,34] were performed to investigate the impact of different geometries and PTO parameters on power efficiency and the LCoE. However, in these prior works, only some predefined geometries were studied, and the results showed that in the cylinder-shaped WEC, an optimal tethers angle depends on the ratio between the buoy height and radius. However, optimisation procedures were not adequately outlined [34]. In another study [12], the performance of a few conventional optimisation methods was investigated in order to maximise the absorbed power and minimise the LCoE.

This paper improves upon previous research by expanding the findings of [12] to include another two state-of-the-art meta-heuristics including the Grey Wolf Optimiser [35] (GWO) and a self-adaptive version of differential evolution (LSHADE-EpSin [36]). Moreover, we propose two novel bi-level optimisation methods consisting of a global search method that works in the upper-level combined with a local search method in the lower-level. In total, nine optimisation methods are applied and compared in order to maximise the absorbed power and minimise the LCoE in a real wave regime from the southern coast of Marettimo (an island in the Mediterranean Sea). We also improve previous research by modelling waves regimes with a higher granularity of wave-directions.

The experimental outcomes show that a bi-level optimisation technique consisting of a self-adaptive differential evolution search (LSHADE-EpSin) interleaved with Nelder–Mead (NM) simplex direct search outperforms previous heuristic methods used in prior works in terms of convergence rate, higher absorbed power output, and lower levelised cost of energy.

The paper is structured as follows. Section 2 outlines the design of the WEC and the model that is applied to simulate both the absorbed power and LCoE. In the next section, the optimisation problem is described, and Section 4 represents the proposed meta-heuristic methods. The optimisation achievements are presented and considered in Section 5. Finally, Section 6 presents the conclusions of this work and canvasses future work.

2. Modelling

2.1. Wave Energy Converter

A wave energy converter chosen for this case study is a fully submerged cylindrical buoy connected to three tethers to absorb wave power from its motion in multiple degrees-of-freedom (or multiple modes), namely surge, heave and pitch. As shown in Figure 1, the geometry of this WEC is determined by the radius a and height H of the cylinder, tethers inclination angle α_t, and the angle α_{ap} that defines the tether attachment point (from the centre of mass of the buoy). The submergence depth (distance from the undisturbed water level to the top of the buoy) is considered fixed and equal to 2 m regardless of the buoy size. The mass of the buoy is taken as half the displaced mass of water $m_b = 0.5\rho_w V$ (the density of water is $\rho_w = 1025\ \mathrm{kg/m^3}$, and the buoy volume is $V = \pi a^2 H$). The hollow buoy houses three direct mechanical drive power take-off units (each connected to the tether). Each PTO acts as a spring-damper system where stiffness and damping coefficients can be adjusted for each sea state.

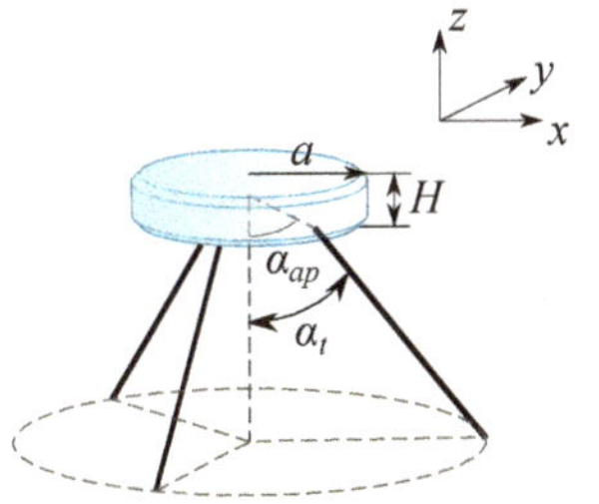

Figure 1. A three-tether wave energy converter.

2.2. Wave Climate

A potential wave energy development site located near the west coast of Marretimo Island (Italy) in the Mediterranean Sea is chosen for this analysis. According to the WXSD classification [20], this wave climate belongs to resource class 1 due to its low energy content (6.4 kW/m). The k-means clustering method has been applied to extract 10 sea states that represent this wave climate as shown in Figure 2 and listed in Table 1. A weighted aggregation of these 10 irregular sea states are used to calculate the annual average power production of the WEC. It is assumed that all waves are unidirectional and propagate in the positive x-direction.

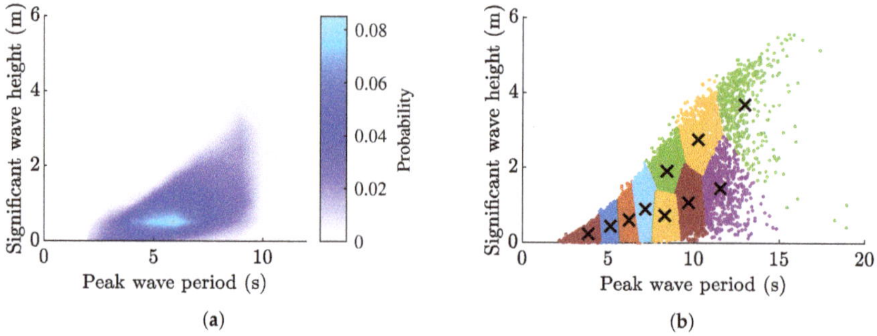

Figure 2. The wave climate at the Marettimo deployment site, Italy (12.04°E, 37.96°N, 6.38 kW/m mean annual wave power resource) [37]: (**a**) wave scatter diagram, and (**b**) clustering of the wave data where crosses correspond to ten representative states.

Table 1. Ten irregular sea states that represent the Marettimo deployment site.

Sea State	T_p, s	H_s, m	Probability O, %
1	3.82	0.24	8.06
2	5.13	0.44	14.62
3	6.20	0.61	17.80
4	7.18	0.90	18.01
5	8.30	0.73	12.10
6	8.43	1.92	9.58
7	9.68	1.08	8.68
8	10.24	2.76	5.78
9	11.56	1.46	3.30
10	12.99	3.69	2.07

2.3. Equations of Motion

The following time-domain model describes the WEC response under the wave and PTO loads:

$$\mathbf{M}\ddot{\mathbf{x}}(t) = \mathbf{F}_{exc}(t) + \mathbf{F}_{rad}(t) + \mathbf{F}_{visc}(t) + \mathbf{F}_{buoy}(t) + \mathbf{F}_{tens}(t), \tag{1}$$

where the $\mathbf{x} \in \mathbb{R}^{6\times1}$ is the buoy position vector in $Oxyz$ coordinate system, $\mathbf{M}$ is a mass matrix, $\mathbf{F}_{exc}$ is the wave excitation force, $\mathbf{F}_{rad}$ is the wave radiation force, $\mathbf{F}_{visc}$ is the viscous drag force, $\mathbf{F}_{buoy}$ is the buoyancy force, $\mathbf{F}_{tens}$ is the tether tension force expressed in the Cartesian space that includes the pre-tension force and control (PTO) forces. The force acting along the k-th tether can be modelled as $F_{t,k} = F_{t0} + K_{pto}\Delta\ell_k + B_{pto}\Delta\dot{\ell}_k$ ($k = 1\ldots3$) being proportional to the tether extension $\Delta\ell$, the rate of change of the tether length $\Delta\dot{\ell}$ and includes the initial tension F_{t0}. The PTO stiffness K_{pto} and damping B_{pto} coefficients take the same values for all three tethers. The transformation between the buoy velocity $\dot{\mathbf{x}}$ and the tether velocity vector $\dot{\mathbf{q}} = [\Delta\dot{\ell}_1 \quad \Delta\dot{\ell}_2 \quad \Delta\dot{\ell}_3]^\top$ has a form of $\dot{\mathbf{q}}(t) = \mathbf{J}^{-1}(\mathbf{x})\dot{\mathbf{x}}(t)$, where $\mathbf{J}^{-1}(\mathbf{x}) \in \mathbb{R}^{3\times6}$ is the inverse kinematic Jacobian that depends on the buoy position at each time instance [34]. So the tether force vector can be converted to the Cartesian space according to $\mathbf{F}_{tens} = -\mathbf{J}^{-\top}\mathbf{F}_t$.

The time-domain model in Equation (1) has a relatively high computation time and may not be suitable for optimisation purposes when a large number of evaluations are required. If to assume that all processes are Gaussian, it is possible to derive a spectral-domain model that can capture all required nonlinear forces using statistical linearisation technique [38,39]. The spectral-domain model approximates the system dynamics in the frequency domain by replacing all nonlinear terms with equivalent linear matrices [40]. The dynamic model in Equation (1) has two sources of nonlinearity: the viscous drag force $\mathbf{F}_{visc}$ and the generalised tether tension force $\mathbf{F}_{tens}$. Due to the fact that geometric nonlinearity contained within $\mathbf{F}_{tens}$ is much weaker than the quadratic nonlinearity in $\mathbf{F}_{visc}$, $\mathbf{F}_{tens}$ can be linearised around the zero position without loss of accuracy for the proposed configuration. If nonlinear effects from tethers become relevant, the equivalent terms can be derived as shown in [38,41,42]. Moreover, it should be noted that other nonlinear forces can be included in the model but omitted in this study, e.g., nonlinear Froude–Krylov force that becomes relevant when the buoy experiences large motion amplitudes [43]. As a result, a nonlinear dynamic Equation (1) is replaced by the equivalent frequency domain model:

$$\left[-\omega^2\left(\mathbf{M}+\mathbf{A}(\omega)\right)+i\omega\left(\mathbf{B}(\omega)+\mathbf{B}_{pto}+\mathbf{B}_{eq}\right)+\mathbf{K}_{pto}\right]\hat{\mathbf{x}}(\omega)=\hat{\mathbf{F}}_{exc}(\omega), \tag{2}$$

where $\mathbf{x}(t) = \Re\mathfrak{e}\{\hat{\mathbf{x}}\, e^{i\omega t}\}$, the radiation force is expressed using the frequency dependent added mass $\mathbf{A}(\omega)$ and radiation damping matrix $\mathbf{B}(\omega)$, $\hat{\mathbf{F}}_{rad}(\omega) = -\left(-\omega^2\mathbf{A}(\omega)+i\omega\mathbf{B}(\omega)\right)\hat{\mathbf{x}}(\omega)$, the tether tension force is linearised as $\hat{\mathbf{F}}_{tens}(\omega) = -(i\omega\mathbf{B}_{pto}+\mathbf{K}_{pto})\hat{\mathbf{x}}(\omega)$ (see [44] for more details), and the viscous drag force is replaced by $\hat{\mathbf{F}}_{visc}(\omega) = -i\omega\mathbf{B}_{eq}\hat{\mathbf{x}}(\omega)$. The equivalent damping term $\mathbf{B}_{eq}$ is unknown and determined iteratively (for each wave condition separately) using the procedure explained in [38]:

$$\mathbf{B}_{eq} = -\left\langle \frac{\partial \mathbf{F}_{visc}}{\partial \dot{\mathbf{x}}} \right\rangle, \tag{3}$$

where $\langle\cdot\rangle$ indicates mathematical expectation, and the viscous force is interpreted as:

$$\mathbf{F}_{visc} = -\frac{1}{2}\rho_w \mathbf{C}_d \mathbf{A}_d (||\dot{\mathbf{x}}|| \odot \dot{\mathbf{x}}), \tag{4}$$

ρ_w is the density of water, $\mathbf{C}_d$ and $\mathbf{A}_d$ are the matrices of the drag coefficients and the cross-section areas of the buoy perpendicular to the direction of motion respectively, and $\odot$ represents the Hadamard product (element-wise multiplication). Note that only the body velocity (not the relative fluid/body velocity) has been considered in the drag force formulation. A detailed methodology of how to incorporate the wave-particle velocity into the spectral-domain model is demonstrated in [45].

The following iterative procedure is used to estimate $\mathbf{B}_{eq}$ and approximate the response of the WEC in irregular waves:

Step 1. Define the sea state and corresponding incident wave spectrum $S_\eta(\omega)$.

Step 2. Compute the power spectral density (PSD) matrix of the excitation force:

$$\mathbf{S}_{\mathbf{F}}(\omega) = S_\eta(\omega)\hat{\mathbf{f}}_{exc}(\omega)\hat{\mathbf{f}}^*_{exc}(\omega), \tag{5}$$

where $\hat{\mathbf{f}}_{exc}$ is the vector of excitation force coefficients, and $()^*$ denotes the conjugate transpose of a vector/matrix.

Step 3. Calculate the WEC response matrix assuming $\mathbf{B}_{eq} = \mathbf{0}_{6\times6}$ in the first iteration:

$$\mathbf{H}(\omega) = \left[-\omega^2\left(\mathbf{M}+\mathbf{A}(\omega)\right)+i\omega\left(\mathbf{B}(\omega)+\mathbf{B}_{pto}+\mathbf{B}_{eq}\right)+\mathbf{K}_{pto}\right]^{-1}. \tag{6}$$

Step 4. Establish the power spectral density matrix of the buoy motion:

$$\mathbf{S}_{\mathbf{x}}(\omega) = \mathbf{H}(\omega)\mathbf{S}_{\mathbf{F}}(\omega)\mathbf{H}^*(\omega). \tag{7}$$

Step 5. Calculate the covariance matrix of the WEC velocity:

$$\sigma_{\dot{\mathbf{x}}}^2 = \text{cov}[\dot{\mathbf{x}}, \dot{\mathbf{x}}] = \int_0^\infty \omega^2 \mathbf{S}_{\mathbf{x}}(\omega) d\omega. \tag{8}$$

Step 6. Estimate the equivalent damping matrix $\mathbf{B}_{eq}$ using the analytical expression from [38]:

$$\mathbf{B}_{eq} = -\left\langle \frac{\partial \mathbf{F}_{visc}}{\partial \dot{\mathbf{x}}} \right\rangle = \frac{1}{2}\sqrt{\frac{8}{\pi}} \rho_w \mathbf{C}_d \mathbf{A}_d \sigma_{\dot{\mathbf{x}}}^2. \tag{9}$$

Step 7. Check the convergence criteria:

$$|\mathbf{B}_{eq}[n] - \mathbf{B}_{eq}[n-1]| < \delta. \tag{10}$$

where n corresponds to the iteration number, and the threshold is set to $\delta = 0.01$. If this condition is not satisfied, go to Step 3.

It can take up to 10 iterations to estimate $\mathbf{B}_{eq}$ and the WEC response in irregular waves. Once calculated, the average power absorbed by each PTO unit $k = 1 \ldots 3$ is calculated as [38]:

$$\bar{P}_k = B_{pto} \sigma_{\dot{q}_k}^2, \tag{11}$$

where $\sigma_{\dot{q}_k}^2$ is the variance of the tether length rate change $\dot{\mathbf{q}}$:

$$\sigma_{\dot{q}_k}^2 = \int_0^\infty \omega^2 S_{q_k}(\omega) d\omega, \tag{12}$$

and the transformation between the Cartesian coordinate system and the tether space is obtained using $\mathbf{S}_{\mathbf{q}}(\omega) = \mathbf{J}_0^{-1} \mathbf{S}_{\mathbf{x}}(\omega) \mathbf{J}_0^{-\mathrm{T}}$, where $\mathbf{J}_0^{-1} = \mathbf{J}^{-1}(\mathbf{x}_0)$ is linearised about the nominal operating position $\mathbf{x}_0 = \mathbf{0}_{6\times 1}$.

The total power generated by three PTO units in an irregular wave with the significant wave height H_s and peak wave period T_p is:

$$\bar{P}(H_s, T_p) = B_{pto} \sum_{k=1}^{3} \sigma_{\dot{q}_k}^2 (H_s, T_p). \tag{13}$$

The expected average annual power production from the WEC for a specific deployment site is estimated as:

$$P_{AAP} = \sum_{H_s} \sum_{T_p} O(H_s, T_p) \cdot \bar{P}(H_s, T_p), \tag{14}$$

where the matrix $\mathbf{O}(H_s, T_p)$ contains the occurrence probability of each sea state within the wave climate.

To demonstrate that the spectral-domain model is an effective tool that can fully capture the nonlinear dynamics of the considered WEC while significantly decreasing the computation time, a comparison of average power estimated using three different models is shown in Figure 3. The frequency-domain model is implemented based on Equation (2) assuming $\mathbf{B}_{eq} = \mathbf{0}$, the spectral-domain model is specified in Equation (2) where $\mathbf{B}_{eq}$ is estimated iteratively for each sea state, and the time-domain model is represented by Equation (1). Good agreement is achieved between the spectral-domain and time-domain models, while the frequency domain model significantly overestimates power generation potential of the WEC.

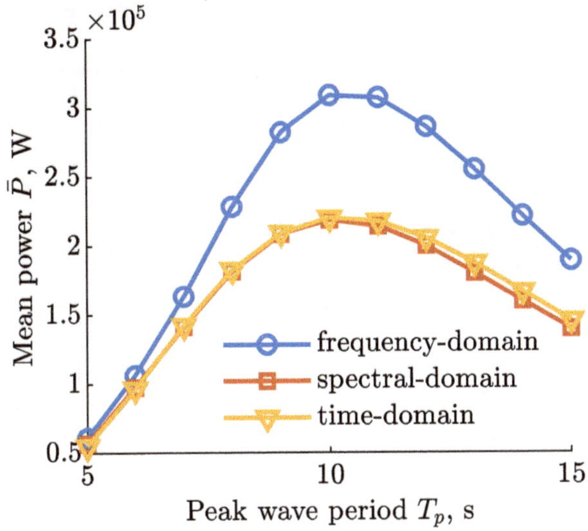

Figure 3. Power production of a three-tether WEC in irregular waves estimated using three different models: frequency-, spectral-, and time-domain. Parameters of the WEC are $a = 5.5$ m, $H = 5.5$ m, $\alpha_{ap} = \alpha_t = 45$ deg, $K_{pto} = 200$ kN/m, $B_{pto} = 150$ kN/(m/s)), irregular waves have the significant wave height of $H_s = 3$ m and modeled using the Pierson–Moskowitz spectrum.

2.4. Economic Model

Levelised cost of energy (LCoE) is used to measure the economic attractiveness of the proposed energy project. Due to the lack of publicly available information of the detailed cost estimations for wave energy technology, [46] proposed to approximate LCoE by the following equation:

$$\text{LCOE}\left(\frac{€}{\text{kWh}}\right) = \text{RDC} \times \left(\frac{\text{Energy (MWh)}}{\text{Mass (kg)}}\right)^{-0.5}, \tag{15}$$

where RDC is a site-specific coefficient that is set to 1 in this study, the characteristic mass of the system includes the mass of the buoy and the anchoring system.

The characteristic mass of the WEC is calculated using the following assumptions:

- The mass of the buoy is calculated based on a given geometry as $m_b = 0.5\rho_w \pi a^2 H$;
- The needed mass of the anchoring system (three piles) relays on the tether tension associated with buoyancy and the wave force, and can be approximated by $m_{as} \approx 0.116 F_t^{peak}$ using case presented in [47] as a reference. The tether peak force (99% $= 2.57\sigma_{F_t}$) is estimated from the spectral-domain model.

As a consequence, the LCoE model applied in this research is:

$$\text{LCOE} = \left(\frac{8760 P_{AAP}}{m_b + m_{as}}\right)^{-0.5}. \tag{16}$$

2.5. Implementation

To estimate the power output and LCoE for any WEC geometry, Equation (2) is solved in MATLAB. The mass matrix has a diagonal form $\mathbf{M} = \text{diag}(m_b, m_b, m_b, I_{xx}, I_{yy}, I_{zz})$ with moments of inertia calculated for the cylindrical body. Hydrodynamic parameters of the WEC, including the added mass $\mathbf{A}(\omega)$, hydrodynamic damping $\mathbf{B}(\omega)$, and excitation force vector $\hat{\mathbf{F}}_{exc}(\omega)$ are estimated using a semi-analytical model [48,49]. $\mathbf{B}_{eq}$ is calculated based on the iterative procedure explained in Section 2.3.

Even though only one geometric shape (vertical cylinder) is used in the study, the magnitude of the viscous drag force, and the corresponding $\mathbf{B}_{eq}$, are highly dependent of the ratio between the cylinder height to its diameter, especially for the heave mode. Therefore, it order to develop an optimisation procedure that can accommodate WEC geometries with various aspect ratios (H/a), the drag coefficient in heave is expressed as a function $C_{d_3} = -0.12(H/a) + 1.2$ based on published data [50] shown in Figure 4. Drag coefficients in other directions are not sensitive to the cylinder aspect ratio and are kept fixed $C_{d_1} = C_{d_2} = 1$ for surge and sway, and $C_{d_4} = C_{d_5} = 0.2$ for roll and pitch. The irregular waves from Table 1 are modelled using the Bretschneider (modified Pierson–Moskowitz) spectrum according to [51].

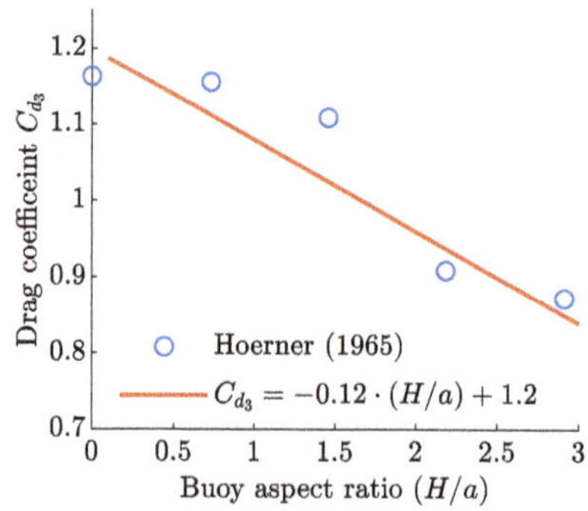

Figure 4. Drag coefficient of the cylindrical body in axial flow as a function of its aspect ratio H/a.

3. Optimisation Configuration Models

In this research, The optimisation decision variables of the cylinder are including the radius of the buoy a, the aspect ratio that is considered as the proportion of the height over the radius of the buoy (H/a), two tether angles (attachment α_{ap} and inclination angle α_t), two vectors of power take-off parameters, damping and stiffness coefficients represented $\mathbf{b}_{pto} = [B^{(1)}_{pto}, B^{(2)}_{pto}, \ldots, B^{(N)}_{pto}]^{\mathrm{T}}$ and $\mathbf{k}_{pto} = [K^{(1)}_{pto}, K^{(2)}_{pto}, \ldots, K^{(N)}_{pto}]^{\mathrm{T}}$, respectively. The length of each PTO vector is $N = 10$. The whole number of decision designs are 24 which should be optimised in the following:

$$\mathbf{z}_1 = [a,\ H,\ \alpha_t,\ \alpha_{ap},\ \mathbf{k}_{pto} \in \mathbb{R}^{N\times 1},\ \mathbf{b}_{pto} \in \mathbb{R}^{N\times 1}]. \tag{17}$$

$$\mathbf{z}_2 = [a,\ (H/a),\ \alpha_t,\ \alpha_{ap},\ \mathbf{k}_{pto} \in \mathbb{R}^{N\times 1},\ \mathbf{b}_{pto} \in \mathbb{R}^{N\times 1}]. \tag{18}$$

We apply two fitness functions in order to maximise the power output and minimise the LCoE.

(i) The average annual produce power output computed utilising Equation (14), that is maximised as

$$f_{O1} = \arg\max_{\mathbf{z}} P_{AAP}(\mathbf{z}), \text{ subject to: } \mathbf{z}_1 \in [\mathbf{z}_{\min}, \mathbf{z}_{\max}] \tag{19}$$

(ii) The LCoE is minimised using the below equation that is specified in Equation (16):

$$f_{O2} = \arg\min_{\mathbf{z}} \mathrm{LCOE}(\mathbf{z}), \text{ subject to: } \mathbf{z}_2 \in [\mathbf{z}_{\min}, \mathbf{z}_{\max}] \tag{20}$$

Table 2 shows the ranges of all design variables which are involved in the optimisation process.

Table 2. Boundary constraints of the cylinder parameters.

Parameter	Unit	Min	Max	Length
radius, a	m	1	20	1
height, H	m	1	30	1
aspect ratio, (H/a)		0.4	2	1
Tether inclination angle, α_t	deg	10	80	1
Tether attachment angle, α_{ap}	deg	10	80	1
PTO stiffness, K_{pto}	N/m	10^3	10^8	10
PTO damping, B_{pto}	N/(m/s)	10^3	10^8	10

4. Optimisation Algorithms

In this paper, we focus on two widespread optimisation strategies in order to maximise harnessed power and minimise the levelised cost of energy (LCoE) of a fully-submerged three-tether WEC. The first approach applies optimisation algorithms to all decision variables simultaneously. These design variables consist of the buoy geometry parameters (radius a, height H and aspect ratio (H/a)), the tether angles (inclination angle α_t and the tether attachment angle α_{ap}), and the PTO parameters (spring stiffness $\mathbf{k}_{pto}$ and damping coefficients $\mathbf{k}_{pto}$). In total, there are 24 parameters that are optimised all-at-once.

The second strategy is to apply bi-level optimisation methods [52], which solve the problem using a two-level optimisation procedure, where one optimisation problem is nested within the other. The outer optimisation task is generally regarded as the upper-level optimisation problem, and the interior one is recognised as the lower-level optimisation problem. A significant characteristic of the bi-level optimisation problem is that the fitness functions of each level may be partly defined by variables advised by other levels. Following this strategy, we propose two bi-level optimisation methods and compare their performance with seven other well-known global search methods. The details of the optimisation algorithms performed for each strategy are outlined in Table 3.

Table 3. The details of the optimisation methods settings. All approaches are restricted to the same evaluation number.

Methods	Settings
Nelder–Mead [53]	Nelder–Mead simplex direct search (NM)
1+1EA [54]	mutation step sizes are $\sigma_a = \xi_1 \times (U_a - L_a)$, $\sigma_H = \xi_1 \times (U_H - L_H)$, $\sigma_{\alpha_t} = \sigma_{\alpha_{ap}} = \xi_1 \times (U_{\alpha_t} - L_{\alpha_t})$, $\sigma_{K_{pto}} = \sigma_{B_{pto}} = \xi_2 \times (U_{K_{pto}} - L_{K_{pto}})$, and Probability mutation rate=$\frac{1}{N}$, $\xi_1 = 0.3$, $\xi_2 = 0.01$
CMA-ES [55]	with the default settings and $\lambda = 13$;
PSO [56]	with $\lambda = 25$, $c_1 = 1.5$, $c_2 = 2$, $\omega = 1$ (decreased with a damping ratio $w_f = 0.99$ exponentially);
GWO [35]	with λ= 25, $\alpha = 2$ (linearly decreased to zero)
DE [57]	with $\lambda = 25$, $F = 0.5$, $P_{cr} = 0.8$
SaDE [58]	with $\lambda = 25$, $LP = 50$, $NumSt = 4$
LSHADE-EpSin [36]	$\lambda = 25$, historical memory size $H = 5$, $Num_{LS} = 10$
Bi-level-1	SaDE +NM, WEC's dimensions and tether angles are optimised in the lower-level, default settings of SaDE
Bi-level-2	LSHADE-EpSin + NM, WEC's dimensions and tether angles are optimised in the lower-level, default settings of LSHADE-EpSin

4.1. All-at-Once Optimisation

Various factors associated with WEC design, tether angles and PTO parameters combined to form a non-convex, dynamic, constrained and large-scale optimisation problem. These challenges serve as our primary motivation for applying the meta-heuristics like evolutionary and swarm optimisation algorithms. We apply and compare the performance of seven well-known meta-heuristics that reliably optimise all decision variables of WECs all-at-once. This optimisation process leads to maximise the produced power and minimise the levelised cost of energy. The optimisation methods applied in this research include 1+1EA [59]; Differential Evolution (DE) [57], Covariance matrix adaptation evolution strategy (CMA-ES) [55], Particle Swarm Optimisation (PSO) [56], Grey Wolf Optimiser (GWO) [35] and two state-of-the-art self-adaptive optimisation methods including SaDE [58] and LSHADE-EpSin [36].

4.1.1. L-SHADE with an Ensemble Pool of Sinusoidal Parameter Adaptation (LSHADE-EpSin)

The Differential Evolution (DE) algorithm, and its adaptive and self-adaptive variants, are simple and robust evolutionary algorithms. Researchers from various fields of science and engineering have applied DE algorithms to various optimisation problems, notwithstanding problems with characteristic such being continuous, multi-modal, combinatorial or mixed variable. DE is able to obtain superior optimisation results across widely encountered real-world engineering problems [60,61]. Among a wide range of self-adaptive DE algorithms, LSHADE-EpSin performs outstandingly in solving different benchmarks and real-world problems [36]. LSHADE-EpSin is a modified version of the L-SHADE algorithm [62] with linear population size reduction and an ensemble pool of sinusoidal parameter adaptations. L-SHADE is a developed version of the SHADE algorithm [63] that practices a history-based parameter adaptation trajectory based on the JADE algorithm [64] which proposed the novel mutation strategy (*current/to/pbest*).

Mutation Strategy with External Archive

In LSHADE-EpSin, one of the best-performing mutation strategies for generating promising mutant vectors during the optimisation process is *current-to-pbest/1* which is initially proposed by JADE. This mutation strategy can be seen in Equation (21).

$$v_{i,g} = x_{i,g} + F_{i,g}(x_{pbest,g} - x_{i,g}) + F_{i,g}(x_{r_1,g} - x_{r_2,g}) \tag{21}$$

where $x_{pbest,g}$ is chosen from the best solutions $N \times p(p \in [0,1])$ of the current parent population (g). $x_{r_1,g}$ is randomly taken from the population and $x_{r_2,g}$ is randomly chosen from a combination of the current population and the external archive (A). The external archive keeps a record of the lower-ranking parents recently replaced by offspring.

Ensemble of Parameter Adaptation

An ensemble of parameter configurations is used in LSHADE-EpSin to control the adaptation of parameters. The adaptive parameters are associated with a combination of two sinusoidal formulas to adjust the scaling factor. Firstly, a non-adaptive sinusoidal adjustment technique is used to adjust the scale factor $(F_{i,g})$ which decreases during the optimisation process. Equation (22) shows this non-adaptive technique.

$$F_{i,g} = \frac{1}{2} \times (sin(2\pi \times freq \times g_{s_1} + \pi) \times \frac{iter_{max} - g_{s_1}}{iter_{max}} + 1) \tag{22}$$

where $freq$ describes a pre-defined frequency for the sinusoidal function and $iter$ denotes the current generation number ($g_{s_1} <= \frac{iter_{max}}{2}$). The second strategy for the adjustment of the scale factor is an adaptive sinusoidal adjustment method. This formulation can be seen in Equation (23).

$$F_{i,g} = \frac{1}{2} \times (sin(2\pi \times freq \times g_{s_1}) \times \frac{g_{s_1}}{iter_{max}} + 1) \tag{23}$$

where $freq$ is an adaptive frequency based on a Cauchy distribution and a successful history-based of settings. $iter$ denotes the current generation number. One of the most effective DE parameter adaptation techniques is recording an archive of both mutation factors and probabilities of crossover based on their success during the optimisation process. The control parameters history-based was proposed by Zhang et al. [64] in JADE. In each generation of JADE, in order to generate an offspring, we have an array of the crossover probability rate that is produced based on a normal distribution of the mean (μ_{CR}) and variance at 0.1. The successful crossover probabilities (S_{CR}) are recorded and updated at each generation. The μ_{CR} is initialised by 0.5 and in the next generation it is updated by Equation (24).

$$\mu_{CR} = (1 - c) \times \mu_{CR} + c \times mean_A(S_{CR}) \tag{24}$$

where c is a constant generated between 0 and 1 randomly and $mean_A$ is a simple arithmetic mean. Likewise, the mutation factor F_i of each x_i is separately generated at each generation, as stated in a Cauchy distribution with the mean μ_F and scale parameter 0.1. (Equation (25))

$$F_i = randc_i(\mu_F, 0.1) \tag{25}$$

where the $randc_i$ is the Cauchy distribution. All successful mutation factors are archived and point out as a set of S_F at the end of each generation. The value of μ_F is updated using Equation (26).

$$\mu_F = (1 - c) \times \mu_F + c \times mean_L(S_F) \tag{26}$$

where $mean_L$ is the Lehmer mean [65] and computed as follows:

$$mean_L(S_F) = \frac{\sum_{F \in S_F} F^2}{\sum_{F \in S_F} F} \tag{27}$$

Linear Population Size Reduction

The LSHADE-EpSin algorithm benefits from a linear reduction in population size to fit the population size (N) iteratively at each generation as exposed in the following equation:

$$N_{g+1} = Round[\left(\frac{N_{min} - N_{max}}{iter_{max}}\right) \times iter + N_{max}] \tag{28}$$

where N_{min} is the minimum population size, and initialised at 4 that is required to make the *current-to-pbest* mutation strategy. The four required solutions are x_i, x^p_{best}, x_{r_1} and x_{r_2}. The mutant vector of this strategy is generated using Equation (29).

$$V_{i,g} = x_{i,g} + F_i \times (x^p_{best,g} - x_{i,g}) + F_i(x_{r_1,g} - x_{r_2,g}) \tag{29}$$

Local Search

In order to extend the exploitation capability of LSHADE-EpSin, a stochastic local search is proposed that works based on Gaussian Walks. The local search is activated when the population size is less than 20 ($N_{ini} = 25$), and 25 random samples are evaluated to exploit the neighbourhood of the best-found design among the current population. The Gaussian walks applied can be seen in Equation (30).

$$y_i = \mathcal{N}(\mu_b, \sigma) + (r_1 \times x_{best} - r_2 \times x_i) \tag{30}$$

where x_{best} is the best-found solution in the local search and μ_b is equal to x_{best}. r_1 and r_2 are two uniform random numbers from the range of $[0, 1]$. Besides, the standard deviation (σ) of this Gaussian Walks is calculated using Equation (31).

$$\sigma = \frac{log(iter)}{iter} \times (x_i - x_{best}) \tag{31}$$

4.2. Bi-Level Optimisation

In this paper, we propose two bi-level optimisation methods, including Bi-level-1 (SaDE+NM) and Bi-level-2 (LSHADE-EpSin+NM). We also provide a general formulation in order to maximise the harnessed power and minimise the LCoE of a cylindrical wave energy converter. These proposed approaches comprise two levels of optimisation tasks where one optimisation process is nested within the other. The exterior optimisation method (which is a global search method) is referred to as the leader's (upper level) optimisation process. In the upper level, we apply two self-adaptive meta-heuristics, including Self-adaptive DE (SaDE) and LSHADE-EpSin. Both methods improve the ability of an adaptive learning strategy to fine-tune the control parameters and mutation strategy and demonstrate a considerable performance in optimising real engineering problems [66,67].

In the second level, the internal method is recognised as the follower's (lower level) optimisation process. In the current study, the inner method is a Nelder–Mead (NM) simplex search method [68]. NM simplex is a downhill local search method, and it is straightforward to hybridise combine with other meta-heuristic methods. The primary reason for such hybridisation (or for using NM as the lower-level in a bi-level method) is to tune a more suitable trade-off between global optimality and computational budgets [69,70].

Figure 5 shows that the proposed bi-level optimisation framework consists of a global search method designed to optimise all decision variables in the upper-level, and both geometry parameters (radius and height) that given from upper-level decision vector are optimising in the lower-level. To adjust the geometry parameters of the cylinder, we use a local search method. The best-found geometry configuration in the lower-level will be replaced in the upper-level decision variables.

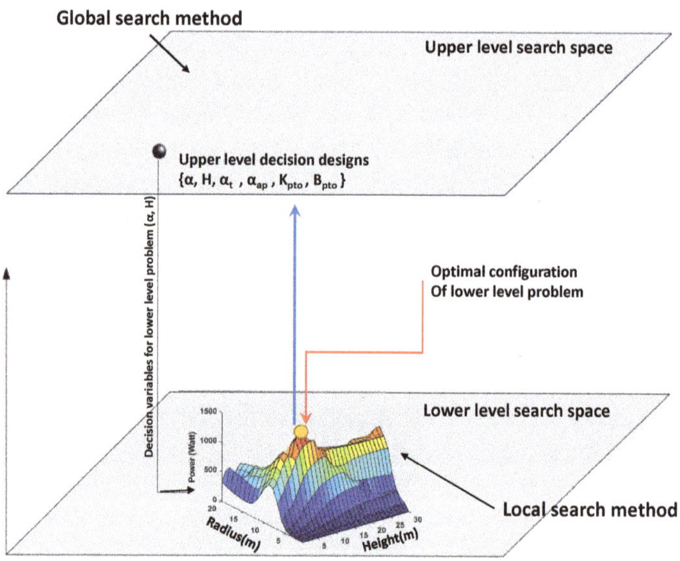

Figure 5. A general sketch of the bi-level optimisation applied in order to maximise the produced power.

The pseudo-code of the proposed Bi-level-2 algorithm is shown in Algorithm 1. It can be seen that the algorithm is divided into two primary sections. At the top level, we have a self-adaptive DE (LSHADE-EpSin) employing two strategies to adjust the control parameters. These strategies are (1) adaptive sinusoidal increasing adjustment and (2) non-adaptive sinusoidal decreasing adjustment. The benefit of this ensemble approach is that it allows the algorithm to converge to a sufficient balance [36] between searching the neighbourhood of current bet-found solutions and the exploration of non-visited search space zones. In the lower-level, there are two nested inner local search methods. The initial local search is used to explore the search space of the cylinder dimensions (radius and height) where other decision variables are fixed. Next, both tether angles (inclination and attachment) are optimised using the second local search. In order to save computational budget, we define a performance criterion for both local search methods. This condition evaluates the local search performance; if the obtained power improvement cannot satisfy the criterion, Bi-level-2 will withdraw the local optimisation process and allocate the remaining budget to the global search method.

Tuning the Local Search

One of the significant parameters of the bi-level optimisation method is the maximum evaluation number (Max_{eval}) of the local search (NM). Tuning this variable plays an important role in obtaining a greater balance between saving on the computational budget and converging to the local optimum as much as possible. In order to tune the Max_{eval}, we perform the local search to optimise the WEC geometry parameters (a, H) and keep the other decision variables fixed. This experiment iterates ten times with different initial solutions. Meanwhile, the same tuning process runs to optimise both tether angles. Figure 6 shows the convergence curves of these experiments. We observe that the local search converges rapidly to a local optimum in the geometry and tether angles optimisation processes after 20 and 40 iterations, respectively on average. Therefore, we set the Max_{eval} of the local search to 20 and 40 iterations.

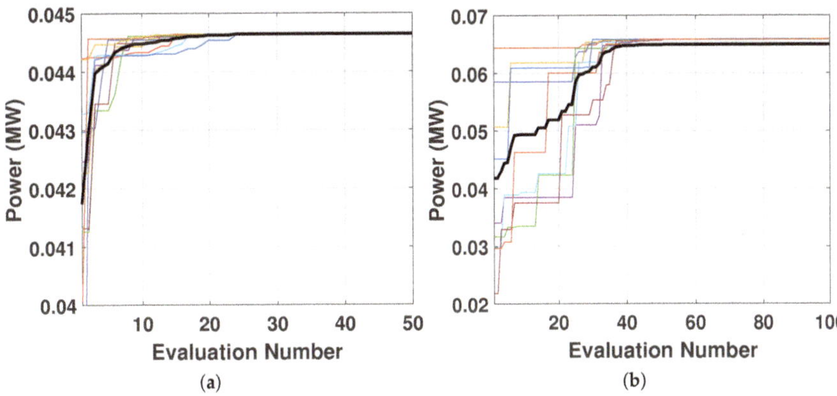

Figure 6. The effect of computational budget on tuning the local search iterations. (**a**) dimension optimisation (a, H) , (**b**) Tether angles optimisation (α_t, α_{ap}).

5. Optimisation Results and Discussions

5.1. Multi-Modality of Search Space

In order to characterise the search space, we perform an experiment using a Nelder–Mead (NM) search method. Twenty random initial configurations are generated and NM is applied to optimise the absorbed power output. Figure 7 shows the trajectory of the NM performance during the optimisation process. It can be seen that the majority of the trajectories in the cylinder dimension (subplot (a)) converged to a specific area of the search space as expected. This is because large WECs can harness more power than small ones. The second observation is that the PTO search space is not uni-modal and each trajectory converged to different configurations (subfigure (c,d,e)).

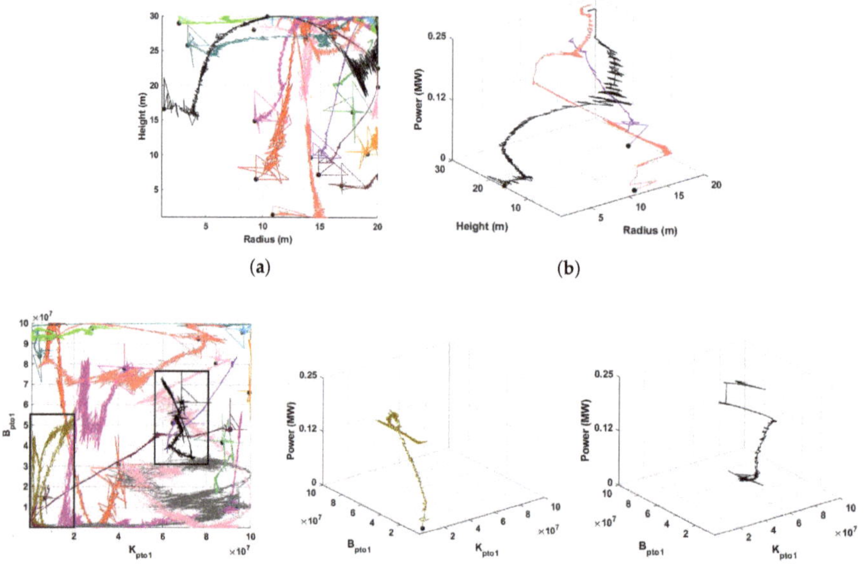

(**a**) (**b**)

(**c**) (**d**) (**e**)

Figure 7. Twenty independent NM runs with the random initial solutions. (**a**) The NM's trajectory in the cylinder's dimension (radius and height) optimisation, (**b**) 3D NM's trajectory in the cylinder's dimension and the absorbed power. (**c**) NM's trajectory in the initial value of the damping (B_{pto}) and spring (K_{pso}) array. (**d**,**e**) two examples of 3D NM's trajectory in B_{pto} and K_{pso}.

5.2. Power Landscape Analysis

With regard to evaluating the impact of each buoy design variable on the level of produced power, we perform a sensitivity analysis experiment. Here, we assume both tether angles are kept fixed at 45°; note that this size is not optimal, because tether angles should be adjusted based on the buoy's dimensions, as recommended by prior works [34]. Moreover, the search space of the K_{pto} and B_{pto} parameters are discretised, where each interval is 10^6. In the next step, for each discrete configuration of PTO parameters, we evaluate the importance of the cylinder dimensions (a, H) using a grid search technique where the discretisation step size is 1 (m).

The results are shown in Figure 8, which includes 400 sub-figures. Each sub-figure represents the relationship of the cylinder radius and height sizes with the absorbed power, where the K_{pto} and B_{pto} are fixed. It is important to note that a variation in the size of the radius has a more substantial effect on the power output than a variation in the cylinder height. In this wide power landscape, we can see that the maximum produced powers are achieved when the PTO parameters are assigned around 10^7, and the buoy radius and height sizes are large. However, it should be noted that the effect of PTO parameters on the absorbed power is more significant than the size of the cylinder dimensions.

Algorithm 1 Bi-level Optimisation method (LSHADE-EpSin+NM)

procedure BI-LEVEL OPTIMISATION METHOD
Initialization
  $P = \{\langle a_1, H_1, \alpha_{t_1}, \alpha_{ap_1}, K_1^1, ..., K_1^{10}, B_1^1, ..., B_1^{10}\rangle, \ldots$
  $\ldots, \langle a_N, H_N, \alpha_{t_N}, \alpha_{ap_N}, K_N^1, ..., K_N^{10}, B_N^1, ..., B_N^{10}\rangle\}$ ▷ initial population
  M:μF =μCR=0.5 ▷ initialise memory of first control settings
  M_{freq}:μfreq = 0.5,$Imp - rate_d = Imp - rate_\alpha = 1$ ▷ initialise memory of second control settings
Upper-Level (Global search method)
  for *iter* in $iter_{max}$ **do** ▷ termination criteria
    if $iter > \frac{iter_{max}}{2}$ **then**
      Call second control parameter settings
      $S_F = S_{CR} = \emptyset$ ▷ Reset successful mean vectors
      $r_i = rand(1, H)$ ▷ Generate a random index, H is memory size
      $F_i = randc(\mu F_{r_i}, 0.1)$,$CR_i = randn(\mu CR_{r_i}, 0.1)$
    end if
    if $iter \le \frac{iter_{max}}{2}$ **then**
      Call first control parameter settings
      $c = rand(0,1)$
      if $c < 0.5$ **then**
        $F_i = \frac{1}{2} \times (sin(2\pi \times freq \times iter + \pi) \times \frac{iter_{max} - iter}{iter_{max}} + 1)$
      else
        $F_i = \frac{1}{2} \times (sin(2\pi \times freq \times iter) \times \frac{iter}{iter_{max}} + 1)$
      end if
      Generate CR_i same as first control parameters (Equation 23)
    end if
    for $i = 1$ to N **do**
      Generate $p = rand(0,1) \times n$, $n = 0.1 \times N$
      $v_i = x_i + F_i \times (x_{pbest} - x_i) + F_i \times (x_{r_1} - x_{r_2})$ ▷ Mutation *current-to-pbest/1*
      $u^j_{i,iter} = \begin{cases} v^j_{i,iter}, & \text{if } (rand < CR_i) \text{ or } (j == j_{rand}) \\ P^j_{i,iter}, & \text{Otherwise} \end{cases}$ ▷ Binomial Crossover
      $P_{i,iter+1} = \begin{cases} u_{i,iter}, & \text{if } (f(u_{i,iter}) > f(P_{i,iter}))\ Maximisation \\ P_{i,iter}, & \text{Otherwise} \end{cases}$ ▷ Selection
      Store successful F_i and CR_i
    end for
    Update the memory according to used settings
    Update the population size by Equation (28)
    $N_{diff} = N_g - N_{g+1}$
    Sort P_{iter} based on the fitness function
    Remove worst solutions N_{diff} from P_{iter} AND Select the best solution P_{best}
Lower-Level (Local search method)
    if $Imp - rate_d$> 0.001% **then** ▷ Optimise Cylinder dimension
      $P_{best}(a, H) = Nelder - Mead(P_{best}(a, H), Max_{eval})$
      Compute improvement rate $Imp - rate_d$
    end if
    if $Imp - rate_\alpha$> 0.001% **then** ▷ Optimise tether angles
      $P_{best}(\alpha_t, \alpha_{ap}) = Nelder - Mead(P_{best}(\alpha_t, \alpha_{ap}), Max_{eval})$
      Compute improvement rate $Imp - rate_\alpha$
    end if
    Update P^{best}_{iter} by the best-found NM configurations
  end for
end procedure

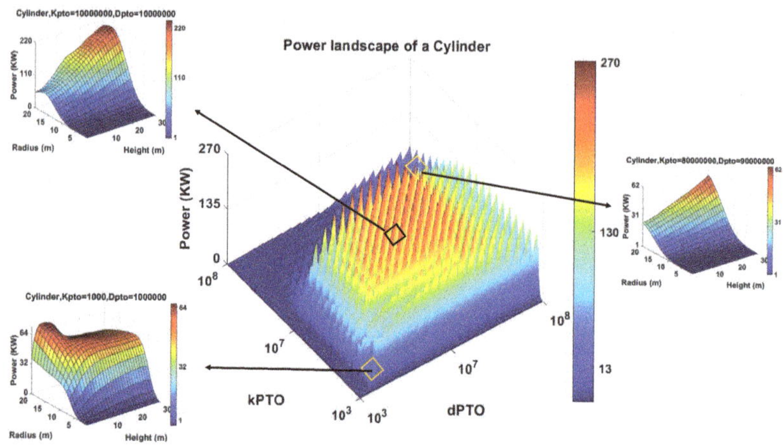

Figure 8. A power landscape of the cylinder with the fixed angles $\alpha_t, \alpha_{ap} = 45$ and various dimensions and PTO parameters.

5.3. *The Annual Average Power Output Maximisation*

In this section, we describe the optimisation results of our cylinder design experiments in order to maximise the annual average power output. Furthermore, we compare the performance of the optimisation algorithms outlined above in terms of best-found designs and speed of convergence.

Table 4 reports the best-found cylinder designs using seven meta-heuristics and two new bi-level optimisation methods that produced the highest power output among all ten runs. Furthermore, it can be seen that Bi-level-2 performs better than other applied optimisation methods and that it can produce a considerable amount of power of 279 kW. The second observation is that almost all (8 out of 9) optimisation methods converged to the cylinder of 15 m radius with the largest possible height of 30 m. However, it should be noted that producing electricity using such large WECs can be expensive, due to the high manufacturing costs. In terms of the angles and PTO settings, a large range of values is proposed by all optimisation methods even though the maximised power output is not dramatically different. This fact proves that it is not straightforward to optimise a multi-mode WEC due to the strong dependencies between angles, PTO parameters, and the hydrodynamic model which dominates the power absorption (heave, surge or pitch).

Table 4. Best-found design parameters in order to maximise the average annual absorbed power.

Parameter	1+1EA	CMA-ES	PSO	GWO	DE	SaDE	LSHADE-EpSin	Bi-Level-1	Bi-Level-2
a [m]	16.62	16.10	19.99	16.68	15.46	15.50	15.49	15.61	14.51
H [m]	30	30	14.80	30	30	30	30	30	30
α_t [deg]	70	26	60	14	48	26	39	50	10
α_{ap} [deg]	10	13	63	28	10	11	29	40	67
$\sum_{i=1}^{N_K} K_{pto}(\times 10^7)$	0.665	0.863	3.796	1.51	1.894	2.883	0.882	0.665	0.514
$\sum_{i=1}^{N_B} B_{pto}(\times 10^7)$	2.765	3.928	4.676	1.51	3.775	4.036	2.479	2.095	1.129
P_{AAP} [kW]	259	248	239	261	259	261	262	265	279

Table 5 presents the average best-found power output per each run for all optimisation methods. Bi-level-2 is not only capable of finding the best design configuration; it also performs the best average power output (Figure 9a) compared with other meta-heuristics. In terms of the convergence rate, Figure 10a depicts the applied optimisation method experiments during the 5000 evaluations. As we can see, GWO and LSHADE-SeSin rapidly converge to considerable settings; however, they could not sustain this upward trajectory and converge near locally optimal designs. Obviously, the fastest convergence rate is allocated to Bi-level-2.

Table 5. Performance comparison of various optimisation methods based on the maximum, minimum and average power output and LCoE of the best-found design per each experiment.

	Power [MW]								
	1+1EA	CMA-ES	PSO	GWO	DE	SaDE	LSHADE-EpSin	Bi-Level-1	Bi-Level-2
Mean	0.2325	0.2329	0.2208	0.2537	0.2501	0.2537	0.2541	0.2551	0.2612
Min	0.1941	0.2121	0.1934	0.2467	0.2327	0.2498	0.2473	0.2526	0.2544
Max	0.2590	0.2476	0.2392	0.2615	0.2589	0.2610	0.2621	0.2610	0.2792
STD	0.0234	0.0117	0.0181	0.0049	0.0087	0.0036	0.0046	0.0032	0.0088
	LCoE								
	1+1EA	CMA-ES	PSO	GWO	DE	SaDE	LSHADE-EpSin	Bi-Level-1	Bi-Level-2
Mean	0.0443	0.0303	0.0678	0.0315	0.0334	0.0309	0.0280	0.0295	0.0268
Min	0.0316	0.0284	0.0556	0.0297	0.0282	0.0277	0.0248	0.0267	0.0243
Max	0.0599	0.0382	0.0794	0.0335	0.0514	0.0329	0.0361	0.0324	0.0285
STD	0.0109	0.0036	0.0071	0.0014	0.0079	0.0019	0.0041	0.0019	0.0012

5.4. LCoE Minimisation

In this section, we describe the second applied objective function related to LCoE and approximated as a ratio of the generated energy to the significant mass of the system. The best-found LCoE values and their relevant cylinder configurations which are obtained using nine meta-heuristic approaches are shown in Table 6. Interestingly, all optimisation methods (except PSO) converged to a narrow range of radii between 5 and 7.3 m, with the smallest possible aspect ratio of 0.4. This geometry leads to the fact that the power generation will be dominated by the heave mode rather than surge. Moreover, this is clearly seen from the optimised values of the tether angles as to absorb power from the vertical motion, the tether angles should be closer to vertical leading to $\alpha_t < 35°$. Another important finding is that the power production of WECs optimised for LCoE is relatively low leading to 28.3 kW.

Figure 9b shows the box-and-whiskers plot for the best configurations of the WEC which deliver the minimum LCoE for each run for nine search heuristics. It can be seen that the performance of Bi-level-2 is more reliable than that of the other meta-heuristic algorithms we applied. Both LSHADE-EpSin and Bi-level-1 show the next best average performances by 0.028 and 0.0295, respectively.

Investigating the convergence trajectories (Figure 10) from this experiment in the real wave model, it is clear that Bi-level-2 converges faster than other optimisation methods. It is noteworthy that among the seven optimisation methods in the all-at-once strategy, the LSHADE-EpSin convergence speed is substantially better than the others due to both adaptive and non-adaptive strategies in order to adjust the control parameters as well as to conduct an embedded local search in the initial iterations. However, it can be seen that the convergence rate of GWO is considerable in the initial 1000 evaluations.

In order to see the convergence performance of Bi-level optimisation algorithms, the search trajectory of the best agent in each generation for all decision variables is shown in Figure 11. Initially, we can see the high convergence ability of Bi-level-2 compared with DE in order to find and converge to the optimal range of both radius and height. Meanwhile, It can be observed that Bi-level-2 tends to explore promising areas of the tether angle search space broadly, and finally, to exploit the best values.

Table 6. Best-found design parameters in order to minimise the LCoE.

Parameter	1+1EA	CMA-ES	PSO	GWO	DE	SaDE	LSHADE-EpSin	Bi-Level-1	Bi-Level-2
a [m]	7.31	6.40	14.32	7.00	7.38	6.57	5.00	6.15	5.00
H/a	0.40	0.40	0.40	0.4	0.40	0.40	0.40	0.40	0.40
α_t [deg]	28	29	10	10	31	25	35	31	34
α_{ap} [deg]	10	11	10	31	14	11	10	12	10
$\sum_{i=1}^{N_K} K_{pto}(\times 10^7)$	0.647	0.919	3.90	0.651	3.50	0.383	2.094	0.77	2.071
$\sum_{i=1}^{N_B} B_{pto}(\times 10^7)$	0.577	0.332	3.52	0.847	1.15	0.481	1.350	0.256	1.914
$LCoE$	0.0316	0.0284	0.0556	0.0297	0.0287	0.0277	0.0248	0.0267	0.0243
P_{AAP} [kW]	53.1	43.6	131	51.4	64.8	50.6	27.1	43.5	28.3

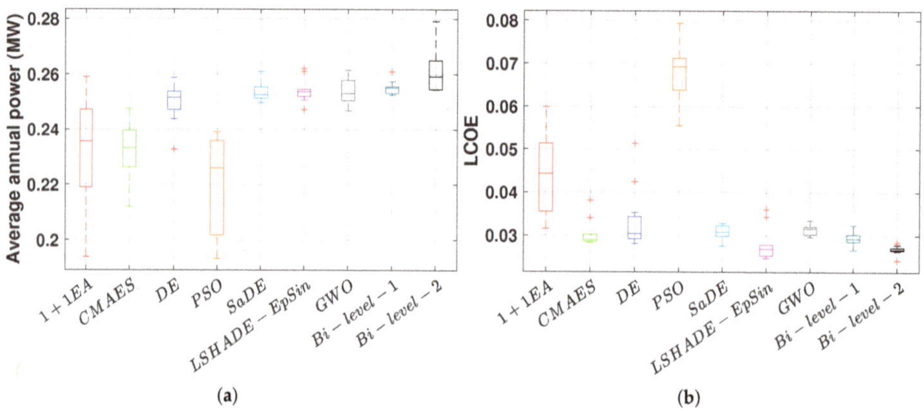

Figure 9. Each method runs 10 times. (**a**) Average annual produced power, (**b**) Levelised cost of energy (LCoE).

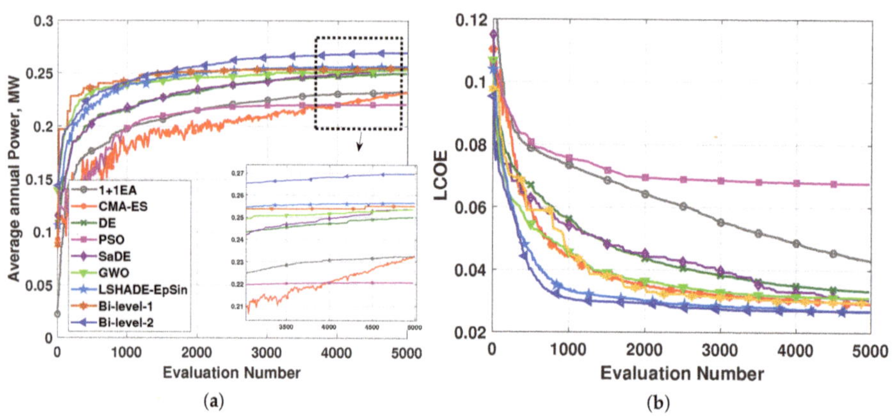

Figure 10. The average convergence rate comparison of the absorbed power and LCoE of the cylinder. Each method runs 10 times. (**a**) Average annual produced power, (**b**) Levelised cost of energy (LCoE).

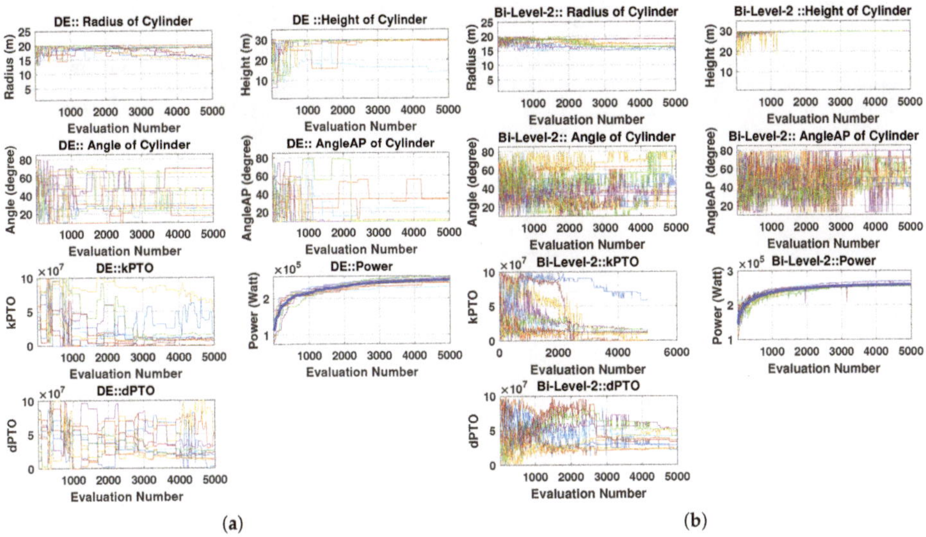

Figure 11. Search history and trajectory of the best solution per each population in all decision variables. (a) the optimisation process (power maximisation) of DE, (b) Bi-level-2.

6. Conclusions

In this paper, two new bi-level optimisation methods are proposed with the aim of maximising the harnessed power output. These methods are also designed to minimise the levelised cost of energy of a fully-submerged, cylindrical WEC with three tethers for the wave climate of a Mediterranean sea site in the west of Sicily, Italy (featuring unidirectional irregular waves). The optimisation of a combination of WEC radius, height, tether inclination and attachment angles, and power take-off parameters is a relatively computationally expensive (5000 evaluations take around 15 h), multi-modal, large-scale and complex problem. These characteristics provided the principal motivation for investigating and proposing a faster and more reliable optimisation technique. With this in mind, we applied a bi-level strategy to optimise the design variables at various levels. A global search method was used at the upper level to optimise the parameters of the whole WEC's. Furthermore, in the lower level, a Nelder–Mead (NM) simplex search method was applied to adjust the geometry settings and tether angles. To systematically compare the effectiveness of the proposed optimisation method, we considered seven state-of-the-art evolutionary and swarm algorithms. The experimental results showed that the bi-level method can outperform other meta-heuristics in terms of both convergence rate and the quality of WEC's configuration. Moreover, according to the best-found configurations, if we focus on maximising the harnessed power output without considering the costs, a large cylindrical buoy is recommended. However, the cheapest energy can be delivered by a relatively small WEC with a radius of 5 m and a height of 2 m.

Author Contributions: Conceptualization, M.N., N.Y.S., B.A. and M.W.; Data curation, N.Y.S. and M.M.N.; Formal analysis, M.N. and N.Y.S.; Methodology, M.N., N.Y.S., M.M.N., E.A., B.A. and M.W.; Resources, M.M.N.; Validation, M.N. and N.Y.S.; Visualization, M.N., N.Y.S., B.A. and M.W.; Writing—original draft, M.N., N.Y.S. and M.M.N.; Writing—review and editing M.N., N.Y.S., D.A.G., B.A. and M.W.; Supervision, D.A.G., B.A. and M.W.; project administration, M.W. All authors have read and agreed to the published version of the manuscript.

Funding: This research was funded by European Union's Horizon 2020 research and innovation programme under grant agreement No 727277 within the project ODYSSEA "Operating a network of integrated observatory systems in the Mediterranean sea" in order to collect data.

Acknowledgments: The authors would like to express their gratitude to Civil Engineering Department of Catania University and Favignana Municipality for their cooperation to provide all data. This research has been carried out within ODYSSEA project that received funding from the European Union's Horizon 2020 research and innovation programme under grant agreement No 727277. Moreover, we would like to show our gratitude to Fabien Voisin from the information technology and digital services, the University of Adelaide due to his valuable participation in the parallel computing services. This research is supported by the supercomputing resources provided by the Phoenix HPC service at the University of Adelaide.

Conflicts of Interest: The authors declare no conflict of interest.

Abbreviations

The following abbreviations are used in this manuscript:

WEC	Wave Energy Converter
PTO	Power Take-off system
PSO	Particle Swarm Optimisation
DE	Differential Evolution
SaDE	Self adaptive Differential Evolution
CMA-ES	Covariance Matrix Adaptation Evolution Strategy
LSHADE	Local Success-history Adaptive Differential Evolution

References

1. Murdock, H.E.; Gibb, D.; André, T.; Appavou, F.; Brown, A.; Epp, B.; Kondev, B.; McCrone, A.; Musolino, E.; Ranalder, L.; et al. *Renewables 2019 Global Status Report*; United Nations Environment Programme: Nairobi, Republic of Kenya, 2019. Available online: http://hdl.handle.net/20.500.11822/28496 (accessed on 19 September 2020).
2. Tronchin, L.; Manfren, M.; Nastasi, B. Energy analytics for supporting built environment decarbonisation. *Energy Procedia* **2019**, *157*, 1486–1493.
3. Mazzoni, S.; Ooi, S.; Nastasi, B.; Romagnoli, A. Energy storage technologies as techno-economic parameters for master-planning and optimal dispatch in smart multi energy systems. *Appl. Energy* **2019**, *254*, 113682.
4. Aderinto, T.; Li, H. Ocean wave energy converters: Status and challenges. *Energies* **2018**, *11*, 1250.
5. Falnes, J. A review of wave-energy extraction. *Mar. Struct.* **2007**, *20*, 185–201.
6. Astariz, S.; Iglesias, G. Wave energy vs. other energy sources: A reassessment of the economics. *Int. J. Green Energy* **2016**, *13*, 747–755.
7. Wen, Y.; Wang, W.; Liu, H.; Mao, L.; Mi, H.; Wang, W.; Zhang, G. A Shape Optimization Method of a Specified Point Absorber Wave Energy Converter for the South China Sea. *Energies* **2018**, *11*, 2645.
8. Alamian, R.; Shafaghat, R.; Safaei, M.R. Multi-Objective Optimization of a Pitch Point Absorber Wave Energy Converter. *Water* **2019**, *11*, 969.
9. Esmaeilzadeh, S.; Alam, M.R. Shape optimization of wave energy converters for broadband directional incident waves. *Ocean Eng.* **2019**, *174*, 186–200, doi:10.1016/j.oceaneng.2019.01.029.
10. Wang, L.; Ringwood, J.V. Geometric optimization of a hinge-barge wave energy converter. In Proceedings of the 13th European Wave and Tidal Energy Conference, Napoli, Italy, 1–6 September 2019; p. 1389.
11. Garcia-Teruel, A.; Forehand, D.I.M.; Jeffrey, H. Metrics for wave energy converter hull geometry optimisation. In Proceedings of the 13th European Wave and Tidal Energy Conference EWTEC, Napoli, Italy, 1–6 September 2019.
12. Sergiienko, N.Y.; Neshat, M.; da Silva, L.S.; Alexander, B.; Wagner, M. Design optimisation of a multi-mode wave energy converter. In Proceedings of the ASME 2020 39th International Conference on Ocean, Offshore and Arctic Engineering (OMAE2020), Fort Lauderdale, FL, USA, 28 June–3 July 2020.
13. Abdelkhalik, O.; Zou, S.; Robinett, R.D.; Bacelli, G.; Wilson, D.; Coe, R.G.; Korde, U.A. Multiresonant Feedback Control of a Three-Degree-of-Freedom Wave Energy Converter. *IEEE Trans. Sustain. Energy* **2017**, *8*, 1518–1527.
14. Neshat, M.; Alexander, B.; Sergiienko, N.; Wagner, M. A Hybrid Evolutionary Algorithm Framework for Optimising Power Take Off and Placements of Wave Energy Converters. *arXiv* **2019**, arXiv:1904.07043
15. Sharp, C.; DuPont, B. Wave energy converter array optimization: A genetic algorithm approach and minimum separation distance study. *Ocean Eng.* **2018**, *163*, 148–156.

16. Fang, H.W.; Feng, Y.Z.; Li, G.P. Optimization of Wave Energy Converter Arrays by an Improved Differential Evolution Algorithm. *Energies* **2018**, *11*, 3522.
17. Neshat, M.; Alexander, B.; Wagner, M.; Xia, Y. A detailed comparison of meta-heuristic methods for optimising wave energy converter placements. In Proceedings of the Genetic and Evolutionary Computation Conference. ACM, Kyoto, Japan, 15–19 July 2018; pp. 1318–1325.
18. Neshat, M.; Alexander, B.; Sergiienko, N.Y.; Wagner, M. Optimisation of Large Wave Farms Using a Multi-Strategy Evolutionary Framework. In *Proceedings of the 2020 Genetic and Evolutionary Computation Conference*; Association for Computing Machinery: New York, NY, USA, 2020; pp. 1150–1158, doi:10.1145/3377930.3390235.
19. Giassi, M.; Castellucci, V.; Göteman, M. Economical layout optimization of wave energy parks clustered in electrical subsystems. *Appl. Ocean Res.* **2020**, *101*, 102274, doi:10.1016/j.apor.2020.102274.
20. Fairley, I.; Lewis, M.; Robertson, B.; Hemer, M.; Masters, I.; Horrillo-Caraballo, J.; Karunarathna, H.; Reeve, D.E. A classification system for global wave energy resources based on multivariate clustering. *Appl. Energy* **2020**, *262*, 114515.
21. Franzitta, V.; Rizzo, G. Renewable energy sources: A mediterranean perspective. In Proceedings of the 2010 2nd International Conference on Chemical, Biological and Environmental Engineering, Cairo, Egypt, 2–4 November 2010; pp. 48–51.
22. Rusu, E.; Onea, F. Estimation of the wave energy conversion efficiency in the Atlantic Ocean close to the European islands. *Renew. Energy* **2016**, *85*, 687–703.
23. Rusu, E. Wave energy assessments in the Black Sea. *J. Mar. Sci. Technol.* **2009**, *14*, 359–372.
24. Bouali, B.; Larbi, S. Contribution to the geometry optimization of an oscillating water column wave energy converter. *Energy Procedia* **2013**, *36*, 565–573.
25. Kramer, M.V.; Frigaard, P. Efficient wave energy amplification with wave reflectors. In *The Twelfth International Offshore and Polar Engineering Conference. International Society of Offshore and Polar Engineers*; International Society of Offshore and Polar Engineers: Mountain View, CA, USA, 2002.
26. Vantorre, M.; Banasiak, R.; Verhoeven, R. Modelling of hydraulic performance and wave energy extraction by a point absorber in heave. *Appl. Ocean Res.* **2004**, *26*, 61–72.
27. Goggins, J.; Finnegan, W. Shape optimisation of floating wave energy converters for a specified wave energy spectrum. *Renew. Energy* **2014**, *71*, 208–220.
28. Hager, R.; Fernandez, N.; Teng, M.H. Experimental study seeking optimal geometry of a heaving body for improved power absorption efficiency. In *the Twenty-second International Offshore and Polar Engineering Conference. International Society of Offshore and Polar Engineers*; International Society of Offshore and Polar Engineers: Mountain View, CA, USA, 2012.
29. McCabe, A. Constrained optimization of the shape of a wave energy collector by genetic algorithm. *Renew. Energy* **2013**, *51*, 274–284.
30. De Andres, A.; MacGillivray, A.; Roberts, O.; Guanche, R.; Jeffrey, H. Beyond LCOE: A study of ocean energy technology development and deployment attractiveness. *Sustain. Energy Technol. Assessments* **2017**, *19*, 1–16.
31. Piscopo, V.; Benassai, G.; Della Morte, R.; Scamardella, A. Cost-based design and selection of point absorber devices for the mediterranean sea. *Energies* **2018**, *11*, 946.
32. Piscopo, V.; Benassai, G.; Cozzolino, L.; Della Morte, R.; Scamardella, A. A new optimization procedure of heaving point absorber hydrodynamic performances. *Ocean Eng.* **2016**, *116*, 242–259.
33. Piscopo, V.; Benassai, G.; Della Morte, R.; Scamardella, A. Towards a cost-based design of heaving point absorbers. *Int. J. Mar. Energy* **2017**, *18*, 15–29.
34. Sergiienko, N.Y.; Cazzolato, B.S.; Ding, B.; Arjomandi, M. An optimal arrangement of mooring lines for the three-tether submerged point-absorbing wave energy converter. *Renew. Energy* **2016**, *93*, 27–37, doi:10.1016/j.renene.2016.02.048.
35. Mirjalili, S.; Mirjalili, S.M.; Lewis, A. Grey wolf optimizer. *Adv. Eng. Softw.* **2014**, *69*, 46–61.
36. Awad, N.H.; Ali, M.Z.; Suganthan, P.N.; Reynolds, R.G. An ensemble sinusoidal parameter adaptation incorporated with L-SHADE for solving CEC2014 benchmark problems. In Proceedings of the 2016 IEEE Congress on Evolutionary Computation (CEC), Vancouver, BC, Canada, 24–29 July 2016.
37. Iuppa, C.; Cavallaro, L.; Vicinanza, D.; Foti, E. Investigation of suitable sites for Wave Energy Converters around Sicily (Italy). *Ocean Sci. Discuss.* **2015**, *12*, 315–354.

38. Silva, L.; Sergiienko, N.; Pesce, C.; Ding, B.; Cazzolato, B.; Morishita, H. Stochastic analysis of nonlinear wave energy converters via statistical linearization. *Appl. Ocean Res.* **2020**, *95*, 102023, doi:10.1016/j.apor.2019.102023.
39. Silva, L.S.P. Nonlinear Stochastic Analysis of Wave Energy Converters Via Statistical Linearization. Master's Thesis, University of São Paulo, São Paulo, Brazil, 2019.
40. Folley, M. *Numerical Modelling of Wave Energy Converters: State-of-the-Art Techniques for Single Devices and Arrays;* Elsevier Science: Saint Louis, MI, USA, 2016.
41. Spanos, P.D.; Arena, F.; Richichi, A.; Malara, G. Efficient dynamic analysis of a nonlinear wave energy harvester model. *J. Offshore Mech. Arct. Eng.* **2016**, *138*, 041901.
42. Silva, L.S.P.; Morishita, H.M.; Pesce, C.P.; Gonçalves, R.T. Nonlinear analysis of a heaving point absorber in frequency domain via statistical linearization. In Proceedings of the ASME 2019 38th International Conference on Ocean, Offshore and Arctic Engineering. American Society of Mechanical Engineers, Glasgow, Scotland, 9–14 June 2019.
43. Penalba, M.; Giorgi, G.; Ringwood, J.V. Mathematical modelling of wave energy converters: A review of nonlinear approaches. *Renew. Sustain. Energy Rev.* **2017**, *78*, 1188–1207.
44. Scruggs, J.T.; Lattanzio, S.M.; Taflanidis, A.A.; Cassidy, I.L. Optimal causal control of a wave energy converter in a random sea. *Appl. Ocean Res.* **2013**, *42*, 1–15, doi:10.1016/j.apor.2013.03.004.
45. Da Silva, L.S.P.; Cazzolato, B.S.; Sergiienko, N.Y.; Ding, B.; Morishita, H.M.; Pesce, C.P. Statistical linearization of the Morison's equation applied to wave energy converters. *J. Ocean. Eng. Mar. Energy* **2020**, 6, 1–13.
46. De Andres, A.; Maillet, J.; Hals Todalshaug, J.; Möller, P.; Bould, D.; Jeffrey, H. Techno-Economic Related Metrics for a Wave Energy Converters Feasibility Assessment. *Sustainability* **2016**, 8, 1109.
47. Sergiienko, N.Y.; Rafiee, A.; Cazzolato, B.S.; Ding, B.; Arjomandi, M. Feasibility study of the three-tether axisymmetric wave energy converter. *Ocean Eng.* **2018**, *150*, 221–233, doi:10.1016/j.oceaneng.2017.12.055.
48. Jiang, S.C.; Gou, Y.; Teng, B. Water wave radiation problem by a submerged cylinder. *J. Eng. Mech.* **2014**, *140*, 6014003, doi:10.1061/(ASCE)EM.1943-7889.0000723.
49. Jiang, S.C.; Gou, Y.; Teng, B.; Ning, D.Z. Analytical solution of a wave diffraction problem on a submerged cylinder. *J. Eng. Mech.* **2014**, *140*, 225–232, doi:10.1061/(ASCE)EM.1943-7889.0000637.
50. Hoerner, S. *Fluid-Dynamic Drag: Practical Information on Aerodynamic Drag and Hydrodynamic Resistance;* Hoerner Fluid Dynamics: Midland Park, NJ, USA, 1965.
51. The Specialist Committee on Waves. Final Report and Recommendations to the 23rd ITTC. In Proceedings of the 23rd International Towing Tank Conference, Venice, Italy, 8–14 September 2002; Volume II, pp. 505–736.
52. Sinha, A.; Malo, P.; Deb, K. A review on bilevel optimization: From classical to evolutionary approaches and applications. *IEEE Trans. Evol. Comput.* **2017**, *22*, 276–295.
53. McKinnon, K.I. Convergence of the Nelder–Mead Simplex Method to a Nonstationary Point. *SIAM J. Optim.* **1998**, *9*, 148–158.
54. Jansen, T.; Wegener, I. On the choice of the mutation probability for the (1+ 1) EA. In *International Conference on Parallel Problem Solving from Nature;* Springer: Cham, Switzerland, 2000; pp. 89–98.
55. Hansen, N. The CMA evolution strategy: A comparing review. *Towards a New Evolutionary Computation;* Springer: Berlin/Heidelberg, Germany, 2006; pp. 75–102.
56. Eberhart, R.; Kennedy, J. A new optimizer using particle swarm theory. In Proceedings of the Symposium on Micro Machine and Human Science (MHS), Nagoya, Japan, 4–6 October 1995; pp. 39–43.
57. Storn, R.; Price, K. Differential evolution–a simple and efficient heuristic for global optimization over continuous spaces. *J. Glob. Optim.* **1997**, *11*, 341–359.
58. Qin, A.K.; Huang, V.L.; Suganthan, P.N. Differential evolution algorithm with strategy adaptation for global numerical optimization. *IEEE Trans. Evol. Comput.* **2008**, *13*, 398–417.
59. Neumann, F.; Wegener, I. Randomized local search, evolutionary algorithms, and the minimum spanning tree problem. *Theor. Comput. Sci.* **2007**, *378*, 32–40.
60. Goudos, S.K.; Deruyck, M.; Plets, D.; Martens, L.; Joseph, W. Optimization of power consumption in 4G LTE networks using a novel barebones self-adaptive differential evolution algorithm. *Telecommun. Syst.* **2017**, *66*, 109–120.
61. Ramli, M.A.; Bouchekara, H.; Alghamdi, A.S. Optimal sizing of PV/wind/diesel hybrid microgrid system using multi-objective self-adaptive differential evolution algorithm. *Renew. Energy* **2018**, *121*, 400–411.

62. Tanabe, R.; Fukunaga, A.S. Improving the search performance of SHADE using linear population size reduction. In Proceedings of the 2014 IEEE Congress on Evolutionary Computation (CEC), Beijing, China, 6–11 July 2014; pp. 1658–1665.
63. Tanabe, R.; Fukunaga, A. Success-history based parameter adaptation for differential evolution. In Proceedings of the 2013 IEEE Congress on Evolutionary Computation, Cancun, Mexico, 20–23 June 2013; pp. 71–78.
64. Zhang, J.; Sanderson, A.C. JADE: Adaptive differential evolution with optional external archive. *IEEE Trans. Evol. Comput.* **2009**, *13*, 945–958.
65. Bullen, P.S. *Handbook of Means and Their Inequalities*; Springer Science & Business Media: Cham, Swtzerland, 2013; Volume 560.
66. Goudos, S.K.; Siakavara, K.; Samaras, T.; Vafiadis, E.E.; Sahalos, J.N. Self-adaptive differential evolution applied to real-valued antenna and microwave design problems. *IEEE Trans. Antennas Propag.* **2011**, *59*, 1286–1298.
67. Rajagopalan, A.; Sengoden, V.; Govindasamy, R. Solving economic load dispatch problems using chaotic self-adaptive differential harmony search algorithm. *Int. Trans. Electr. Energy Syst.* **2015**, *25*, 845–858.
68. Nelder, J.A.; Mead, R. A simplex method for function minimization. *Comput. J.* **1965**, *7*, 308–313.
69. Ghasemi, M.; Ghavidel, S.; Ghanbarian, M.M.; Habibi, A. A new hybrid algorithm for optimal reactive power dispatch problem with discrete and continuous control variables. *Appl. Soft Comput.* **2014**, *22*, 126–140.
70. Rajan, A.; Malakar, T. Optimal reactive power dispatch using hybrid Nelder–Mead simplex based firefly algorithm. *Int. J. Electr. Power Energy Syst.* **2015**, *66*, 9–24.

energies

Article

Numerical Simulation and Experimental Validation of an Oil Free Scroll Compressor

Massimo Cardone * and **Bonaventura Gargiulo**

Department of Chemical, Materials and Production Engineering, University of Naples Federico II, 80125 Naples, Italy; bonaventura.gargiulo@unina.it

* Correspondence: massimo.cardone@unina.it; Tel.: +39-081-7683675; Fax: +39-081-2394165

Received: 15 October 2020; Accepted: 6 November 2020; Published: 10 November 2020

Abstract: This paper presents a virtual model of a scroll compressor developed on the one-dimensional analysis software Simcenter Amesim®. The model is semi-empirical: it needs some physical details of the modelled machine (e.g., the cubic capacity), but, on the other hand, it does not require the geometrical features of the spirals, so it needs experimental data to calibrate it. The model also requires rotational speed and the outlet temperature as boundary conditions. The model predicts the power consumption and the mass flow rate and considers leakages and mechanical losses. After the model presentation, this paper describes the test bench and the obtained data used to calibrate and validate the model. At last, the calibration process is described, and the results are discussed. The calculated values fit the experimental data also in extrapolation, despite the model is simple and performs calculations within 7 s. Due to these characteristics, the model is suitable for being used in a larger model as a sub-component.

Keywords: scroll-compressor; experimental validation; numerical model

1. Introduction

Scroll compressors are widely used in applications where noise pollution and low vibrations are relevant factors, such as domestic refrigeration and domestic climatic control. Due to its unique properties, these machines are given much attention by scientific and industrial researchers. Some works, especially the oldest ones [1,2], are mainly focused on the theoretical functioning of those machines, even though the final aim has always been the improvement of the scroll efficiency. The scroll compressors efficiency-enhancing is pursued using different approaches, such as researching on the design of rotor profiles to reduce volumetric losses [3,4] or investigating the cooling effects on the compression work [5,6].

The latest most attractive research branches on scroll machines concern studies on the machine behaviour into thermodynamics cycles [7], works on the performance of injecting water (or vapour) compressors [8,9] and studies on the scroll expanders [10].

Many works on scroll compressors or expanders use a mathematical model that is generally virtualised with a low-level programming language. Among them, some use a geometrical approach [11–15] while others use semi-empirical methods [16–18]. Some details of one work for each group are briefly illustrated below. Blunier et al. [14] presented a model written in very-high-speed-integrated-circuits hardware description language (VHDL) code including the scroll's geometrical features in it; the model does not need any calibration. Winandy et al. [18] used the Engineering Equation Solver (EES) software to virtualise the model equations. The model needs seven parameters to be obtained through a calibration process on mass flow rate, power output and outlet temperature. The required data points are obtained by themselves through several experimental runs.

The use of commercial 0-1D fluid-dynamic code was not frequent for machines models, while they are often used to simulate a whole thermodynamic cycle [18–20]. Ziliani et al. [19] used the commercial software Amesim® (Siemens PLM Software, Plano, TX, USA)to model an entire Organic Rankine Cycle (ORC) plant where it is supposed to be used a screw or a scroll expander. The machines' behaviour was deduced by a Computational Fluid Dynamics (CFD) simulation (for the scroll compressor) and a geometry-based simulation (for the screw compressor). Bracco et al. [20] used a similar approach for their ORC plant simulation: their machine model needs four parameters, that are based on a combination of some manufacturer information, experimental data and a self-made scroll simulation tool.

However, Bell et al. [21] recently (2020) developed an open-source platform (named PDSim) specific for modelling positive-displacement compressors. Tanveer et al. [22] compared different software on a reciprocating compressor finding that the PDSim suit has good potential. On the other hand, Rak et al. [23] relied on a CFD analysis to model in detail a cooled scroll compressor, thus considering heat transfer issues.

The main purpose of this work was the development of a fluid-dynamic model of a scroll compressor and its experimental validation. The model does not consider the internal geometry of the scroll, nor its kinematic behaviour. The model aim is to perform calculations in a few seconds so that it can be used as a sub-component of a whole plant model. In this paper, an oil-less commercial scroll compressor is modelled by a zero-dimensional semi-empirical model developed using the commercial software Simcenter Amesim® (version 19.2). The compressor is viewed as a pneumatic system made up by a series of peculiar reciprocating compressors with driven valves. Leakages and a keyed fan power consumption are considered. The experimental activities are performed on a commercial oil-less scroll compressor, at four rotational speed levels and six pressure levels.

2. The Scroll Compressor

The same timing scheme characterises all rotary machines. The rotors uncover ports and intercept cells carved in the stator, carrying the working fluid from the inlet side to the outlet. As the cells have a decreasing volume, a design pressure ratio β_i is generated by the internal volume reduction (for roots compressors $\beta_i = 1$). The compression ratio β, required by the application, may overlap or not with the compression ratio β_i, that the machine can produce due to volume change of cells. Usually, regarding the scroll compressors, there is a moving spiral and a fixed spiral. The moving spiral describes an orbit around the base circle centre of the fixed. The spirals are thick circle involutes, and, in most cases, they are equals. Many authors described the scroll geometry and kinematic characteristics in details (e.g., Chen et al. [13]). The proposed model is not based on the geometry of the spirals, so this work will not examine these issues.

Figure 1a shows a scroll compressor working scheme, while Figure 1b shows a rotary volumetric compressor's ideal cycle diagram. Referring to Figure 1, these machines make a first internal compression (from state 0 to state 1) through volume reduction. Then, at the opening of the last internal contacts of the machine, two volumes are put into communication: the last cell filled with gas in state 1 (p_1, T_1) and the discharge volume filled with gas in state 2 (p_2, T_2). Therefore, an instantaneous mixing phase (at constant volume) starts: the state of the whole gas becomes an intermediate state (p_x, T_x). In the next step of the machine rotors, while the volume of the last cell decreases, state 2 in the volume V_m is restored.

An authors' previous work [24] theoretically analysed the machine $\beta \geq \beta_i$ field. It illustrated an alternative representation of a generic ideal rotary compressor's working scheme (see Figure 1). The theoretical analysis confirmed that the two theoretical models are equivalent. It was demonstrated that the proposed representation leads to a simpler but rigorous equation to calculate the ideal specific work consumption (see Equation (1)).

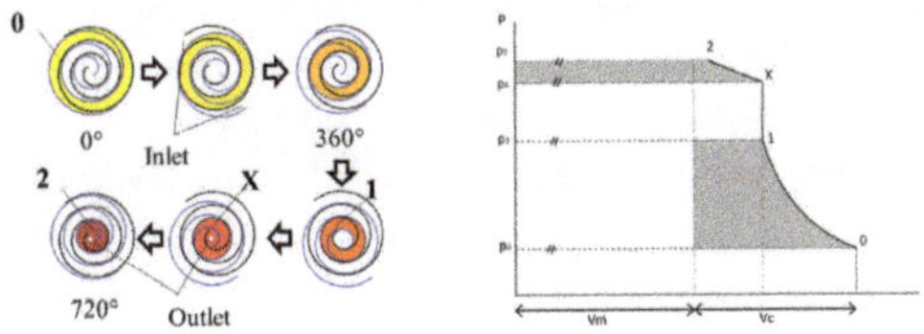

(a) (b)

Figure 1. (**a**) Scroll functioning scheme; (**b**) rotary compressors' diagram.

In this work, Equation (1) is used to give a theoretical validation to the proposed model, before implementing real losses and calibrating it on experimental data.

$$w_{0-2} = \frac{k}{k-1} R T_0 \left(\beta_i^{\frac{k-1}{k}} - 1 \right) + \frac{k}{k-1} R T_X \left(\beta_{X2}^{\frac{k-1}{k}} - 1 \right) \left(\frac{m_m + m_C}{m_C} \right) \tag{1}$$

where k is the isentropic index, R is the gas constant (for air), T_0 is the inlet temperature, T_X is the temperature in the state X, m_c is the mass elaborated per cycle, m_m is the mass in the discharge volume. The other abbreviations are collected in Equation (2).

$$\beta_i = \frac{p_1}{p_0} \quad ; \quad \beta_{X2} = \frac{p_2}{p_X} \; ; \; T_X = \frac{p_1 V_1 + p_2 V_m}{\frac{p_1 V_1}{T_1} + \frac{p_2 V_m}{T_2}} \tag{2}$$

3. Numerical Model

The numerical model is developed on Simcenter Amesim® software.

Referring to Figure 2, two pairs (A, B) of variable volumetric chambers are used to simulate the scroll compressor. The volumetric variation does not follow the actual scroll chambers variation. The model uses a simple sinusoidal function of the shaft rotation. The compression process is modelled through the subsequent steps: the suction phase of chambers A1 and B1 stands for the scroll suction. They are in phase opposition so that the system has the suction phase 360° long and the sum of their cubic capacity is the scroll capacity. Three controlled valves for each pair of chambers are used to simulate the openings of the scroll contacts. These valves are opened every 180°. Focus on one pair of chambers (e.g., chambers A1 and A2). The first valve stands for the contacts that enclose half scroll suction chamber when it reaches its maximum capacity; the second valve separates the first chamber from the second and it stands for the contacts that separate the scroll suction chambers from the other scroll chambers. Then, the third valve stands for the contacts that separate the scroll compression chamber from the discharge chamber. Chambers A2 and B2 are smaller than the others and they are in phase opposition respective to chamber A1 and B1. When the second valve opens, the compression phase begins due to the total volumetric decreasing achieved after the second valve opening.

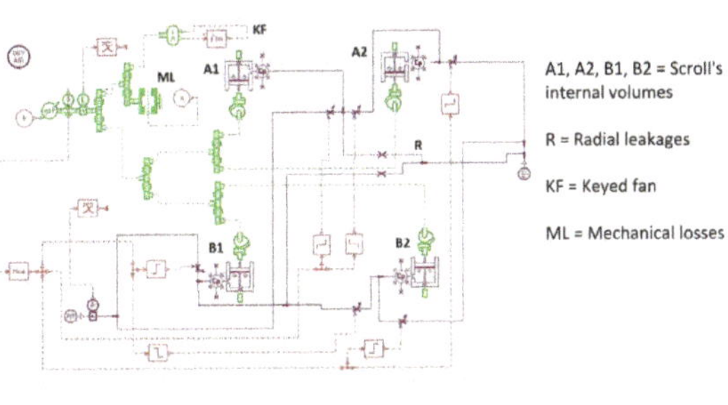

Figure 2. Model code in Amesim environment.

An ideal electric engine imposes a constant speed to the whole model. In this work, electric losses are neglected, because the experimental work input is measured downstream of the electric motor.

After the compression phase, the third valve opens and the second close. The discharge phase and the suction phase start respectively in chambers 2 and 1. The other pair of chambers (e.g., B1 and B2) are just the same, but they are in opposition of phase. Therefore, there is an entire suction and discharge that lasts 2π radians long in a revolution. An ideal infinite volume simulates the compressor discharge tank.

In this model, the heat losses are neglected, so the chambers are adiabatic. Therefore, the outlet temperature, as well as the inlet temperature, are set based on the experimental data. According to Yu Chen et al. [13] work there are two types of leakages: the flank leakages and the radial leakages. The flank leakages are simulated by an imperfect closing of the valves (the valves V1, V2, V3 that simulate the internal contacts). Therefore, when they should be closed, they are slightly opened. The radial leakages are simulated by internal by-pass (R). The mass flow rate through the valves is calculated via Equation (3) [25].

$$\dot{m} = AC_qC_m\frac{P_{up}}{\sqrt{T_{up}}} \tag{3}$$

where A is the actual orifice area, C_q is Perry's coefficient from his correlation [26], C_m is the flow coefficient as also described by Szente et al. [25], P_{up} and T_{up} are the upstream pressure and temperature.

The mechanical losses (ML) are modelled as a fixed frictional loss (one constant to be estimated). The keyed fan (KF) is simulated by a torque load function of the square of the rotational speed, so there are three parameters (A, B, C) to be estimated (see also Equation (4)).

$$Torque\ load = A\omega^2 + B\omega + C \tag{4}$$

At first, the model is tested on an ideal case, suppressing the elements KF, ML and R. The model results are compared with the theoretical results of Equation (1), at different internal and total pressure ratios. Figure 3 shows that there are no significant deviations between the scroll ideal performances and the simplified numerical model. The maximum deviation obtained is always under 0.8% of the theoretical values (the model calculation is always greater than the theoretical value). These differences are caused by some small pressure losses still present in the model, and, secondary, by the model outlet tank, that the code treats as an infinite volume (the theoretical calculation is under the assumption of an outlet tank of 10^5 times the scroll cubic capacity). Overall, the proposed model is congruent with the theory, so, enabling KF, ML and R, it should be able to simulate real scrolls. The values of these elements' parameters are estimated through a calibration process based on experimental data.

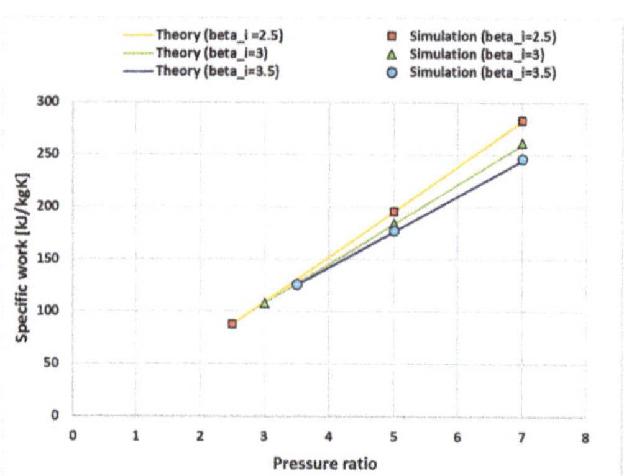

Figure 3. Comparison between theory and the proposed model.

4. Results

4.1. Experimental Activities

4.1.1. Experimental Setup

Figure 4a shows the experimental test bench used, Figure 4b illustrates its scheme and Table 1 collects the details on its main components. The machine tested is a 2.1 kW oil-free scroll compressor (C) taken from the ATLAS COPCO SF2 clean air generator (ATLAS COPCO, Nacka, Sweden). The required power is given by a three-phase oscillating-casing electrical asynchronous motor (EM). The motor is connected to an inverter that controls the motor speed modifying the current frequency. A load cell measures the required torque (Tq): the motor oscillating-casing is constrained by the load cell through an arm of a known length, so the product of the sensed force and the arm length is equal to the torque given by motor. A trapezoidal belt connects the compressor to the motor with a unitary transmission ratio (TB). The encoder (RPM) is integral with the compressor axis, measuring the compressor speed. The compressor original cooling fan is a centrifugal fan (KF), and it is keyed on the scroll axis. A turbine flow meter (V) measures the air volumetric flow rate at the suction of the compressor. Two K-type thermocouples (T) (mounted through a T-joint) measure the temperatures, both at the inlet and outlet pipe of the machine. The outlet thermocouple is mounted at 15 cm from the outlet port, due to the compressor built-in external fins. A piezo-resistive sensor (p) is linked to the calm reservoir (TANK) at the outlet to measure the required average pressure imposed on the compressor. The circuit ends with a regulator valve (RV) to control the outlet pressure. The data are digitalized by data acquisition system made up by a NI-DAQ USB 6259 (National Instrument, Austin, TX, USA) and a NI-FieldPoint cFP 1808 coupled with a cFP-TC-120 module that provides built-in cold joint correction for the thermocouples. All data are processed by a self-made software realized in LabVIEW ™ (version 15.5, National Instrument, Austin, TX, USA) code.

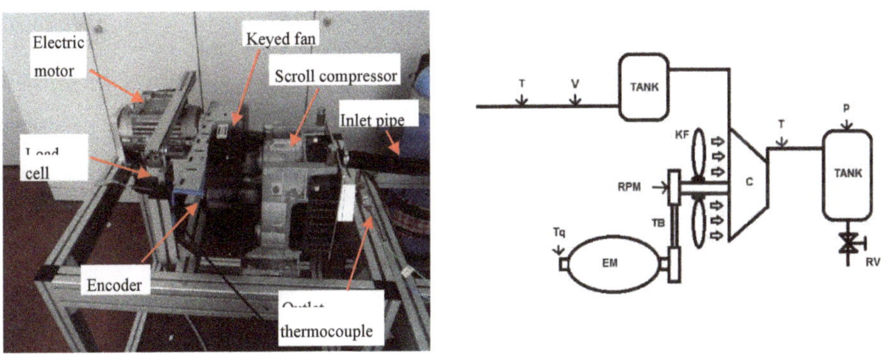

(a) (b)

Figure 4. (**a**) Test bench; (**b**) test bench scheme.

Table 1. Test bench elements.

Symbol	Element	Details	Accuracy
EM	Electric engine		
C	Compressor	2.1 kW scroll compressor	
TB	Transmission belt	Trapezoidal transmission belt. Transmission ratio = 1:1	
p	Pressure sensor	Piezo-resistive sensor	±4%
Tq	Torque sensor	Strain-gauge force sensor applied to the electric engine's known length arm	±0.011 Nm
RPM	Encoder	500 pulse per round	
RV	Regulation valve		
T	Temperature sensor	K-type thermocouple	±1.5 K
V	Volumetric flow rate sensor	Turbine flow meter	±3%
KF	Keyed fan	Centrifugal cooling fan	
TANK	Dumping tank		

4.1.2. Experimental Plan

The test campaign considered in this work consists of four runs, each performed at total pressure ratios (β) from 2 to 7 and then from 7 to 2. Each run is performed at a different constant speed: 1000, 1500, 2000 and 2500 RPM. Once the compressor speed is set, the pressure in the outlet tank is controlled through the regulation valve. Starting from the lowest pressure ratio, when the desired value is reached, the acquisition system waits until the outlet temperature is stationary and then saves the data. Then, the regulation valve is tightened to achieve the next desired pressure ratio. The process is repeated until the pressure ratio reaches 7, and the system saves the data. Then, compressor state is modified reaching a pressure ratio of 7.5 (approaching the limits of the test bench). After 5 min, the process is repeated from pressure ratio 7 to 2. Despite the different thermal dynamics between the rising and the falling part of the run, the next paragraphs show that the measured data are close to each other.

4.1.3. Experimental Results

Figure 5 presents the working fluid (air) temperature versus the total pressure ratio and for various rotational speed. The experimental reproducibility is average: the higher error is below 5%. It is shown that the temperature is an increasing function of both compression ratio and rotational speed. The temperature is still increasing due to both the rising of the compressed mass flow rate and the rise of frictional losses, despite a higher rotational speed causes a higher cooling flow rate (generated by the keyed fan).

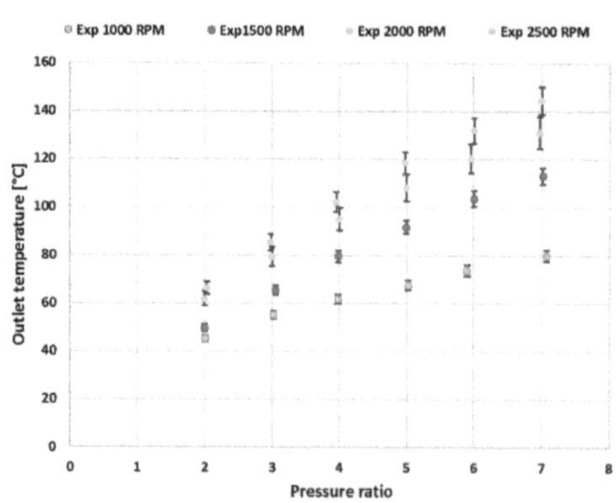

Figure 5. Experimental data: outlet temperature.

The mass flow rate and power consumption experimental data are shown together with the simulated data (the model results will be discussed in the next paragraphs): Figure 6 shows the compressor mass flow rate and Figure 7 shows the power consumption with the fan. According to Figure 6, the higher the pressure in the delivery tank, the lower the mass flow rate. This effect is heavier for low rotation speed. This tendency is caused by both the temperature effect (the whole machine temperature is higher at higher compression ratios) and the leakages (a higher-pressure gradient across the gaps leads to more air leaked). The experimental reproducibility is good: the higher error is below 3%, and the absolute errors are under 0.1 g/s.

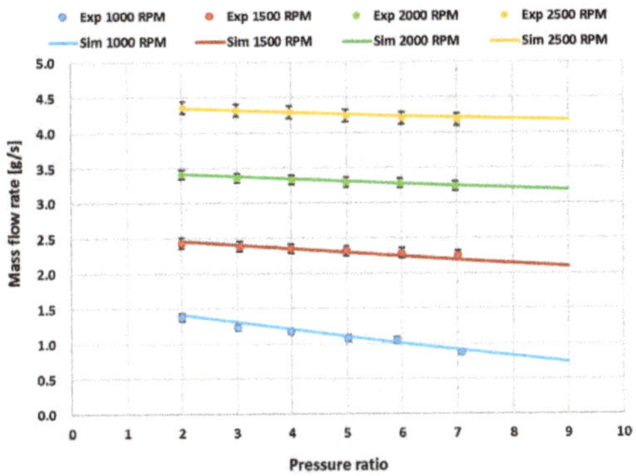

Figure 6. Model validation: mass flow rate.

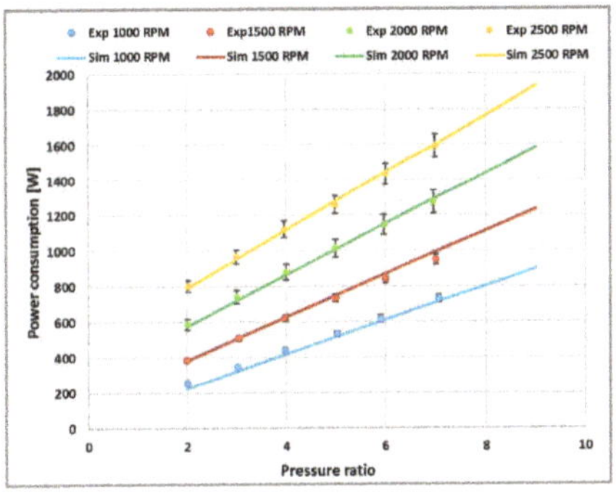

Figure 7. Complete model validation: power consumption.

Figure 7 shows the power consumption data of compressor for all experimental tests, calculated via Equation (5), where M_{exp} and ω_{exp} are the measured torque and the measured rotational speed. The experimental reproducibility is average: the higher error is below 5%

$$P_{exp} = M_{exp}\omega_{exp} \tag{5}$$

Both the power and the temperature are increasing functions of both compression ratio and rotational speed.

4.2. Model Results

4.2.1. Model Calibration and Validation: Leakages

In the model, the adiabatic process simplified hypothesis is done, supposing that the fan cooling affects only the final stage of the compression process (i.e., the curve X-2 shown in Figure 1). The first step of the calibration process is to determine the leakages. In the model (see Figure 3), the flank gaps are modelled as an imperfect closing of the internal contacts (the valves V1–V3) while the radial leakages are modelled as a by-pass. The model requires the orifice area for each of them. These areas are not the physical areas of the scroll gaps, but their model representation. It is possible to determine them using the experimental data on the mass flow rate. It can be postulated that both the centrifugal force and the increasing temperature would reduce the internal clearances. Therefore, the influence of the rotor speed over the leakages must be considered.

The calibration is performed using experimental data at $\beta = 2$ and $\beta = 6$ at 1000, 2000 and 2500 RPM. All the other experimental points are used to validate the model. In particular, the $\beta = 7$ points are used to verify the model consistency outside the calibration data field, while the 1500 RPM experimental points are used to evaluate the model consistency in simulating the scroll running at another rotational speed. The calibration points temperatures are set based on the experimental data. In other cases, the temperatures are based on a linear regression of the values measured at $\beta = 2$ and $\beta = 6$. As for the 1500 RPM temperatures, they are based on a linear regression of the values determined at 1000, 2000 and 2500 RPM. Similarly, the orifice areas are estimated anew for each rotor speed. For the validation at 1500 RPM, the orifice areas are calculated based on a linear regression of the values determined at 1000, 2000 and 2500 RPM. Figure 6 shows the model compliance with the real scroll in terms of mass flow rate after the calibration process (coloured dots for the experimental data and coloured lines the calculated one).

4.2.2. Model Calibration and Validation: Mechanical Losses

A new set of experimental runs were performed to calculate the mechanical losses (modelled by the ML block). This experimental set-up is characterised by the absence of the keyed fan. In this way, we get the difference between the experimental power consumption and the simulated one (once the KF block is suppressed, see Equation (6)).

$$M_{ML\ block} = \frac{P_{experimental}}{\omega} - M_{model\ scroll} \tag{6}$$

The machine temperatures were higher than the previous case, so the higher pressure ratio tested is 4. As the thermal conditions were altered, the spirals thermal deformation is different; therefore, a new leakages calibration was needed. Figure 8 shows the accordance between the model and the new experimental set (the one without the fan) after the new calibration. It is possible to determine the value of the ML torque confronting the power consumption experimental data with the model output. The torque can be assumed as a constant independent by the rotational speed as the mechanical losses are just simple friction losses (there are no other auxiliaries nor inertial forces).

4.2.3. Model Calibration and Validation: Keyed Fan

The speed effect on the torque losses (ML) was neglected, so the mechanical loss is set to constant. As previously said, the keyed fan (KF) power consumption is modelled by a torque load function of the square of the rotational speed, so there are three parameters (A–C) to be estimated (see also Equation (2)). Once these parameters are determined, they are set as constant, independent from both the speed and the pressure ratio. As previously, the data at $\beta = 2$ and $\beta = 6$ at 1000, 2000 and 2500 RPM are used to calibrate, the others are used to validate. Figure 7 shows the comparison between the experimental power consumption (coloured dots) and the calculated one (coloured lines), after the calibration process. The figure shows the power consumption as a function of the pressure ratio and the rotational speed. In addition, the 1500 RPM and all $\beta = 7$ experimental data are well predicted by the model. The model complies with experimental data, but some deviations remain. These deviations are higher for lower rotational speed (that are further from the real compressor nominal speed and further from adiabatic behaviour).

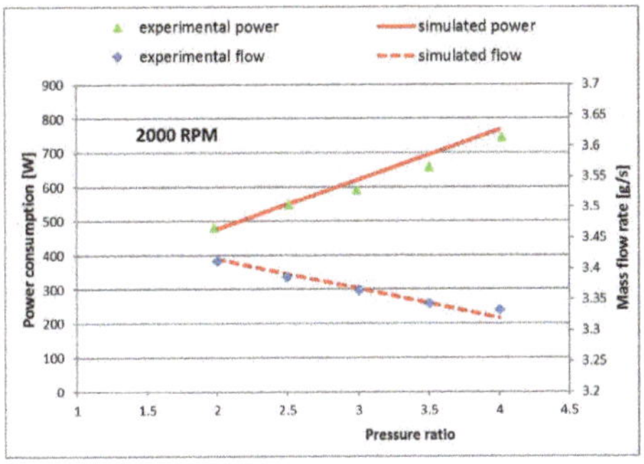

Figure 8. Comparison of mass flow rates and power consumption between the model and the real compressor without the keyed fan.

5. Conclusions

In the first part of the paper, the key features of a scroll compressor are presented. Then a numerical model developed in Simcenter Amesim® is presented: the model includes leakages and mechanical losses. Despite that the integrated keyed cooling fan power consumption is considered, its cooling effect is neglected, and the compressor is considered adiabatic. To partially overcome this limitation, the outlet temperature is set as a boundary condition using experimentally derived values. The model was preliminary tested on ideal cases (no losses or leakages). Then, the experimental activity is presented: a series of experimental runs were performed on a commercial scroll compressor. Four levels of rotational speed (1000, 1500, 2000, 2500 RPM) and six levels of the pressure ratio (2–7) were considered. The experimental data are reproducible as the errors are under 5%. Finally, the model is calibrated on the experimental data. Six points are used to calibrate the model (at $\beta = 2$ and $\beta = 6$ at 1000, 2000 and 2500 RPM), while the other 26 are used to validate. After the proper calibration of the leakages, mechanical losses and fan power consumption, the model can follow the scroll's real behaviour.

The model does not consider the actual scroll geometry and kinematic. This feature could be both an advantage or a limitation: it is a limitation because it is not possible to calculate any inner quantity nor instantaneous quantities (e.g., instantaneous torque); it is an advantage when the geometrical features are unknown. Moreover, the calculations last less than seven seconds. Overall, the model is not suitable as a machine design aid, but, on the other hand, it can be used as a component for a whole plant simulation.

Author Contributions: The conceptualization, methodology and experimental design have been developed by M.C.; the data acquisition, software model development and model calibration have been performed by B.G. All authors have read and agreed to the published version of the manuscript.

Funding: This research received no external funding.

Acknowledgments: The authors would like to thank Salvatore De Cristofaro for his technical assistance in laboratory activities.

Conflicts of Interest: The authors declare no conflict of interest.

Abbreviations

c_p	Isobaric specific heat capacity [kJ/kgK]
c_v	Isochoric specific heat capacity [kJ/kgK]
H	Enthalpy [kJ]
h	Specific enthalpy [kJ/kg]
k	Isentropic index
M	Torque [Nm]
m_c	Inlet air mass [kg]
m_m	Mass inside discharge volume [kg]
p	Pressure [Pa]
R	Gas constant [kJ/kgK]
T	Temperature [K]
V	Volume [m^3]
v	Specific volume [m^3/kg]
V_c	Inlet volume [cm^3]
V_c'	Last closed cell volume [cm^3]
V_m	Discharge volume [cm^3]
W	Work [kJ]
w	Specific work [kJ/kg]
0	Inlet state point
1	End of internal compression state point
X	End of isochoric compression state point
2	End of global compression state point
β	Total compression ratio p_2/p_0
βi	Internal compression ratio p_1/p_0
β_{X2}	p_2/p_X
ρ_i	Internal volumetric compression ratio
ω	Rotational speed [rad/s]

References

1. Stone, C. The efficiency of Roots compressors and compressors with fixed internal compression. *Proc. Instit. Mech. Eng. Part A Power Proc. Eng.* **1988**, *202*, 199–205. [CrossRef]
2. Yanagisawa, T.; Cheng, M.D.; Fukuta, M.; Shimizu, T. Optimum operating pressure ratio for scroll compressors. In Proceedings of the International Compressor Engineering Conference, Purdue, IN, USA, 17–20 July 1990.
3. Ishii, N.; Sakai, M.; Sano, K.; Yamamoto, S.; Otokura, T. A fundamental optimum design for high mechanical and volumetric efficiency of compact scroll compressors. In Proceedings of the International Compressor Engineering Conference, Purdue, IN, USA, 23–26 July 1996; p. 1176.
4. Lee, Y.R.; Wu, W.F. On the profile design of a scroll compressor. *Int. J. Refrig.* **1995**, *18*, 308–317. [CrossRef]
5. Xudong, W.; Yunho, H.; Reinhard, R. Investigation of potential benefits of compressor cooling. *Appl. Therm. Eng.* **2008**, *28*, 1791–1797.
6. Shuaihui, S.; YuanYang, Z.; Pengcheng, S. Simulation research on scroll refrigeration compressor with external cooling. *Int. J. Refrig.* **2010**, *33*, 897–906.
7. Yang, S.C.; Hung, T.C.; Feng, Y.Q.; Wu, C.J.; Wong, K.W.; Huang, K.C. Experimental investigation on a 3 kW organic Rankine cycle for low-grade waste heat under different operation parameters. *Appl. Therm. Eng.* **2017**, *113*, 756–764. [CrossRef]
8. Mendoza, L.C.; Sylvain, L.; Jürg, S. Testing and modelling of a novel oil-free co-rotating scroll machine with water injection. *Appl. Energy* **2017**, *185*, 201–213. [CrossRef]
9. Tello-Oquendo, F.M.; Navarro-Peris, E.; Gonzálvez-Maciá, J. New characterization methodology for vapor-injection scroll compressors. *Int. J. Refrig.* **2017**, *74*, 526–537. [CrossRef]
10. Zhang, X.; Xu, Y.; Xu, J.; Sheng, Y.; Zuo, Z.; Liu, J.; Chen, H.; Wang, Y.; Huang, Y. Study on the performance and optimization of a scroll expander driven by compressed air. *Appl. Energy* **2017**, *186*, 347–358. [CrossRef]
11. Cho, I.; Yong, S.; Bin, K.; Yongchan, K. Optimization of injection holes in symmetric and asymmetric scroll compressors with vapor injection. *Int. J. Refrig.* **2012**, *35.4*, 850–860. [CrossRef]

12. Lemort, V.; Sylvain, Q.; Jean, L. Numerical simulation of a scroll expander for use in a Rankine Cycle. In Proceedings of the International Compressor Engineering Conference, Purdue, IN, USA, 14–17 July 2008.
13. Chen, Y.; Halm, N.P.; Groll, E.A.; Braun, J.E. Mathematical modeling of scroll compressors—Part I: Compression process modelling. *Int. J. Refrig.* **2002**, *25*, 731–750. [CrossRef]
14. Blunier, B.; Cirrincione, G.; Herve, Y.; Miraoui, A. A new analytical and dynamical model of a scroll compressor with experimental validation. *Int. J. Refrig.* **2009**, *32*, 874–891. [CrossRef]
15. Sung, J.P.; Joon, H.B.; Eui, G.J. Transient thermodynamic modeling of a scroll compressor using R22 refrigerant. *Energies* **2020**, *13*, 3911. [CrossRef]
16. Winandy, E.L.; Lebrun, J. Scroll compressors using gas and liquid injection: Experimental analysis and modelling. *Int. J. Refrig.* **2002**, *25*, 1143–1156. [CrossRef]
17. Dardenne, L.; Fraccari, E.; Maggioni, A.; Molinaroli, L.; Proserpio, L.; Winandy, E. Semi-empirical modelling of a variable speed scroll compressor with vapour injection. *Int. J. Refrig.* **2015**, *54*, 76–87. [CrossRef]
18. Winandy, E.; Saavedra, C.; Lebrun, J. Experimental analysis and simplified modelling of a hermetic scroll refrigeration compressor. *Appl. Therm. Eng.* **2002**, *22*, 107–120. [CrossRef]
19. Ziviani, D.; Suman, A.; Lecompte, S.; De Paepe, M.; van de Broek, M.; Spina, P.R.; Pinelli, M.; Venturini, M.; Beyene, A. Comparison of a single-screw and a scroll expander under part-load conditions for low-grade heat recovery ORC systems. *Energy Proc.* **2014**, *61*, 117–120. [CrossRef]
20. Bracco, R.; Clemente, S.; Micheli, D.; Reini, M. Experimental tests and modelization of a domestic-scale ORC (Organic Rankine Cycle). *Energy* **2013**, *58*, 107–116. [CrossRef]
21. Bell, I.H.; Ziviani, D.; Lemort, V.; Bradshaw, C.R.; Mathison, M.; Horton, W.T.; Braun, J.E.; Groll, E.A. PDSim: A general quasi-steady modeling approach for positive displacement compressors and expanders. *Int. J. Refrig.* **2020**, *110*, 310–322. [CrossRef]
22. Tanveer, M.M.; Bradshaw, C.R. Quantitative and qualitative evaluation of various positive-displacement compressor modeling platforms. *Int. J. Refrig.* **2020**, *119*, 48–63. [CrossRef]
23. Rak, J.; Pietrowicz, S. Internal flow field and heat transfer investigation inside the working chamber of a scroll compressor. *Energy* **2020**, *202*, 117700. [CrossRef]
24. Abagnale, C.; Cardone, M.; Gargiulo, B.; Marialto, R. Ideal specific work of rotary compressors: A new approach. *Energy Procedia* **2016**, *101*, 710–717. [CrossRef]
25. Szente, V.; Vad, J. A semi-empirical model for characterisation of flow coefficient for pneumatic solenoid valves. *Period. Polytech. Mech. Eng.* **2003**, *47*, 131–142.
26. Perry, J.A. Critical flow through sharp-edged orifices. *Trans. ASME* **1949**, *71*.

energies

MDPI

Article

A Parametric Study of Wave Energy Converter Layouts in Real Wave Models

Erfan Amini [1], Danial Golbaz [1], Fereidoun Amini [2], Meysam Majidi Nezhad [3], Mehdi Neshat [4] and Davide Astiaso Garcia [5,*]

1 Coastal and Offshore Structures Engineering Group, School of Civil Engineering, University of Tehran, Tehran 13145-1384, Iran; erfan.amini@ut.ac.ir (E.A.); Dgolbaz@ut.ac.ir (D.G.)
2 School of Civil Engineering, Iran University of Science and Technology, Tehran 13114-16864, Iran; Famini@iust.ac.ir
3 Department of Astronautics, Electrical and Energy Engineering (DIAEE), Sapienza University of Rome, 00184 Rome, Italy; meysam.majidinezhad@uniroma1.it
4 Optimization and Logistics Group, School of Computer Science, The University of Adelaide, Adelaide 5005, Australia; mehdi.neshat@adelaide.edu.au
5 Department of Planning, Design, and Technology of Architecture, Sapienza University of Rome, 00197 Rome, Italy
* Correspondence: davide.astiasogarcia@uniroma1.it

Received: 2 October 2020; Accepted: 18 November 2020; Published: 20 November 2020

Abstract: Ocean wave energy is a broadly accessible renewable energy source; however, it is not fully developed. Further studies on wave energy converter (WEC) technologies are required in order to achieve more commercial developments. In this study, four CETO6 spherical WEC arrangements have been investigated, in which a fully submerged spherical converter is modelled. The numerical model is applied using linear potential theory, frequency-domain analysis, and irregular wave scenario. We investigate a parametric study of the distance influence between WECs and the effect of rotation regarding significant wave direction in each arrangement compared to the pre-defined layout. Moreover, we perform a numerical landscape analysis using a grid search technique to validate the best-found power output of the layout in real wave models of four locations on the southern Australian coast. The results specify the prominent role of the distance between WECs, along with the relative angle of the layout to dominant wave direction, in harnessing more power from the waves. Furthermore, it is observed that a rise in the number of WECs contributed to an increase in the optimum distance between converters. Consequently, the maximum exploited power from each buoy array has been found, indicating the optimum values of the distance between buoys in different real wave scenarios and the relative angle of the designed layout with respect to the dominant in-site wave direction.

Keywords: layout assessment; wave energy conversion; renewable energy; real wave model

1. Introduction

Wave energy is expected to contribute towards the development of a carbon-free electricity generation. The theoretical computation of wave energy potential over the oceans is projected to be in the order of 1–10 TW [1], which can cover the current global energy demand [2]. This tremendous potential has attracted attention from research societies, which have proved that harnessing electric power from ocean waves is possible [3,4]. Wave energy converters (WEC) are planned to be stationed in an array constituted of many converters -similar to offshore wind turbines. The initial developments of the analytical modeling of hydrodynamic forces on submerged buoys can be found in [5]. However, it has been further developed since then. The next stage of studies was focused on enhancing the design and power take off system of a single buoy [6]. The next studies bring the idea of WEC's array by conducting a comparative study on different configurations [7]. The proceeding research phase concentrated on finding the optimal value for WEC's array parameters (such as optimal position or layout) using either numerical, parametric or optimisation-based solutions [8–11], as this story falls in this category. The position of converters in the array which is scattered through the array has a direct relationship with the performance of the array because hydrodynamic interactions between them can be constructive or destructive. These interactions depend on the configuration of the array. Consequently, this is the main reason to investigate these interactions in order to apply them to reinforce the total power output. There are many relevant publications with this subject by several R&D units across Europe in the past by the pioneering works [12–16], and it is still an interesting research field, as several investigations have been published recently [17–20] . Furthermore, the identification of techno-economically feasible decarbonisation paths and sustainability transitions have been investigated by [21–23]. Some of the related research projects that considered the performance of arrays or converters' distance were undertaken by [9,24–26] and the effects of nonlinear mooring forces via a time-domain analysis and the influence of interactions between WECs are well described in [27,28], respectively. Table 1 demonstrates a brief survey of some of the recent literature on the various aspects of WECs including layouts, PTO and design optimisation. Some of the mentioned research has used hindcast wave models; however, different layout configurations were considered regarding real wave scenarios in this study.

Table 1. A briefly survey some of the recent literature on the layout, Power Take-Off (PTO) parameters and design optimisation of wave energy converters.

Objective	WECs Type	WECs Number	Method	Year	References
Design & PTOs	submerged	2	Experimental observations	2020	[17]
Layout & PTOs	fully-submerged	4, 16	Cooperative EAs	2020	[18]
Design & PTOs	fully-submerged	1	Hybrid EAs	2020	[29]
Layout	fully-submerged	50, 100	Multi-strategy EAs	2020	[30]
Design & PTOs	heaving WEC	1	Evolutionary and GA	2020	[19]
PTOs	oscillating wave surge converter	1	GA	2020	[20,31]
Design	sloped-motion WEC	1	Heuristic optimization	2020	[32]
PTOs	oscillating water column-based	1	Water cycle algorithm	2020	[33]
PTOs	hinged-type WECs	1	Experimental observations	2020	[34]
PTOs	oscillating wave surge converter	1	GA and ML	2020	[35]
Layout	submerged	25	PSO	2020	[36]
Design	submerged flat plate	1	GA	2019	[37]
Design & Layout	cylindrical heaving WECs	3, 5, 7	GA	2019	[38]
Design	submerged	2	GA	2019	[39]
Layout	fully-submerged	4, 16	Smart heuristic	2019	[40]
Layout	fully-submerged	4, 16	Nuro-adaptive EA	2019	[41]
PTOs	freely floating	2	EAs	2019	[42]
Design	hinge-barge WEC	2	gradient-based method	2019	[43]
Design	fully-submerged	1, 2, 3	GA, PSO	2019	[44]
Layout & PTOs	fully-submerged	16	Hybrid EAs	2019	[45]
Layout & PTOs	fully-submerged	4, 9	Heuristics	2019	[46]
Feasibility Study&Design	oscillating wave surge converter	3	Numerical and GWO	2019	[47]
Layout	heaving WEC	1	GWO	2019	[48]
Layout	heave-constrained cylinder	5	improved GA	2018	[49]

Table 1. *Cont.*

Objective	WECs Type	WECs Number	Method	Year	References
Layout	fully-submerged	4, 16	Local search	2018	[50]
Layout	oscillating WEC	3, 5, 8	improved DE	2018	[51]
Layout &LCoE	fully-submerged	4, 9, 36	Multi-objective EAs	2018	[52]
PTOs	submerged	1	Hidden GA	2018	[53]
Layout & PTOs	submerged	4, 7, 9, 14	hybrid GA	2018	[54]
Layout	semi-submerged	1000	approximate analytical method	2015	[9]
Layout	submerged	32	randomized geometries	2013	[26]
Layout	floating + partially submerged	4	sensitivity analysis	2014	[25]
Design & Layout	submerged	4	sensitivity analysis	2017	[24]
Design	point-absorbing WECs	100	analytical multiple scattering	2015	[55]
Design& Layout	floating over-topping WECs	9	Down-scaling techniques	2018	[56]
Design& PTOs	heaving WEC	9, 16, 25	sensitivity analysis	2012	[57]

The CETO6 is a fully-submerged point absorber wave energy converter that is manufactured, installed, and updated by Carnegie Clean Energy Ltd. The prospective location of the WECs array is off the coast of Albany due to its exposure to open ocean wave conditions [58]. This study has been conducted based on the numerical simulation of this converter's array. Our concentration is particularly on the arrangement optimisation of WEC arrays and shows the effectiveness of the inter-distance among WECs to produce more power. In order to establish an array of WECs, an optimal layout is chosen to maximise the power conversion; however, the number of WECs is a significant factor. We evaluate various numbers of WECs as an array, arrangements and separations, and report the performance of the layouts using q-factor, power of each converter and total power output. The distances between the WECs, and the array size are constrained, which is a more realistic approach for studying WEC arrays. Finally, a landscape numerical analysis is performed with regard to evaluating the position effect of each WEC in the array's power output using a grid search approach.

It should be noticed that such research has not been investigated in the mentioned real wave scenarios (Perth, Adelaide, Sydney, and Tasmania) regarding this parametric study. Therefore, the main motivation of this study is to evaluate the output performance of the simulated CETO6 arrays to find a suitable layout with optimal distance, and the rotation angle to the dominant wave direction in these specific case studies. Moreover, the investigation coverage is more comprehensive than in other research studies by exploiting wave power using a ten-degree resolution covering the whole area of study, compared to, Bozzi et al. [24] presented the implementation and evaluation of a few numbers of WEC separation distances (5, 10, 20 and 30 buoy diameters) and incident wave directions (30° apart).

This paper is structured into five sections. Section 2 presents a brief description of the hydrodynamic WEC array interaction model, modeling the wave climate and the equations used to compute the produced power. Section 3 expresses the layout assessment routine and presents the strategy to explore the optimal position of the WECs in the array. Section 4 discusses the array layout investigation results in terms of performance and optimal array layout solutions. Subsequently, Section 5 summarises the principal finding of the paper.

2. Numerical Modelling

2.1. Wave Energy Converter

In this study, a CETO6 wave energy converter with a three-tethered mooring system is considered which has a fully submerged spherical buoy attached to the seabed by the tethers, as shown in Figure 1. This model is developed in MATLAB and was modified in 2020 [59]. The WEC details are:

buoy radius = 5 (m), submergence depth = 3 (m), water depth = 50 (m), buoy mass = 376 (t), buoy volume = 523.6 (m^3), tether angle = 55 (degree), PTO stiffness = 2.7×10^5 (N/m), PTO damping= 1.3×10^5 (Ns/m).

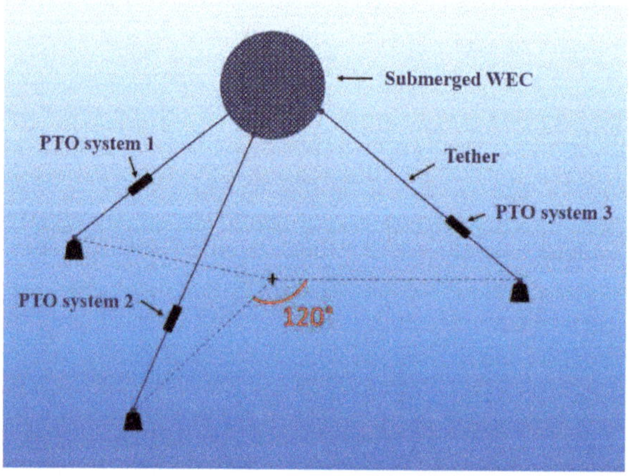

Figure 1. Schematic representation of the CETO6 modelled point absorber wave energy converter (adapted from [60]).

This buoy, which is floating at sea, moves in six degrees of motion. However, due to the converter's spherical shape, its displacement is in three degrees of freedom which are surge, heave, and sway. Based on these degrees, the motion equation can be written on the frequency domain.

$$\begin{aligned} \Sigma F &= m\ddot{z}, \\ &= F_m + F_{hs} + W + F_R + F_{PTO} + F_{W_k} + F_{VD} \end{aligned} \tag{1}$$

where F_m is the mooring force, F_{hs} is the hydro-static force resulting from buoyancy, W is the body weight, F_R represents added mass and wave damping forces, force resulted by PTO system is F_{PTO}, F_{W_k} represents the vertical components of the wave exciting force and F_{VD} is the vertical viscous drag force [61]. This equation is used in order to describe a time-domain response of the WECs in waves, and can be rewritten as:

$$(m + A_\infty)\ddot{z} + \int_0^t K_{rad}(t-\tau)\dot{z}(\tau)\mathrm{d}\tau + Cz = F_{exc} + F_{pto} + F_{hs} \tag{2}$$

where m is a buoy mass, A_∞ is the infinite-frequency added mass coefficient, C is the hydro-static stiffness, $K_{rad}(t)$ is the radiation impulse response function, F_{exc} is the wave excitation force, F_{pto} is the load force exerted on the buoy from the power take-off system [62]. Free surface elevation height results from a linear superposition consisting of some wave characteristics in irregular waves. This is usually determined by a wave spectrum which describes the distribution of energy in a vast number of wave frequencies. Significant wave height and peak period are utilized as the basic identification of the wave in the spectrum. The irregular excitation force can be calculated as the real part of an integral term across all wave frequencies as follows.

$$F_{exc} = \mathbf{R}\left[\int_0^\infty \sqrt{2S(\omega_r)}F_x\, e^{i(\omega_r t+\phi)}\mathrm{d}\omega_r\right] = \int_{-\infty}^{+\infty} \eta_{(\tau)} f_e(t-\tau)\mathrm{d}_{(\tau)} \tag{3}$$

where **R** denotes the real part of the equation, F_x is the excitation vector consists of amplitude and phase of the wave, S is the wave spectrum, ϕ is the stochastic phase angle, η_T represents water elevation and f_e is the element of force vector [48]. The load force of PTO is modeled as a linear spring-damper system.

$$F_{pto} = -B_{pto}\dot{z} - K_{pto}z \tag{4}$$

$$\begin{aligned} F_{hs} = &- K_{hs,\min}(z - z_{\min})u(z_{\min} - z) \\ &- K_{hs,\max}(z - z_{\max})u(z - z_{\max}) \end{aligned} \tag{5}$$

where in Equation (4) K_{pto} and B_{pto} are control parameters which represent stiffness and damping of PTO and in Equation (5) u is the Heaviside step function, $K_{hs,\min}$ and $K_{hs,\max}$ are the hard stop spring coefficients, and $z_{\min}$ and $z_{\max}$ are the stroke limits which are related to the nominal position of the converter. It is important to note that, for computing useful absorbed energy, the effect of this force is not considered [63].

In order to calculate the energy produced by each buoy, the sum of three forces is necessary: wave excitation($F_{exc,p}(t)$), force of radiation ($F_{rad,p}(t)$), and power take off force ($F_{pto,p}(t)$). The scattered irregular waves are included in the wave field when computing the excitation force. Furthermore, the stiffness and damping parameters of the PTO system at the end of each tether along with hydrodynamical parameters are taken in order to compute the total power output of an array. To calculate the average power absorbed by the array, several variables have to be taken in to account, as follows.

$$P_n(H,T) = \int_0^{2\pi} \int_0^{\infty} 2S_n(\omega)D(\beta)p(\beta,\omega)\mathrm{d}\omega\mathrm{d}\beta \tag{6}$$

where $P_n(H,T)$ is the average power absorbed by the array in a regular wave of unit amplitude, $S_n(\omega)$ is the irregular wave spectrum which is calculated with the Bretschneider spectrum and $D(\beta)$ represents the directional spreading spectrum, particularly for this site which is come from the wave rose [64]. ω is the wave frequency and $p(\beta,\omega)$ is the power function of each submerged buoy defined by Equation (7).

$$p(\beta,\omega) = \frac{1}{2}D_{pto}\omega^2\Gamma(\beta,\omega)^2 \tag{7}$$

where $\Gamma(\beta,\omega)$ is the response amplitude operator (RAO) of the productive degree of freedom of the buoy obtained by solving the equation of motion from Equation (2), and D_{pto} is the Power Take-Off (PTO) damping. The wave angle is based on the $z(\beta,\omega)$ [40], which can be calculated by equation (2) at the beginning of this section. The array at a certain test site is generated by total mean annual power P_{array}, and to calculate that, the contribution of energy absorption from a wave climate in each state can be summarized as:

$$P_{array} = \sum_{n=1}^{N_s} O_n(H_s,T_p)P_n(H,T) \tag{8}$$

where N_s is a number of chosen sea state, H_s is the significant wave height and T_p is the peak wave period for each sea state, $O_n(H_s,T_P)$ represents the probability of occurrence of sea state which stems from the wave scatter diagram and $P_n(H_s,T_p)$ is a power which the array produces in the nth sea state [40]. Significant wave height and peak wave period are statistics of a sea state which can refer to the condition of the ocean/sea surface. To calculate $P_n(H_s,T_p)$ in irregular waves, it is necessary to sum all power contributions in each frequency and significant wave direction.

2.2. Wave Resource

According to previous works [40,45], four different sea sites were chosen for this study. The wave height directional distribution (wave rose) can be seen in Figure 2, at the chosen locations (as an example, the wave condition at the Sydney site is shown). The wave rose shows that the significant wave directions are from 15 degrees to 190 degrees, where 90 percent of the incident waves travelled. Consequently, the dominant wave direction is from the south.

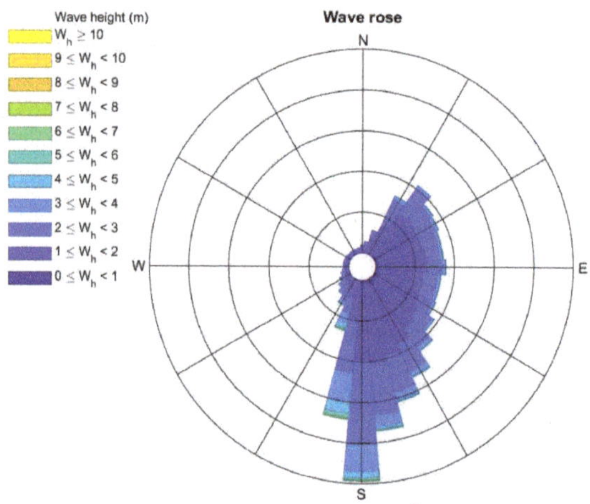

Figure 2. The wave rose plot at Sydney.

Each array is constrained by the maximum area and the minimum distance between WECs. Firstly the minimum separation between buoys (R') has to be 50 m to provide a safe pass for vessels. Secondly, although the area grows by increasing the number of buoys, it has to be constrained within the area Ω, where $\Omega = l \times \omega, \quad l = \omega = \sqrt{N \times 120{,}000}$ m [45].

The Bretschneider spectrum is used for modeling irregular waves in this study. This spectrum is a modified Pierson-Moskowitz spectrum which is based on significant wave height and peak period. These two parameters are highly dependent on wind speed and its direction [65].

$$S(f) = \frac{H_{m0}^2}{4}(1.057f_p)^4 f^{-5} \exp\left[-\frac{5}{4}(\frac{f_p}{f})^4\right] \tag{9}$$

where H_{m0} and f_p are the significant wave height and the frequency of the peak wave period, respectively.

2.3. Array Interaction Criteria

The optimal designs of the array for four different locations in Australia use power matrices of various configurations (i.e., different layout geometry, WEC distance and relative angles regarding dominant wave direction wave directions). To be more precise, the goal is to select the best site for each layout configuration with the optimal separation among WECs and rotation angle, namely, the one that provides the highest annual energy output per each converter. For this aim, based on the number of WECs, different layouts can be deployed with various orientations and separation among converters. Note that there are certain constraints for distances, and this depends on the number of WECs. Similarly, the number of converters in an array allows configurations to be chosen. Hence, it is crucial to measure the effectiveness of interactions between converters by the q-factor coefficient. The q-factor is shown to be an important evaluation criterium such that if $q > 1$, then it has a positive effect on the total energy of an array; otherwise, the interactions are destructive.

$$q = \frac{P_{array}}{NP_{isolated}} \tag{10}$$

where $P_{isolated}$ is the power that an isolated WEC generates, N is the number of converters [41]. As Equation (10) indicates, there is a direct relationship between q-factor and power output of each array; however, both of them have to be studied separately, due to different objectives of finding optimal values for each parameter. The maximum feasible amount of q-factor is investigated to achieve the best constructive effects of interactions of buoys in an array. On the flip side, since the power output of a WEC array plays a significant role in the assessment of the system's response to energy consumption needs in coastal areas, this parameter is considered along with the q-factor. The equation below calculates the mean q-factor by considering the number of converters, variety of wave directions and allowable distance within a 5 m interval.

$$\textbf{mean q-factor}_{\text{(each wave scenario)}} = \frac{\sum_{i=0}^{max\alpha} \sum_{j=50}^{l} \mathbf{q}_{factor_{(i,j)}}}{total\ number\ of\ cases} \tag{11}$$

where α is the direction of wave n 10-degrees resolution, except when N is 5, the interval changes to 9 degrees ranging from 0–63 degrees and l is the maximum allowable distance between buoys within the area. This mean q-factor coefficient will be considered to find the best location for the max q-factor over different angles and distances.

3. Layout Assessment Routine

According to the mentioned equations in Section 2, the following outcomes are obtained. Four different layouts are considered regarding the number of buoys, and they are thoroughly described in detail. By looking at the mentioned literature, it is evident that using more converters results in more potential destructive interaction. Specifically, the q-factor may decrease when the number of buoys rises to greater than five [54]. Therefore, we decided to choose the five buoy layout as the maximum complexity for the model; however, evaluating the larger wave farm characteristics is a part of our future research plan. Furthermore, the symmetric design is proposed for the layouts based on the following reasons: (i) To find a single variable to handle, the buoy-buoy distance must remain constant for each configuration. Thus, the symmetric configuration fulfills this requirement in our assessment, like in previous studies [66]. (ii) To cover the whole area of study in power absorption assessment process, the array rotates 10 degrees, regarding the dominant wave direction, in each evaluation. The asymmetric effect of the array configuration rules out the duplicate assessment, resulting in less computational cost.

In the first step, There are two buoys in this array, and one line connecting them. The dominant wave direction indicates the direction in which most waves travel. However, in the calculation of power output from each buoy or an array, the significant wave directions are used. These are directions from where 90 percent of the waves are traveling. Furthermore, the resolution size of the evaluation is 15 degrees, by which we detect the dominant direction of the waves.

The dominant wave direction is obtained by considering one-third of the maximum waves in the wave rose. The angle between dominant wave direction and a hypothetical line is considered to be alpha (α), which is clearly illustrated in Figure 3. The interval of alpha is chosen to be tested every 10 degrees; therefore, there would be 18 different alpha ranging from 0 to 170 degrees. This range has been considered to prevent the extra calculation of results that have already been calculated. When there are three buoys to consider in an array, one of the most common geometries is the equilateral triangle. If some lines are used for connecting these buoys, angles between vertexes of the triangle will be 60. There is a line from this converter perpendicular to the line, which connects the two other buoys. The angle between the dominant wave direction and this perpendicular line is alpha, shown in Figure 3, and this parameter has twelve degrees from 0 to 110, which changes every 10 degrees. A regular quadrilateral is taken into account to configure four buoys. To describe alpha in this layout, firstly, the dominant wave direction needs to be determined. Secondly, a hypothetical line should be drawn from one converter to the furthest one. For example, if the converters are numbered clockwise and the closest buoy to the front wave is buoy number one, the line should be drawn from 1 to 3, exactly like in Figure 3. Finally, the angle between this line and dominant wave direction is alpha, and the range of this is from 0 to 80 degrees, which has nine different amounts with equal intervals.

In this layout, five similar converters form an array are shown in the shape of a regular pentagon. The dominant wave direction is illustrated in Figure 3 with a blue arrow. Converters are numbered clockwise, and the first number starts from the closest buoy to the front wave. As shown in Figure 3, each converter has the longest distance with two buoys. In this case, the furthest converter to buoy number 1 is number 3 and 4. If a perpendicular line is drawn from the first converter to the connecting line between furthest converters, the angle between the dominant wave direction and the perpendicular line represents alpha. The range for alpha is from 0 to 63 in 9-degree intervals, so there are eight alphas to test in this layout. Distances are also assumed to change every five meters between the allowable period. In the end, three measurements, which are the power output of each buoy, array power, and q-factor, are taken in each step for all layouts, separately. The details of all results are discussed comprehensively in Section 5.

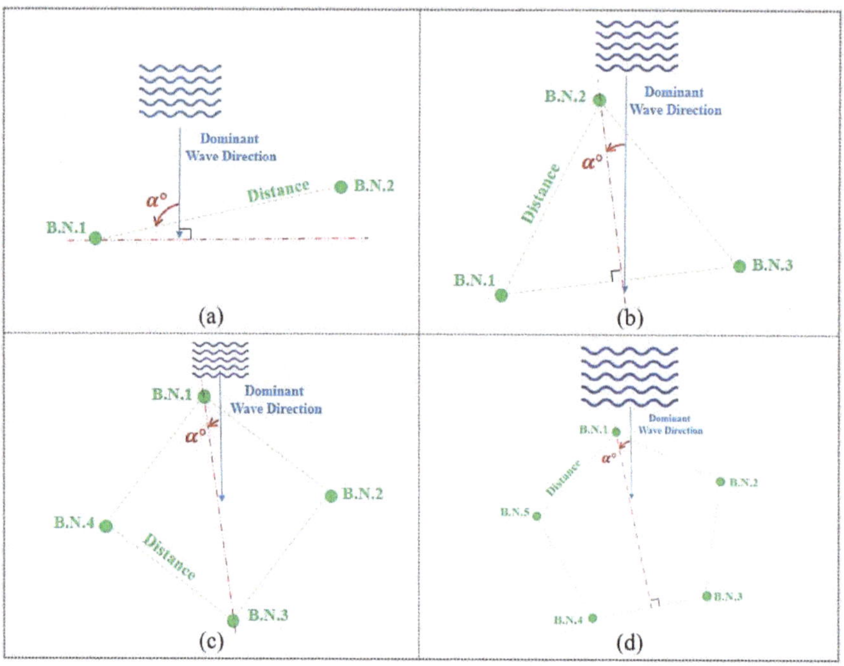

Figure 3. Layout Setup of (**a**) 2 buoys array (linear), (**b**) 3 buoys array (triangle-shape), (**c**) 4 buoys array (square-shape) and (**d**) 5 buoys array (pentagon-shape), with regard to dominant wave direction.

4. Results and Discussions

This section represents the results of different array layouts when the number of buoys rises from 2 to 5 in considered locations on the Australian coast. The results demonstrate the sensitivity of the array power and the q-factor due to the changes in buoy-buoy distance and rotation angle. It is worth mentioning that there are 16 conditions in this study, which will be discussed in detail as follows. To choose the optimal degree in this section, one of the most important variables is alpha, whose optimum value leads to the average maximum amount of the power output.

4.1. Sensitivity of Two-Buoy Array Performance To Distance

Figure 4 shows the sensitivity analysis of the array power output to the different buoy-buoy distances. It can be seen that Tasmania has the most wave array power, which is almost 0.534 Mw where the α is 80 degrees and the buoy-buoy distance is 160 m. The 60-degree angle line, which has the most average power, rises with a sinusoidal trend from the beginning to 200 m. Then, it increases gradually. The second location is Sydney, which has a considerable array power. Although its array power is 0.218 Mw, which is far less than Tasmania, the maximum average array power occurs in 130 degrees with a similar trend to the mentioned location. The maximum power that Sydney's layout determines is achievable when the distance is around 400 m. Adelaide and Perth are similar in terms of the array power range, which is roughly from 0.18 to 0.196 Mw. As the green line in these figures shows, when the rotation angle and buoy-buoy distance are 40 degrees and 165 m, respectively, both reach the highest array power. The maximum average of array power can be witnessed in 20 degrees in Adelaide and 30 in Perth. In the two mentioned sites, it is apparent that figure lines follow different trends. To compare, when α is 20 degrees, and buoy-buoy distance is 100 m, the first peak of the array power is observed in Adelaide. Next, it falls until the distance goes over 150 m. The first peak in Perth happens when the distance is near 60 m. Then, it remains unchanged for the next 60 m. After significant growth, it reaches around 0.194 Mw with 160 m buoy-buoy distance. Overall, it is remarkable that the maximum power can be harnessed in three locations when the distance is 160 m.

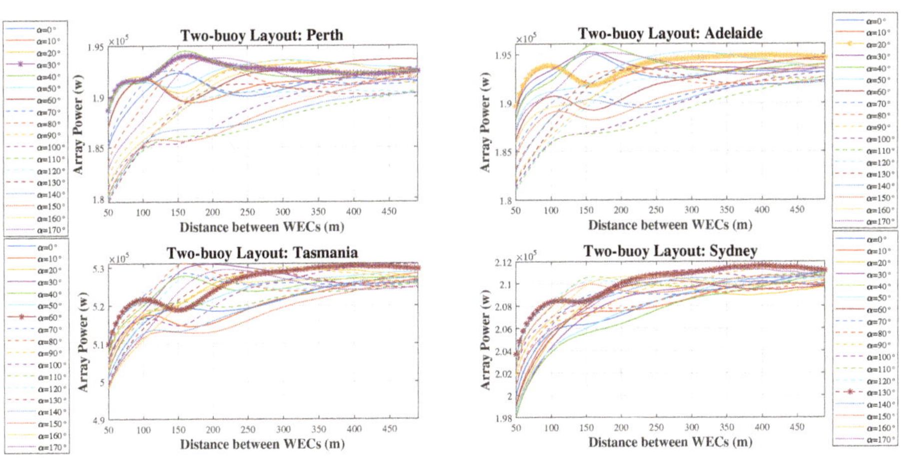

Figure 4. Array power of the two-buoy layout over different distances in four wave models.

4.2. Sensitivity of Three-Buoy Array Performance to Distance

It can be seen that in Figure 5, the most obvious inferred outcome is that, the longer distance between WECs leads to more extracted power output. However, widening the area might not be proficient because the line only rises 0.2 Mw by increasing the distance from 250 to 500 m. The maximum harnessed power output can be seen when α is 10 degrees, except in Sydney, which is 110. The range of array power is quite narrow in the mentioned locations, where only a 0.05 Mw gap can be witnessed among 12 tested angles. The main reason for obtaining the similar results among different angles' experiments can be explained as follows. Where the three converters are placed in equilateral triangle geometry, there would always be two buoys in the zone of radiation. Therefore, the changes in α cannot produce considerable effects. By comparing the power output of WECs over the changes of distances, we can see the same overall trend has been followed in all studied locations. A sharp rise in array power can be achieved by increasing the distance up to 100 m, followed by a gradual rise by increasing the distance up to the maximum allowed size. It can be mentioned that differences between maximum distance in each layout relate to the area constraints, which have already been discussed in Section 2.

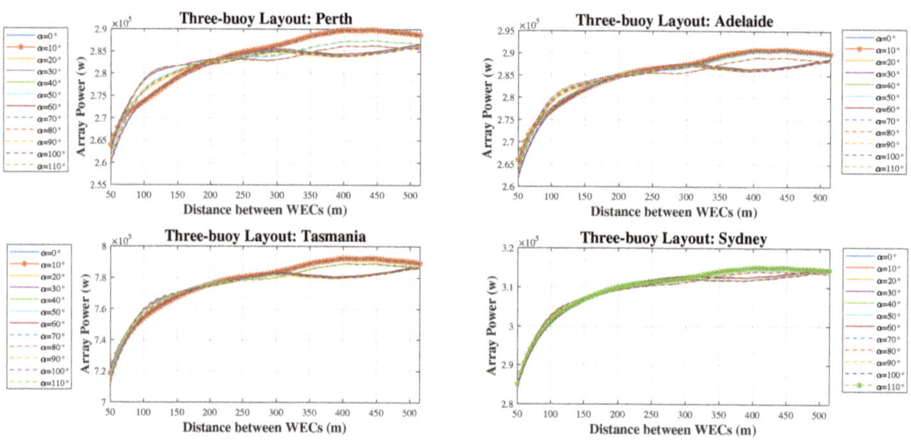

Figure 5. Array power of the three-buoy layout over different distances in four wave models.

4.3. Sensitivity of Four-Buoy Array Performance to Distance

The geometry chosen for four converters is a square shape. α values in this layout range from 0 to 80 degrees with 10-degree intervals. The highest harnessed array power is 1.05 Mw in Tasmania, while the lowest one observed in the Perth layout is around 0.387 Mw. Turning to the rotation angle, for Perth and Sydney, the angle is 40 degrees to extract the maximum average array power; however, for Tasmania and Adelaide, α is 0 and 80 degrees in order of appearance. Considering the distances between WECs, it is interesting that where the buoy-buoy distance is between 150 and 200 m, the maximum wave array power can be exploited in Perth, Adelaide, and Tasmania. However, in Sydney, it seems that in this α, the array power evens off after reaching the highest amount. Hence, the minimum distance between WECs is more cost beneficial for considering layout design 160 m of distance takes into account. In contrast, in Adelaide and Tasmania, a 160 or 170-m distance seems to be the best distance between converters. This amount is a bit greater in Perth, where the highest array power is firstly seen in the 180-m distance (Figure 6).

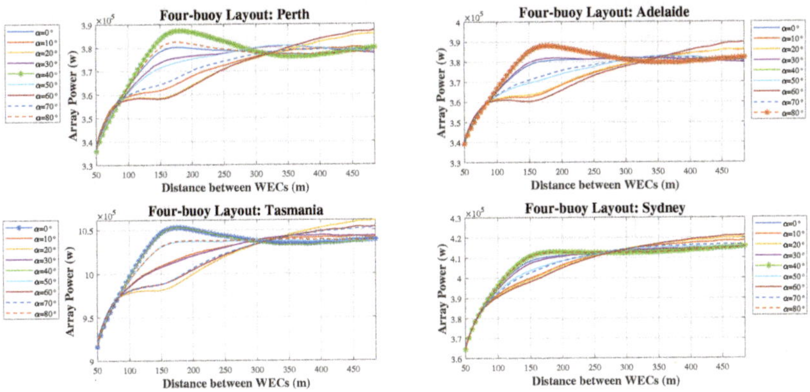

Figure 6. Array power of the four-buoy layout over different distances in four wave models.

4.4. *Sensitivity of Five-Buoy Array Performance to Distance*

In this study, the configuration of five converters is chosen as a regular pentagon array. This is because there is no difference between each WEC, and the range of rotation angles is restricted to be between 0 and 63 with eight different angles. The maximum averaged array power in all four case studies is witnessed when the α is either 18 or 63 degrees. To be more precise, in Tasmania and Perth, the α is 18 degrees, and for the other two, it is 63 degrees. Also, it is evident that when the distance is between 200 and 250 m, the maximum power output is harnessed in all wave scenarios; and the optimal choice can be found in the mentioned range consequently. As Figure 7 shows, the trend of all case studies are similar, except in the 18-degree's line in Tasmania and Perth, where the array power reduces gradually after the peak, instead of leveling at the peaks power. By comparing this result with recent similar studies, we can see the same trend of absorbed power by raising the distance between WECs in each layout up to an optimal value, after which the results were roughly stable [54,67].

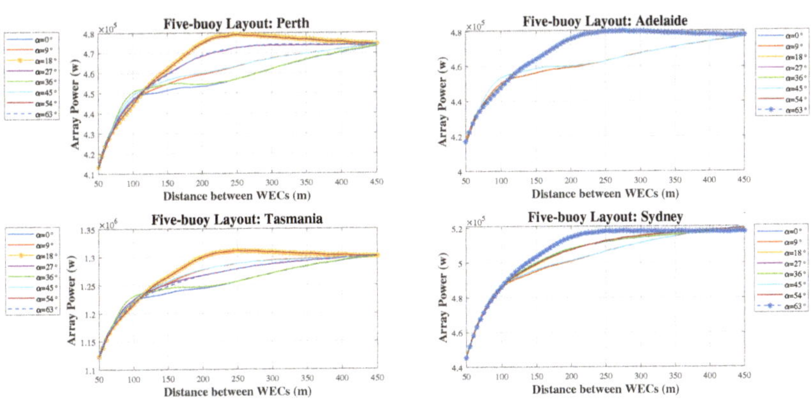

Figure 7. Array power of the five-buoy layout over different distances in four wave models.

4.5. Sensitivity Analysis of q-Factor to the Relative Angle of Rotation

The interaction of converters is measured with a well-known parameter called q-factor. A sensitivity analysis has done by monitoring the q-factor distribution over different rotation angles of the WECs in each layout. Since the interaction of the buoys in each layout could be constructive or destructive, the q-factor is calculated per 5 m of distance between converters. Figures 8–11 show the distribution of results of calculated q-factors over different relative angles (α) in a box chart, describing the values as they spread across the entire range. In each angle, there is a box that reveals the amount of fifty percent of q-factor results. Also, the middle line indicates the mean value of all results. The other amounts of q-factor, which are far from the mean values (i.e., the greatest 25 percent and the least 25 percent of the results), are shown by two lines located above and below the rectangular box.

There is a lot of similarity between Adelaide and Perth in terms of their q-factors, but Sydney and Tasmania have different trends. In both Adelaide and Perth, as Figure 8 and 9 show in two-buoy layout, a fluctuating pattern is witnessed which indicates the importance of the rotation angle of the array power with respect to the dominant wave direction. Thus, when α is between 30 and 40 degrees, the highest q-factor is achieved, and the layout design process should be followed by choosing the best buoy-buoy distance in the mentioned angle. The distinction between q-factors in the three-buoy layout is negligible due to the effects of dominant wave direction on the equilateral triangle layout, in which one converter affects two others by radiated waves. The maximum q-factor can be seen in the 30 and 40 degrees area. In the four-buoy layout, when α is 40 degrees, the q-factor is around 1 in both locations. Among the 9 discussed dominant wave directions, the highest q-factors are seen when α is 20, 40, or 60 degrees, and the average q-factor in each direction is a bit over 0.96. In the five-buoy layout, it is clear that the average q-factors are between 0.95 and 0.98, and the highest q-factors happen when α is either 20, 30, 60, or 70 degrees, and the range of q-factors is from 0.85 to 0.98. It is important to note that in the mentioned degrees, q-factors are mostly close to the highest amount because the average line is on top of each box.

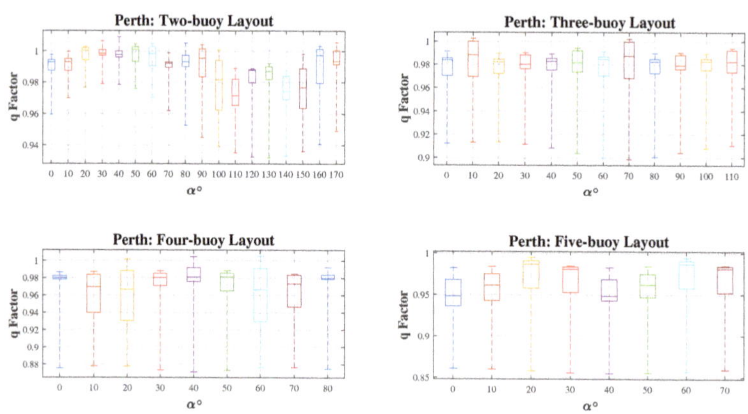

Figure 8. q-factor results distribution and mean value per five meters of the Wave Energy Converters (WECs)' distance over rotation angle due to significant wave direction in the Perth wave model. (Fifty percent of results near the mean value are plotted in a box, the range of other results is shown by a dashed line).

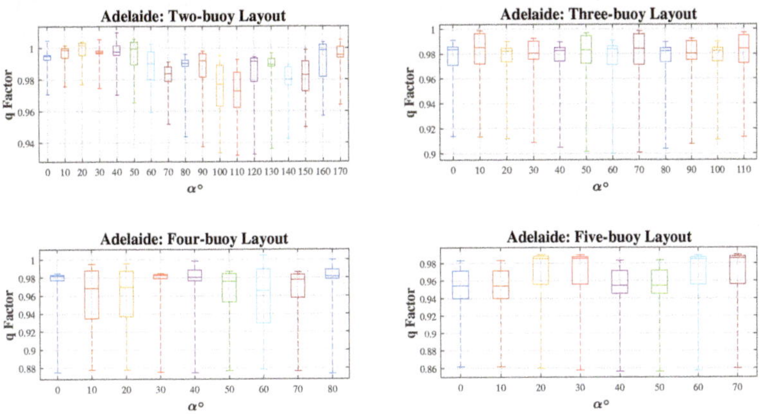

Figure 9. The q-factor results distribution and mean value every five meters of WECs' distance over rotation angle due to the significant wave direction in the Adelaide wave model. (Fifty percent of results near the mean value are plotted in a box, the range of the other results is shown by a dashed line).

In Tasmania, due to the symmetry between WECs and small effects of changing α in q-factor for the three-buoy and five-buoy layout, changes are not considerable. The average q-factor in each rotation angle is approximately 0.98 and 0.96, respectively. When α is between 40 and 90 degrees in the two-buoy layout, q-factors in each distance are between 0.95 and 1.005. The maximum averaged array power occurs at 20 degrees. In the four-buoy layout, maximum q-factors are observed in four rotation angles, which are 0, 20, 40, and 60 degrees. Further details can be found in Figure 10.

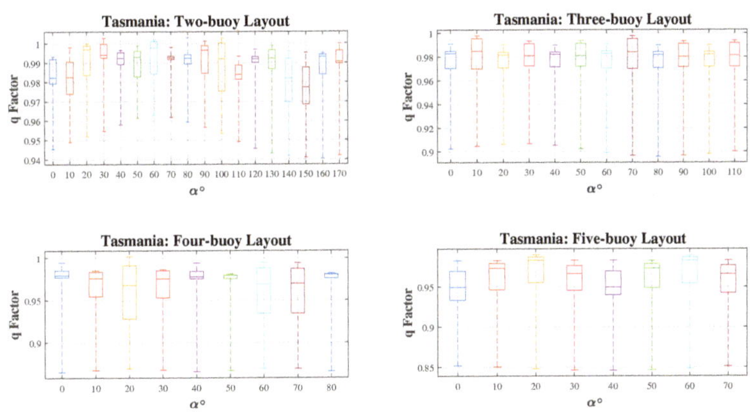

Figure 10. The q-factor results distribution and mean value per five meters of WECs' distance over rotation angle due to significant wave direction in Tasmania wave model. (Fifty percent of results near the mean value are plotted in a box, the range of other results are shown by a dashed line).

Looking at Sydney in Figure 11, in the two-buoy layout, it is evident that maximum q-factors happen when α is either between 110 and 120 degrees or 130 and 140. One of the distinctions compared to the mentioned locations is that the lowest q-factor is seen at 40 degrees. In the three-buoy and four-buoy layout, the average q-factor in each α is around 0.98 and 0.97, respectively, and its changes are not recognizable in all 12 tested angles. The q-factors in the five-buoy layout ranged from 0.84 to 0.98. The closest q-factor values to 1 are found at 10 and 50 degrees. Moreover, it can be inferred that when α is 30 or 70 degrees, the related q-factors are near 0.98. These results would help further feasibility studies of the WECs' array analysis by presenting the possible range of achievable q-factors in Perth, Adelaide, Tasmania, and Sydney ports.

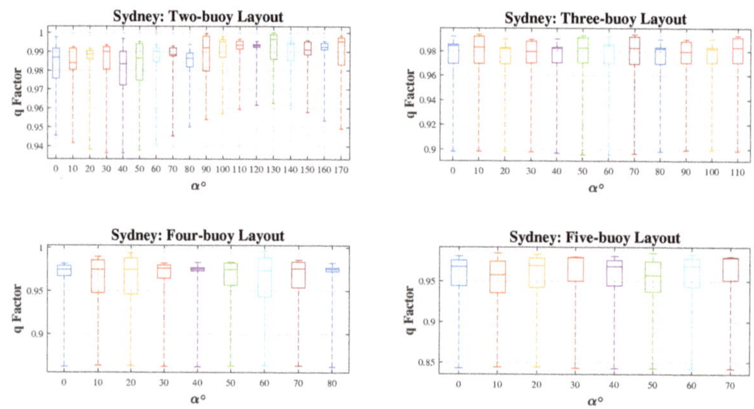

Figure 11. q-factor results distribution and mean value per five meters of WECs' distance over rotation angle due to significant wave direction in the Sydney wave model. (Fifty percent of results near the mean value are plotted in a box, the range of the other results is shown by a dashed line).

4.6. Landscape Analysis

Figures 12–15 reveal the power for each buoy in four different layouts. Overall, it is inferred that the asymmetry in arrays makes the power more predictable in each distance and rotation angle. The illustrated plots in this section indicate four kinds of buoy layout for each area in Adelaide, Tasmania, Sydney, and Perth. For each layout, the Colour-bar presents the amount of total power per buoy. This amount has been shown in the form of a contour. In general, there is a similar pattern for studied locations by considering the configuration of the arrays. The maximum amount of extracted energy is more likely to be found in the 5-buoy layout. Moreover, the power of each converter in the center area is not considerable, and the maximum amounts of the exploited energy found for higher buoy-buoy distances. The asymmetry of these energy distributions is high as well, but Perth is an exception. To observe abrupt changes, an increase in the resolution of distance and angles are needed. Take the 2-buoy layout in Adelaide as another example of non-asymmetric layout; the distribution of the incident waves over different angles implied such non-asymmetric contour of exploited energy. The reason behind the asymmetry in other layouts would be that by increasing the number of buoys, possible shadowing effects proportional to each rotation angle occur in every layout. In some placements of the 2-buoy layout, as the array experiences different rotation angles, the buoys may see the same dominant wave direction; however, in some angles, the shadowing effect of one buoy over the other can play a crucial role in reducing the energy. This shadowing effect of buoys to each other becomes more drastic, as the number of buoys increases because, in each angle, there is more chance of interaction between at least two buoys. To describe blue spots it should be noted that in this research, the minimum separation between buoys is considered to be 50 m. Then, by assessing the different angles, the configuration of the array rotates over the center point, such that the lowest amounts of energy may be witnessed in the middle of the figures by interpolation.

Taking a look at Figure 12, it is interesting that the order of maximum power for each layout is around 0.1 to 0.105 Mw in most areas. However, this range increases in the two-buoy layout, so more areas with dark red colour can be seen. In this figure, asymmetry in each layout is more evident than in the other locations.

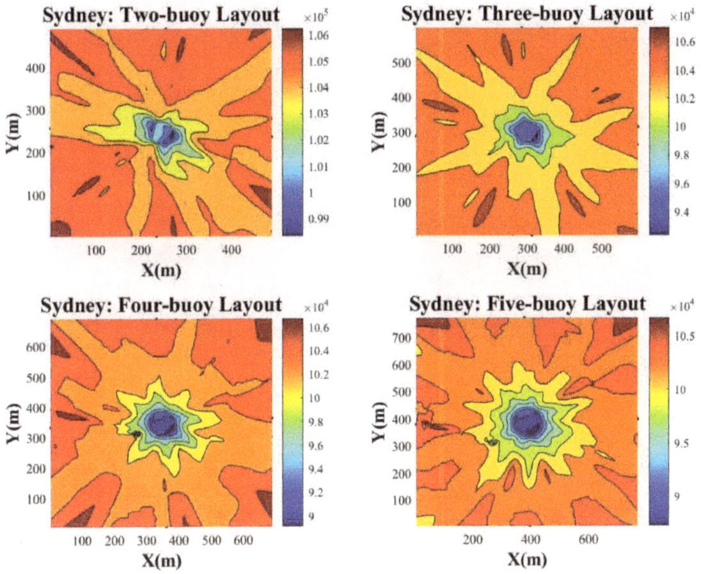

Figure 12. Exploited energy distribution of the WECs array over entire area in Sydeny wave scenario.

There are many similarities between Adelaide and Perth in Figures 13 and 14. For instance, in a two-buoy layout, their contour has resemblance, and the range of power is identical. Furthermore, more power is extracted in Perth based on these plots. To compare the four-buoy layout, Adelaide has more symmetrical power distribution, and the chance of reaching 1 Mw power is more in Adelaide in general. Finally, there is a small difference between choosing Perth or Adelaide as the installation site among the other surveyed sites.

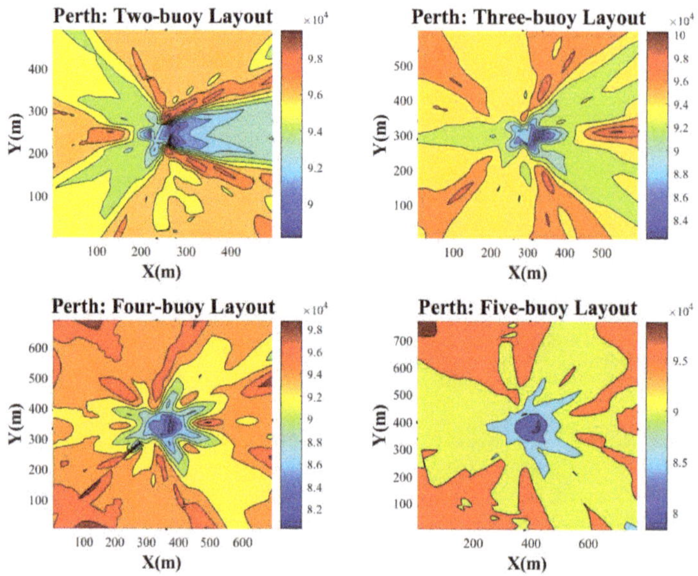

Figure 13. Exploited energy distribution of the WECs array over the entire area in the Perth wave scenario.

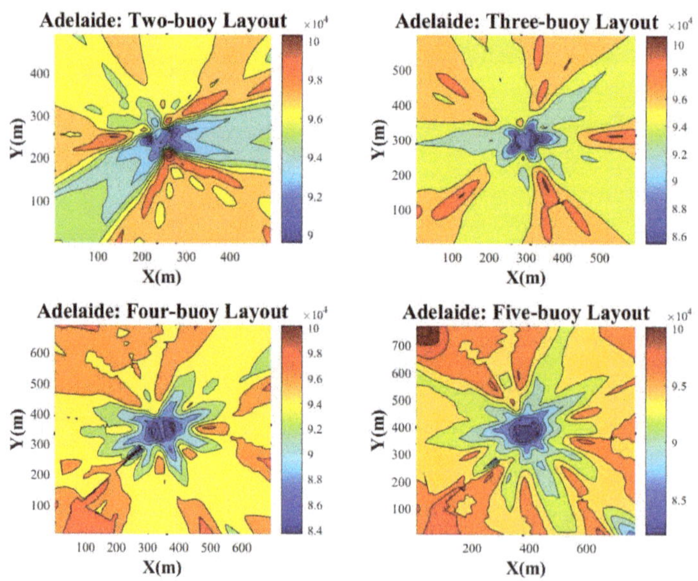

Figure 14. Exploited energy distribution of the WECs array over the entire area in the Adelaide wave scenario.

The highest amount of power can be exploited from converters in Tasmania, which is obvious in Figure 15. Although seemingly green areas in the two-buoy layout cover the majority of the zone, the power that belongs to the green areas is very close to the other layouts of the orange ones. The chance of extracting over 0.27 Mw power is seen in the five-buoy and four-buoy layouts. It is worth considering the three-buoy layout power when the X axis is over 250 m, which reveals the potential of this layout under certain conditions. Likewise, when the Y axis is less than 150 or more than 350, this potential is met.

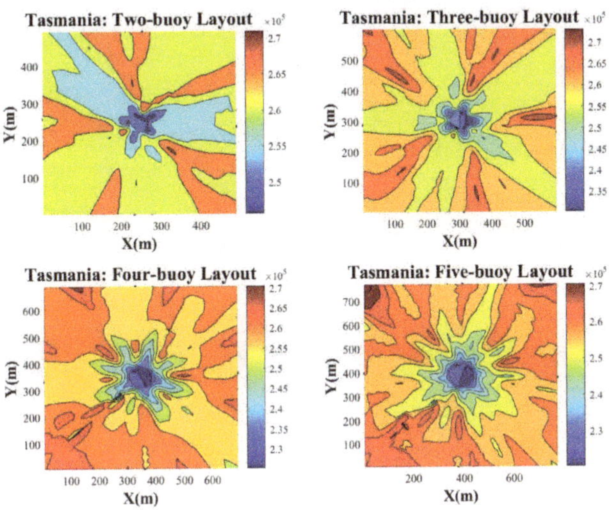

Figure 15. Exploited energy distribution of the WECs array over the entire area in the Tasmania wave scenario.

4.7. Interaction Based Layout Selection

Figure 16 shows the maximum and mean value of the q-factors in a given number of buoys in each wave model. The most significant observations inferred from Figure 16 are addressed as follows. The maximum q-factor in Tasmania and Sydney is less than in the other locations because of the lack of constructive interactions to compare to the other layouts. The mean q-factors are also higher in Perth and Adelaide in all locations for the same reason. Turning to the maximum q-factor, it is apparent that the highest constructive interactions in the two-buoy layout occur in Adelaide. However, in Sydney's wave scenario, installing buoys, whether separately or in an array, is almost the same because the q-factor equals 1 in the best-case scenario. Although this amount is over 1 in all locations of the four buoy layout, Sydney is an exception. In the three-buoy layout and five-buoy layout, only the q-factor of Perth is more than 1. It is worth considering that the latter has the least maximum q-factor. These results confirm results from previous studies on the decrease of q-factor after increasing the number of WECs arrays, specifically after incorporating more than five buoys [54]. Turning to the mean q-factor, it is evident that by increasing the number of buoys, this variable decreases, and a reduction of almost 0.07 is seen by adding a buoy. Also, this measure is observed to be a trend because constructive interactions are more likely to be seen in Perth and Adelaide. These interactions will occur if the α, buoy-buoy distance, and the geometry of layout are appropriately chosen.

Finally, it has to be noticed that further details and numbers are written in Table 2, which enables a comparison between each location. The bold numbers in Table 2 represent constructive interactions between converters. Therefore, in those cases, installing an array is more efficient.

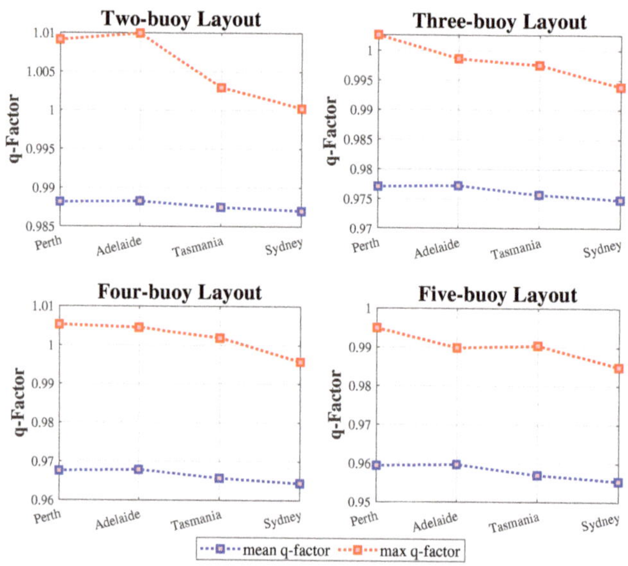

Figure 16. Comparison of maximum and mean q-factor in different wave scenarios in each layout.

Table 2. best solutions related to maximum q-factor in each wave scenario.

Parameter	Perth	Adelaide	Sydney	Tasmania
Two-buoy layout maximum q-factor	**1.0091**	**1.0163**	**1.0003**	**1.003**
$\alpha_{\text{(degrees)}}$	40	0.00	130	80
$distance_{\text{(meter)}}$	160	165	400	160
Three-buoy layout maximum q-factor	**1.0026**	0.9987	0.9939	0.9976
$\alpha_{\text{(degrees)}}$	10	10	10	10
$distance_{\text{(meter)}}$	445	445	445	405
Four-buoy layout maximum q-factor	**1.0053**	**1.0046**	0.9958	**1.0019**
$\alpha_{\text{(degrees)}}$	60	60	60	20
$distance_{\text{(meter)}}$	485	485	485	485
Five-buoy layout maximum q-factor	0.9949	0.9899	0.9849	0.9905
$\alpha_{\text{(degrees)}}$	18	63	45	18
$distance_{\text{(meter)}}$	250	275	450	250

5. Conclusions

Investigating for an appropriate arrangement an array layout constitutes a complicated problem in wave energy projects. Wave energy converters can reinforce each other to provide more power output in the form of an array if the distance among them is efficiently adjusted and the arrangement of the layout appropriately defined. In this paper, we analyzed the CETO6-project WECs separation in an array with different numbers of devices and arrangements. In order to assess the impact of various wave models, we perform and compare all numerical analyses in four real wave scenarios including the Sydney, Perth, Adelaide, and Tasmania sea sites. According to the numerical analysis, there is a direct relationship between the number of converters and optimal inter-distance among them and also relative angle to the significant wave direction. Greater separation between converters leads to more array harnessed power output. However, the most exploited energy can be achieved in 2 buoy layout with a 165 m buoy-buoy distance. A sensitivity analysis has revealed that the q-factor distribution differed due to different rotation angles of the WECs array. Moreover, the maximum q-factor output analysis showed that results in a two-buoy layout in all scenarios, tree-buoy layout in Perth and four-buoy layout in all scenarios (excluding Tasmania) are far higher than the other locations' q-factor, and this parameter is almost the same in the five-buoy layout sea sites. However, the landscape analysis-approved maximum amount in terms of the extracted net power output was found in the 5-buoy layout in the Tasmania wave scenario.

Author Contributions: Conceptualization, E.A.; Data curation, D.G., M.M.N., M.N.; Formal analysis, E.A., D.G., F.A.; Investigation, E.A., D.G., F.A., M.M.N. and M.N.; Methodology, E.A., D.G., F.A., M.M.N. and M.N.; Resources, M.N.; Supervision, D.A.G., F.A.; Validation, E.A., D.G., F.A., and M.M.N.; Visualization, E.A., D.G., and F.A.; Writing—original draft, E.A., D.G., M.M.N. and M.N.; Writing—review & editing, M.N. and D.A.G. All authors have read and agreed to the published version of the manuscript.

Funding: This research received no external funding.

Acknowledgments: The authors would like to express their gratitude to ODYSSEA project that received funding from the European Union's Horizon 2020 research and innovation programme under grant agreement No. 727277. Furthermore, the authors would like to appreciate Nataliia Surgiienko from the University of Adelaide due to publishing the MATLAB source code of the wave energy simulator. This work has been supported by the High Performance Computing Research Center (HPCRC)—Amirkabir University of Technology under Contract No. ISI-DCE-DOD-Cloud-700101-4504.

Conflicts of Interest: The authors declare no conflict of interest.

Abbreviations

The following abbreviations are used in this manuscript:

WEC	Wave Energy Converter
PTO	Power Take-off
PSO	Particle Swarm Optimisation
GA	Genetic Algorithm
EA	Evolutionary Algorithms
DE	Differential Evolution
GWO	Gray Wolf Optimiser
ML	Machine Learning

References

1. Barstow, S.; Mørk, G.; Mollison, D.; Cruz, J. The wave energy resource. In *Ocean Wave Energy*; Springer: Berlin/Heidelberg, Germany, 2008; pp. 93–132.

2. (US) EIA; Government Printing Office. *International Energy Outlook 2016, with Projections to 2040*; Government Printing Office: Washington, DC, USA, 2016.
3. Clément, A.; McCullen, P.; Falcão, A.; Fiorentino, A.; Gardner, F.; Hammarlund, K.; Lemonis, G.; Lewis, T.; Nielsen, K.; Petroncini, S.; et al. Wave energy in Europe: Current status and perspectives. *Renew. Sustain. Energy Rev.* **2002**, *6*, 405–431. [CrossRef]
4. Antonio, F.d.O. Wave energy utilization: A review of the technologies. *Renew. Sustain. Energy Rev.* **2010**, *14*, 899–918.
5. Linton, C. Radiation and diffraction of water waves by a submerged sphere in finite depth. *Ocean Eng.* **1991**, *18*, 61–74. [CrossRef]
6. Babarit, A. On the park effect in arrays of oscillating wave energy converters. *Renew. Energy* **2013**, *58*, 68–78. [CrossRef]
7. Babarit, A.; Hals, J.; Muliawan, M.; Kurniawan, A.; Moan, T.; Krokstad, J. *Numerical Estimation of Energy Delivery from a Selection of Wave Energy Converters—Final Report*; Report; Ecole Centrale de Nantes & Norges Teknisk-Naturvitenskapelige Universitet: Nantes, France, 2011.
8. Göteman, M.; Engström, J.; Eriksson, M.; Isberg, J.; Leijon, M. Methods of reducing power fluctuations in wave energy parks. *J. Renew. Sustain. Energy* **2014**, *6*, 043103. [CrossRef]
9. Göteman, M.; Engström, J.; Eriksson, M.; Isberg, J. Optimizing wave energy parks with over 1000 interacting point-absorbers using an approximate analytical method. *Int. J. Mar. Energy* **2015**, *10*, 113–126. [CrossRef]
10. Giassi, M.; Castellucci, V.; Göteman, M. Economical layout optimization of wave energy parks clustered in electrical subsystems. *Appl. Ocean Res.* **2020**, *101*, 102274. [CrossRef]
11. Neshat, M.; Sergiienko, N.Y.; Amini, E.; Majidi Nezhad, M.; Astiaso Garcia, D.; Alexander, B.; Wagner, M. A New Bi-Level Optimisation Framework for Optimising a Multi-Mode Wave Energy Converter Design: A Case Study for the Marettimo Island, Mediterranean Sea. *Energies* **2020**, *13*, 5498. [CrossRef]
12. Budal, K. Theory for absorption of wave power by a system of interacting bodies. *J. Ship Res.* **1977**, *21*, 248–254.
13. Evans, D. A theory for wave-power absorption by oscillating bodies. *J. Fluid Mech.* **1976**, *77*, 1–25. [CrossRef]
14. Budal, K. *Interacting Point Absorbers with Controlled Motion*; Pascal Academic Press: London, UK, 1980.
15. Thomas, G.; Evans, D. Arrays of three-dimensional wave-energy absorbers. *J. Fluid Mech.* **1981**, *108*, 67–88. [CrossRef]
16. Simon, M. Multiple scattering in arrays of axisymmetric wave-energy devices. Part 1. A matrix method using a plane-wave approximation. *J. Fluid Mech.* **1982**, *120*, 1–25. [CrossRef]
17. Castro, F.A.; Chiang, L.E. Design optimization and experimental validation of a two-body Wave Energy Converter with adjustable Power Take-Off parameters. *Energy Sustain. Dev.* **2020**, *56*, 19–32. [CrossRef]
18. Neshat, M.; Alexander, B.; Wagner, M. A hybrid cooperative co-evolution algorithm framework for optimising power take off and placements of wave energy converters. *Inf. Sci.* **2020**, *534*, 218–244. [CrossRef]
19. Bonovas, M.I.; Anagnostopoulos, I.S. Modelling of operation and optimum design of a wave power take-off system with energy storage. *Renew. Energy* **2020**, *147*, 502–514. [CrossRef]
20. Calvário, M.; Gaspar, J.; Kamarlouei, M.; Hallak, T.; Soares, C.G. Oil-hydraulic power take-off concept for an oscillating wave surge converter. *Renew. Energy* **2020**, *159*, 1297–1309. [CrossRef]
21. Tronchin, L.; Manfren, M.; Nastasi, B. Energy analytics for supporting built environment decarbonisation. *Energy Procedia* **2019**, *157*, 1486–1493. [CrossRef]
22. Mazzoni, S.; Ooi, S.; Nastasi, B.; Romagnoli, A. Energy storage technologies as techno-economic parameters for master-planning and optimal dispatch in smart multi energy systems. *Appl. Energy* **2019**, *254*, 113682. [CrossRef]
23. Nastasi, B. Hydrogen policy, market, and R&D projects. In *Solar Hydrogen Production*; Elsevier: Amsterdam, The Netherlands, 2019; pp. 31–44.
24. Bozzi, S.; Giassi, M.; Miquel, A.M.; Antonini, A.; Bizzozero, F.; Gruosso, G.; Archetti, R.; Passoni, G. Wave energy farm design in real wave climates: The Italian offshore. *Energy* **2017**, *122*, 378–389. [CrossRef]
25. De Andrés, A.; Guanche, R.; Meneses, L.; Vidal, C.; Losada, I. Factors that influence array layout on wave energy farms. *Ocean Eng.* **2014**, *82*, 32–41. [CrossRef]
26. Engström, J.; Eriksson, M.; Göteman, M.; Isberg, J.; Leijon, M. Performance of large arrays of point absorbing direct-driven wave energy converters. *J. Appl. Phys.* **2013**, *114*, 204502. [CrossRef]

27. Vicente, P.C.; Falcão, A.d.O.; Justino, P.A. A time domain analysis of arrays of floating point-absorber wave energy converters including the effect of nonlinear mooring forces. In Proceedings of the ICOE'2010-3rd International Conference on Ocean Energy, Bilbao, Spain, 6–8 October 2010.
28. Yang, S.; Ringsberg, J.; Johnson, E. Analysis of interaction effects between WECs in four types of wave farms. In Proceedings of the 3rd International Conference on Renewable Energies Offshore (RENEW 2018), Lisbon, Portugal, 8–10 October 2018; pp. 647–658.
29. Sergiienko, N.Y.; Neshat, M.; da Silva, L.S.; Alexander, B.; Wagner, M. Design optimisation of a multi-mode wave energy converter. *arXiv* **2020**, arXiv:2001.08966.
30. Neshat, M.; Alexander, B.; Sergiienko, N.Y.; Wagner, M. Optimisation of Large Wave Farms Using a Multi-Strategy Evolutionary Framework. In Proceedings of the 2020 Genetic and Evolutionary Computation Conference, GECCO'20, Cancún, Mexico, 14 June 2020; Association for Computing Machinery: New York, NY, USA, 2020; pp. 1150–1158. [CrossRef]
31. Jusoh, M.; Ibrahim, M.; Daud, M.; Yusop, Z.; Albani, A.; Rahman, S.; Mohad, S. Parameters estimation of hydraulic power take-off system for wave energy conversion system using genetic algorithm. In Proceedings of the International Conference on Sustainable Energy and Green Technology, Bangkok, Thailand, 11–14 December 2019; Volume 463, p. 012129.
32. Rodríguez, C.A.; Rosa-Santos, P.; Taveira-Pinto, F. Hydrodynamic optimization of the geometry of a sloped-motion wave energy converter. *Ocean Eng.* **2020**, *199*, 107046. [CrossRef]
33. M'zoughi, F.; Bouallègue, S.; Garrido, A.J.; Garrido, I.; Ayadi, M. Water cycle algorithm–based airflow control for oscillating water column—Based wave energy converters. *Proc. Inst. Mech. Eng. Part I J. Syst. Control Eng.* **2020**, *234*, 118–133. [CrossRef]
34. Chen, M. Hydrodynamic Analysis and Optimization of a Hinged-Type Wave Energy Converter-SeaWEED. Ph.D. Thesis, Memorial University of Newfoundland, St. John's, NL, Canada, 2020.
35. Liu, Z.; Wang, Y.; Hua, X. Prediction and optimization of oscillating wave surge converter using machine learning techniques. *Energy Convers. Manag.* **2020**, *210*, 112677. [CrossRef]
36. Izquierdo-Pérez, J.; Brentan, B.M.; Izquierdo, J.; Clausen, N.E.; Pegalajar-Jurado, A.; Ebsen, N. Layout Optimization Process to Minimize the Cost of Energy of an Offshore Floating Hybrid Wind–Wave Farm. *Processes* **2020**, *8*, 139. [CrossRef]
37. Esmaeilzadeh, S.; Alam, M.R. Shape optimization of wave energy converters for broadband directional incident waves. *Ocean Eng.* **2019**, *174*, 186–200. [CrossRef]
38. Lyu, J.; Abdelkhalik, O.; Gauchia, L. Optimization of dimensions and layout of an array of wave energy converters. *Ocean Eng.* **2019**, *192*, 106543. [CrossRef]
39. Kelly, M.; Alam, M.R. Shape Optimization of a Submerged Pressure Differential Wave Energy Converter for Load Reductions. In Proceedings of the International Conference on Offshore Mechanics and Arctic Engineering, Glasgow, UK, 9–14 June 2019; American Society of Mechanical Engineers: New York, NY, USA, 2019; Volume 58899, p. V010T09A030.
40. Neshat, M.; Alexander, B.; Sergiienko, N.; Wagner, M. New insights into position optimisation of wave energy converters using hybrid local search. *Swarm Evol. Comput.* **2020**, 59, 100744. [CrossRef]
41. Neshat, M.; Abbasnejad, E.; Shi, Q.; Alexander, B.; Wagner, M. Adaptive neuro-surrogate-based optimisation method for wave energy converters placement optimisation. In Proceedings of the International Conference on Neural Information Processing, Sydney, Australia, 12–15 December 2019; Springer: Berlin/Heidelberg, Germany, 2019; pp. 353–366.
42. Jabrali, A.; Khatyr, R.; Naciri, J.K. Viscous effects and energy recovery optimization for freely floating and bottom fixed wave energy converters. *Int. J. Renew. Energy Res.* **2019**, *9*, 290–300.
43. Wang, L.; Ringwood, J.V. Geometric optimization of a hinge-barge wave energy converter. In Proceedings of the 13th European Wave and Tidal Energy Conference, Naples, Italy, 1–6 September 2019; p. 1389.

44. Faraggiana, E.; Masters, I.; Chapman, J. Design of an optimization scheme for the WaveSub array. In *Advances in Renewable Energies Offshore, Proceedings of the 3rd International Conference on Renewable Energies Offshore (RENEW), Lisbon, Portugal, 8–10 October 2018*; Soares, G., Ed.; Taylor & Francis Group: London, UK, 10 October 2019; pp. 633–638.
45. Neshat, M.; Alexander, B.; Sergiienko, N.Y.; Wagner, M. A Hybrid Evolutionary Algorithm Framework for Optimising Power Take off and Placements of Wave Energy Converters. In Proceedings of the Genetic and Evolutionary Computation Conference, GECCO'19; Prague, Czech Republic, 14 July 2019; Association for Computing Machinery: New York, NY, USA, 2019; p. 1293–1301. [CrossRef]
46. Vatchavayi, S.R. Heuristic Optimization of Wave Energy Converter Arrays. Ph.D. Thesis, University of Minnesota, Minneapolis, MN, USA, 2019.
47. Amini, E. Locating and Evaluating the Oscillating Surge Wave Energy Converter Using Grey Wolf Optimizer Algorithm and WEC-Sim Toolbox. Ph.D. Thesis, University of Tehran, Tehran, Iran, 2019.
48. Amini, E.; Naeeni, S.T.O.; Ghaderi, P. Investigating Wave Energy Potential in Southern Coasts of the Caspian Sea and Evaluating the Application of Gray Wolf Optimizer Algorithm. *arXiv* **2019**, arXiv:1912.13201.
49. Sharp, C.; DuPont, B. Wave energy converter array optimization: A genetic algorithm approach and minimum separation distance study. *Ocean Eng.* **2018**, *163*, 148–156. [CrossRef]
50. Neshat, M.; Alexander, B.; Wagner, M.; Xia, Y. A detailed comparison of meta-heuristic methods for optimising wave energy converter placements. In Proceedings of the Genetic and Evolutionary Computation Conference, Kyoto, Japan, 15–19 July 2018; ACM: New York, NY, USA, 2018; pp. 1318–1325.
51. Fang, H.W.; Feng, Y.Z.; Li, G.P. Optimization of Wave Energy Converter Arrays by an Improved Differential Evolution Algorithm. *Energies* **2018**, *11*, 3522. [CrossRef]
52. Arbonès, D.R.; Sergiienko, N.Y.; Ding, B.; Krause, O.; Igel, C.; Wagner, M. Sparse incomplete lu-decomposition for wave farm designs under realistic conditions. In Proceedings of the International Conference on Parallel Problem Solving from Nature, Coimbra, Portugal, 8–12 September 2018; Springer: Berlin/Heidelberg, Germany, 2018; pp. 512–524.
53. Abdelkhalik, O.; Darani, S. Optimization of nonlinear wave energy converters. *Ocean Eng.* **2018**, *162*, 187–195. [CrossRef]
54. Giassi, M.; Göteman, M. Layout design of wave energy parks by a genetic algorithm. *Ocean Eng.* **2018**, *154*, 252–261. [CrossRef]
55. Göteman, M.; Engström, J.; Eriksson, M.; Isberg, J. Fast modeling of large wave energy farms using interaction distance cut-off. *Energies* **2015**, *8*, 13741–13757. [CrossRef]
56. López-Ruiz, A.; Bergillos, R.J.; Raffo-Caballero, J.M.; Ortega-Sánchez, M. Towards an optimum design of wave energy converter arrays through an integrated approach of life cycle performance and operational capacity. *Appl. Energy* **2018**, *209*, 20–32. [CrossRef]
57. Borgarino, B.; Babarit, A.; Ferrant, P. Impact of wave interactions effects on energy absorption in large arrays of wave energy converters. *Ocean Eng.* **2012**, *41*, 79–88. [CrossRef]
58. Carnegie Clean Energy Limited (formerly Carnegie Wave Energy). Carnegie CETO 6 Technology. Available online: https://arena.gov.au/projects/carnegie-ceto-6-technology/ (accessed on 15 July 2020).
59. Sergiienko, N. Wave Energy Converter (WEC) Array Simulator. 2020, Available online: https://www.mathworks.com/matlabcentral/fileexchange/71840-wave-energy-converter-wec-array-simulator (accessed on 27 July 2020).
60. Sergiienko, N.Y.; Cazzolato, B.S.; Ding, B.; Arjomandi, M. An optimal arrangement of mooring lines for the three-tether submerged point-absorbing wave energy converter. *Renew. Energy* **2016**, *93*, 27–37. [CrossRef]
61. Nolte, J.D.; Ertekin, R. Wave power calculations for a wave energy conversion device connected to a drogue. *J. Renew. Sustain. Energy* **2014**, *6*, 013117. [CrossRef]
62. Cummins, W.E. *The Impulse Response Function and Ship Motions*; Report; DTIC Document: Rockville, MD, USA, 1962.
63. Sergiienko, N.; Cazzolato, B.; Ding, B.; Hardy, P.; Arjomandi, M. Performance comparison of the floating and fully submerged quasi-point absorber wave energy converters. *Renew. Energy* **2017**, *108*, 425–437. [CrossRef]
64. Flavià, F.F.; Babarit, A.; Clément, A.H. On the numerical modeling and optimization of a bottom-referenced heave-buoy array of wave energy converters. *Int. J. Mar. Energy* **2017**, *19*, 1–15. [CrossRef]

65. Zwolan, P.; Czaplewski, K. Sea waves models used in maritime simulators. *Zesz. Nauk. Morska W Szczecinie* **2012**, *104*, 186–190.
66. Chen, W.; Gao, F.; Meng, X.; Fu, J. Design of the wave energy converter array to achieve constructive effects. *Ocean Eng.* **2016**, *124*, 13–20. [CrossRef]
67. De Andres, A.; Maillet, J.; Hals Todalshaug, J.; Möller, P.; Bould, D.; Jeffrey, H. Techno-Economic Related Metrics for a Wave Energy Converters Feasibility Assessment. *Sustainability* **2016**, *8*, 1109. [CrossRef]

energies

MDPI

Article

A Data Analytics-Based Energy Information System (EIS) Tool to Perform Meter-Level Anomaly Detection and Diagnosis in Buildings

Roberto Chiosa, Marco Savino Piscitelli and Alfonso Capozzoli *

Department of Energy "Galileo Ferraris", TEBE Research Group, BAEDA Lab, Politecnico di Torino, Corso Duca degli Abruzzi 24, 10129 Turin, Italy; roberto.chiosa@polito.it (R.C.); marco.piscitelli@polito.it (M.S.P.)
* Correspondence: alfonso.capozzoli@polito.it

Abstract: Recently, the spread of smart metering infrastructures has enabled the easier collection of building-related data. It has been proven that a proper analysis of such data can bring significant benefits for the characterization of building performance and spotting valuable saving opportunities. More and more researchers worldwide are focused on the development of more robust frameworks of analysis capable of extracting from meter-level data useful information to enhance the process of energy management in buildings, for instance, by detecting inefficiencies or anomalous energy behavior during operation. This paper proposes an innovative anomaly detection and diagnosis (ADD) methodology to automatically detect at whole-building meter level anomalous energy consumption and then perform a diagnosis on the sub-loads responsible for anomalous patterns. The process consists of multiple steps combining data analytics techniques. A set of evolutionary classification trees is developed to discover frequent and infrequent aggregated energy patterns, properly transformed through an adaptive symbolic aggregate approximation (aSAX) process. Then a post-mining analysis based on association rule mining (ARM) is performed to discover the main sub-loads which mostly affect the anomaly detected at the whole-building level. The methodology is developed and tested on monitored data of a medium voltage/low voltage (MV/LV) transformation cabin of a university campus.

Keywords: building energy management; energy information systems; anomaly detection and diagnosis; classification tree; symbolic aggregate approximation; association rule mining

Citation: Chiosa, R.; Piscitelli, M.S.; Capozzoli, A. A Data Analytics-Based Energy Information System (EIS) Tool to Perform Meter-Level Anomaly Detection and Diagnosis in Buildings. *Energies* **2021**, *14*, 237. https://doi.org/10.3390/en14010237

Received: 8 December 2020
Accepted: 30 December 2020
Published: 5 January 2021

1. Introduction

The building sector is globally recognized as one of the most energy-intensive, and its energy demand continues to increase as a result of a combination of various factors such as extreme climatic events, increased demand for energy services, and in particular those related to air conditioning and quality of the built environment. According to the International Energy Agency (IEA) for the EU member states, buildings are responsible for around 21% of primary energy consumption [1].

As a result, this sector is currently among the most strategic ones for reducing global energy demand, improving energy efficiency, and achieving specific decarbonization targets. In the last years, the great focus on buildings has also been encouraged by the introduction of a robust regulatory framework that puts in evidence the importance of a more responsible building energy management. In this perspective, the technological advancements that characterized the world of IoT (Internet of Things) and ICT (information and communication technology) has played a fundamental role in determining an ever-increasing spread of advanced monitoring and automation infrastructures in buildings, making it is possible to collect a huge amount of data and information related to the real performance in the operation of such complex systems.

The analysis of data collected represents a huge opportunity to identify and define effective energy-saving strategies and to optimize building performance in operation [2,3]. This process can be considered as the starting point of all the activities that are aimed at reducing the gap between the actual and expected building energy performance that is often generated by incorrect occupant behavior, equipment faults, and wrong or ineffective control strategies of energy systems [4].

Moreover, thanks to the growing availability of open access building data sets [5–7], analysts can quantitatively compare different processes of analysis, evaluating algorithm performance and assessing building energy performance in a more objective and transparent way [8].

Nonetheless, professional figures involved in the energy management process of buildings are now facing great difficulties in managing these large amounts of data and setting their analyses in a systematic way in order to extract useful knowledge and consequently the desired value.

For this purpose, energy management and information systems (EMISs) can be employed. EMISs belong to the rapidly evolving family of tools that monitor, analyze, and control building energy use and system performance, often leveraging advanced data analytics-based technologies. According to [9], the first classification of EMISs distinguishes such systems considering if their functionalities are enabled at the meter or system-level. The first category of EMISs considers data measurements at a high level (e.g., data related to the total load or of the main sub-loads) while system-level EMISs are focused on more detailed data related to the operation of specific systems or components. energy information systems (EISs) are part of EMIS and integrate software solutions conceived for the analysis of meter-level monitored data of buildings that are not usually collected through building automation systems (BAS). EISs typically enable predictive pattern recognition analysis for performing essential tasks in building energy management such as energy consumption forecasting, anomaly detection and diagnosis, advanced benchmarking, load profiling, and schedule optimization of building energy systems [4].

Among these tasks, anomaly detection and diagnosis has been the most underdeveloped for application on meter-level data.

Anomaly detection and diagnosis (ADD) in buildings is often related to fault detection and diagnosis (FDD) analysis conducted at system/component-level where the scale of analysis is small (e.g., air handling unit components). However, in most real cases, just a few aggregate variables related to the total energy consumption of the building are monitored and collected. Improving the building energy performance by analyzing aggregate data is challenging, especially if several factors such as occupant behavior, comfort levels, operational schedules of systems may generate different energy consumption patterns not always easily inferable. In this context, an EIS tool capable to automatically detect anomalous energy trends in building energy consumption allows energy managers to be promptly informed when the building is not behaving as expected and to avoid inefficient energy management procedures.

In the process of ADD, pattern recognition techniques play a key role in the analysis of patterns and trends in high-dimensional time series of building energy consumption [10]. There are three main expected goals behind ADD analysis in buildings that can be summarised as follows:

- Identification of typical load patterns in whole-building energy consumption time series.
- Detection of anomalous load patterns when typical ones are violated over time.
- Diagnosis of the detected anomalies by means of inference analysis performed on the main sub-loads.

According to the aforementioned objectives, this work proposes an EIS tool capable of performing ADD analysis in buildings by exploiting meter-level data. ADD procedures are usually performed offline and on small subsets of historical data, but more and more interest is growing in developing an automatic framework of analysis for online implementations. For this purpose in this paper, an innovative ADD methodology conceived for application in a real testbed (i.e., the university campus of Politecnico di Torino) is presented. The proposed methodology enables the automatic detection of energy anomalies at the whole-building level and their diagnosis at the sub-load level, revealing which sub-load/sub-loads are responsible for the anomalies detected. According to the objective of this paper, the next paragraph reports and discusses the literature concerning the implementation of ADD processes in buildings and presents the main contributions introduced in this work.

Related Work and Contribution of the Paper

ADD is extremely valuable for improving building energy performance and promising in terms of cost reduction potential if implemented in currently adopted EISs [11]. Despite the great potential offered by ADD at different levels of investigation in buildings, the implementation of this kind of analysis has been majorly focused at the system/component-level (e.g., heating, ventilation, and air conditioning (HVAC) systems), often neglecting applications at whole-building. This trend has been justified by the great availability of system-level data collected by building automation systems (BASs) in buildings. However, extracting any kind of meaningful information from BASs (especially from the outdated ones) can be a complicated task usually characterized by limitations on the data availability. Conversely, the collection of meter-level data in buildings is often performed by means of modern IoT devices that make monitored data easily available as never before. In this context, EIS tools focused on the analysis of meter-level data (especially ADD analysis) are becoming a very fast-growing market in the context of building analytics technologies.

According to the literature, the field of ADD in buildings is progressively leveraging on the application of data analytics techniques [12] for addressing both detection and diagnosis tasks.

The first task is often accomplished through the use of classification, regression, and pattern recognition techniques capable of providing estimations of the building energy consumption in normal operation according to specific boundary conditions (e.g., outdoor climatic conditions). The estimations are then used as a reference baseline for detecting the occurrence of abnormal patterns in the time series that significantly differs from the majority of processed data and/or from the expected trend [13].

For what concerns the implementation of supervised techniques for anomaly detection, in [14] the building energy consumption anomalies are identified comparing the actual consumption with the prediction of a hybrid artificial neural network (ANN) model. A similar approach is adopted in [15], where a deep neural network autoencoder was used to create a prediction model able to successfully detect abnormal energy patterns in the building operational data of an educational building in Hong Kong. Similarly, a general anomaly detection process is also proposed in [16], where the authors employed a variational recurrent autoencoder. Among supervised techniques also classification algorithms proved to be effective in anomaly detection. A robust methodology based on classification trees (CT) was proposed in [17]. In more detail, in that study, a set of classifiers were used for predicting the occurrence of categorical patterns in the time series of the total building electrical load, making it possible to detect a potential anomaly in the case of misclassification (i.e., the same concept of residual analysis in the case of regression models). The study underlined the prediction capabilities of CT algorithms and, most of all, the possibility of exploiting their interpretable nature in anomaly detection problems by extracting useful "if-then" decision rules.

In the context of unsupervised learning for anomaly detection, clustering, and association rule mining (ARM) are the most used techniques [18,19]. In [20], the authors used k-means clustering to automatically discover anomalies in whole-building energy consumption among daily load profiles characterized by an infrequent trend. In [21], an agglomerative hierarchical clustering-based strategy and three different dissimilarity measures were used to identify typical electrical usage profiles that enabled the detection of the abnormal ones.

As previously stated, the use of decision rules in the form of "if-then" implications is extremely valuable in anomaly detection. Following an unsupervised approach, this can be achieved by extracting association rules from building an operational dataset. association rules mining (ARM) algorithms have been widely used to discover abnormal patterns in the energy consumption of buildings and systems and then to enhance their performance. ARM allows discovering causal relationships between events also in the time domain [22]. This kind of algorithm is particularly suitable in extracting hidden knowledge from large databases, as it is reported in [23], where an extensive rules extraction is performed to detect energy wastes in the operation of a lighting system. Similarly, in [10], an improved ARM-based method was employed to discover and detect abnormal operational patterns of HVAC systems installed in a commercial building in Shenzhen (China).

More sophisticated approaches for anomaly detection consist of combining several techniques to maximize the amount of knowledge discovered and automatize the process of analysis.

The study conducted in [24] introduced the concept of collective anomaly detection, described as an event that is considered anomalous only if considered in relation to other events. In the proposed framework, ARM, performed through the Apriori algorithm, was used to extract the most frequent items from a time series related to smart grid operation. Then, anomalous behavior was identified through clustering analysis, considering silhouette indicator as a quality metric. Also, in [25], an anomaly detection process based on an ensembling technique was proposed. In detail, typical building operational patterns were identified by means of clustering analysis, and then an ARM algorithm was used to discover an anomalous load of a cooling chiller system installed in a building in Hong Kong. In [26], a multi-step clustering analysis was performed for removing anomalous daily load profiles from the energy consumption time series of a university campus. Then a regression model was developed on the anomaly-free dataset, combining artificial neural network (ANN) and regression tree (RT), to be used in online applications for detecting the occurrence of anomalous trends in the electrical energy consumption.

Another crucial aspect that arises from the literature review deals with the use of data reduction and transformation methods for (i) reducing the computational cost of the analysis, (ii) easily extracting the main patterns from time series, (iii) improving the effectiveness of supervised and unsupervised algorithms in detecting anomalies. In fact, directly analyzing raw data of time series could be extremely onerous, making difficult the handling and the characterization of the data under investigation. In this perspective, dimensionality reduction can be used with a low computational cost, for example, for removing irrelevant patterns and redundancy from energy consumption datasets. As reviewed in [12], various techniques were explored to enable the classification of data as normal or anomalous, such as principal component analysis (PCA) [27], linear discriminant analysis (LDA) [28], singular variable decomposition (SVD) [29].

In this context, symbolic aggregate approximation (SAX) [30] is one of the most promising techniques available to reduce the size of a time series without losing key information [31]. The SAX algorithm is conceived for the reduction of the time series through a piecewise technique and on its transformation into symbolic strings. Frequent symbolic sub-sequences in the whole sequence can then be extracted and defined as motifs (i.e., normal patterns), while infrequent ones can be isolated and labeled as discords (i.e., potential anomalies). In [31], SAX was used to discover patterns in time series related to the energy consumption of the International Commerce Centre (ICC) in Hong Kong and to recognize inefficient operating conditions that could cause energy wastes. Also, in [20], SAX was used for enabling the extraction of infrequent operating patterns in the energy consumption time series of a school campus and an office building. In particular, discords were detected, setting a minimum frequency threshold to the occurrence of SAX symbol sub-sequences representative of the original daily load profiles. In [17], an enhanced version of SAX called adaptive SAX (aSAX) was used for minimizing the information loss due to the reduction and transformation of energy consumption time series and recognizing motif and discord symbolic patterns by means of classification models.

Once the detection of anomalies in energy consumption is performed, a diagnosis analysis makes it possible to identify the main causes associated with them. The field of anomaly diagnosis has been widely explored in buildings but with a greater focus on system-level applications rather than whole-building level. Also, this research field largely benefits from the use of data analytics techniques following both supervised and unsupervised approaches. The study in [32] proposed a process based on the development of a CT for diagnosing anomalies in the operation of air handling unit (AHU) components. Moreover, in [22], a CT was used for diagnosing up to 11 typical faults in AHU with an accuracy higher than 90%. Indeed, similarly to previously presented studies focused on anomaly detection, also the diagnosis analysis often exploits algorithms that allow the extraction of decision rules. Such a condition is particularly favorable for the final user due to the high interpretability of the diagnosis process, which meaningfulness can then be easily validated by domain expertise. To this aim also ARM algorithms can be employed as reported in [33–36].

On the basis of the literature review, in most of the cases, only meter level anomaly detection is performed at the whole-building level without any further analysis for identifying anomaly causes among sub-loads at a lower level.

The work presented in this paper aims to bridge this literature gap by introducing a novel hierarchical multi-level approach in the ADD process. The proposed methodology allows to perform the anomaly detection phase at the whole-building level, and only if an anomalous pattern is detected, an event-based diagnostic process is activated for finding root causes at the sub-load level. The event-based hierarchical approach in anomaly diagnosis makes it possible to reduce the computational cost of the analysis and also to rationalize the number and the quality of feedback generated by the ADD tool during operation. Indeed, the final user is not required to visually inspect the trends of all sub-loads in real-time, but he/she is alerted only when interesting events occur, i.e., when specific anomalous conditions among the sub-loads trends generate a divergence of the total load from the expected pattern.

This work combines different advanced data analytics techniques with the aim of maintaining the output of the ADD process human-readable and interpretable while providing accurate results.

The paper considers as a case study the energy consumption data gathered from a monitoring infrastructure installed in the university campus of Politecnico di Torino. The data refer to the electrical energy consumption of a medium voltage/low voltage (MV/LV) transformation cabin that serves different buildings/zones of the campus. In particular, about ten sub-loads of the cabin are available for developing the introduced hierarchical ADD process. The methodology leverages the reduction and transformation of the analyzed time series through an enhanced and adaptive process based on symbolic aggregate approximation (aSAX) as presented in [17]. The aSAX transformation enabled a reduction of the dataset and an effective identification of unexpected operational energy consumption patterns at the sub-daily time windows level. Furthermore, the diagnosis of the abnormal patterns detected at the total load level (i.e., MV/LV transformation cabin) was provided by implementing an association rule mining (ARM) algorithm on the sub-load time series. In this context, the main innovative aspects introduced by the present paper can be summarised as follows:

- In order to further enhance the pattern recognition process enabled by the aSAX-based process introduced in [17], different features of the energy consumption time series were encoded in symbols in addition to the mean value evaluated in each time window for data reduction purposes. In particular, the encoding of trend features of the time series was performed, allowing an improved characterization of energy consumption behavior and making it possible to reduce the information loss that is always related to the application of temporal abstraction processes such as aSAX. In addition, both the number of time windows and alphabet size for the encoding of the time series in symbols were tuned during the analysis through a fully automatic process.

- The identification of the normal energy consumption pattern is evaluated for specific time periods during the day (i.e., aSAX time windows) by means of classification models capable of estimating the most probable symbol encoded through the aSAX-based process. In particular, globally optimal evolutionary trees were used to accomplish this task. The use of evolutionary trees introduce a twofold advantage in the classification task: (i) the results obtained from their application are fully interpretable as they can be translated in "if-then" decision rules, (ii) the achievable accuracy in high-dimensional problems can be significantly higher than the performance of standard decision trees (e.g., locally optimal classification trees [37]).
- The anomaly diagnosis is performed at the sub-load level by implementing an unsupervised data analytics technique based on an ARM algorithm. The diagnostic process is capable of automatically updating an anomaly library in the form of "if-then" association rules extracted from historical data. This opportunity allows the developed ADD tool to evolve during building operation, significantly increasing its generalizability.
- The whole methodology was conceived for being applied in a real testbed paying attention to its generalizability and scalability to other buildings. In this perspective, the developed ADD process is capable of self-tuning its hyper-parameters ensuring a robust performance in online implementations. As a reference, the algorithms for both detection and diagnosis of the energy anomalies can be easily retrained periodically or considering an event-based approach (e.g., the occurrence of a not pre-identified anomaly).

The rest of the paper is organized as follows. Section 2 provides an overview and a brief theoretical description of the data analytics methods used for conducting ADD analysis. Section 3 presents and describes the case study considered for the analysis. Section 4 introduces the methodological framework on the basis of the ADD analysis performed. Eventually, Sections 5 and 6 presents and discusses the results obtained, while in Section 7, the concluding remarks and future research perspectives are reported.

2. Description of the Data Analysis Methods

In this section, the data analytics methods employed in this work are briefly described. The method description is not intended to be exhaustive, but it is aimed to underline the usefulness in the framework of this study and building energy data exploitation.

2.1. Adaptive Symbolic Aggregate Approximation (aSAX)

Meter-level data measurements are collected in the so-called time series: a two-dimensional matrix where each row corresponds to a single observation in time and the column to a measured variable [31]. The sampling frequency determines the time interval between two consecutive observations, and for building applications, it is usually in the order of minutes. As a consequence, the resulting high-dimensional time series is often computationally expensive to be stored and analyzed in its original form. In this context, many dimensionality reductions and transformation techniques were proposed in the literature; one of the most widely used is the symbolic aggregate approximation (SAX), which makes it possible to compress time series while preserving its fundamental characteristics [30]. This process segments the original time series in sub-sequences, each of them is summarised with a single numerical value (e.g., mean value) that is then encoded into the alphabetic symbol and finally combined into a string. The resulting string is much shorter than the original time series and enables the application of various pattern recognition techniques while reducing the computational cost. In the last years, some variations to the original algorithm have been proposed in the literature, especially with the aim of generalizing some initial assumptions (e.g., data distribution) and facing information loss issues always generated from the reduction and transformation of time series. In the author's opinion, one of the greatest improvements to the SAX was introduced through the so-called adaptive symbolic aggregate approximation (aSAX) [38]. In the following, the main steps of aSAX process are presented, with specific reference to their implementation in the present work.

- **Chunking**: The original time series ($y(t) = \{y_1, \ldots y_n\}$) of length n is divided into N non-overlapping sub-sequences ($T = \{T_1, \ldots T_N\}$) chosen for the specific context. In the case of energy consumption time series, the selection of the length of the sub-sequences is influenced by the periodicity of the energy pattern observed, and for building applications, it is usually set to 24 h. Each sub-sequence is further divided into W segments called time windows ($\tau = \{\tau_1, \ldots \tau_W\}$). The parameter W is word size. During this process, it is possible to choose time windows with equal or different length, based on user preference [17,39];
- **Feature extraction**: In this step, an aggregated numerical feature is calculated starting from the sub-sequence of the original time series that falls in the generic time window $\tau_{i,}$ and this value is considered as representative of all the data points included in that window. Aggregated features can extract some important characteristics of the time series while losing some other information. The analyst chooses which feature is the most significant and whether one or more features are needed for the purpose of the study. The most used and known approach is called piecewise aggregate approximation (PAA), which performs a constant approximation of the original time series y(t) by replacing the values that fall into the same time window τ with their mean [40]. Many other statistical features can be extracted (mean, variance, kurtosis, skewness) not only from the time domain but even from other domains such as the frequency one [41]. A feature representing important characteristics of time series is, for example, the trend angle [42]. This feature is particularly effective in describing the time series trend, and it was employed in this study. In detail, given a time series $y(t) = \{y_1, \ldots y_n\}$ of length n in a given time window $\tau = \{t_1, \ldots t_n\}$, defined $\Delta q(t_1)$ and $\Delta q(t_n)$ the first order distance between the initial and final point with the time series mean, can be defined a trend triangle as shown as in Figure 1. The trend angle feature θ, green in Figure 1, is defined with the following equation:

$$\theta = \text{atan}\left(\frac{\Delta q(t_n) - \Delta q(t_1)}{n}\right) \tag{1}$$

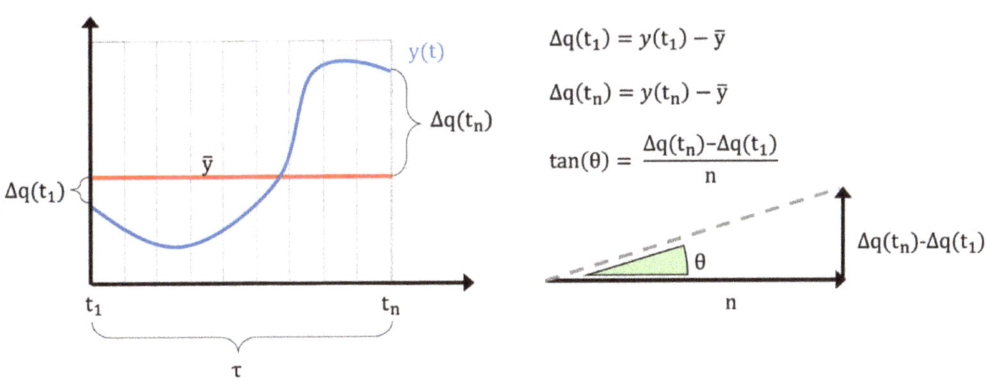

Figure 1. Definition of trend feature triangle and trend angle for a generic time series (y(t)).

The trend angle domain ranges continuously from $-90°$ to $90°$. If the trend angle value is approximately zero ($\theta \approx 0$), the trend is stationary; if it is positive ($\theta > 0$), the trend is rising; vice versa if it is negative ($\theta < 0$), the trend is descending.

- **Encoding**: this step consists of setting an alphabet size α and assigning an alphabetic character to each time window, according to where the extracted numerical feature lies within a set of breakpoints ($\beta = \{\beta_1, \ldots \beta_{\alpha\text{-}1}\}$) identified according to the shape of the feature distribution. The aSAX algorithm [38] finds the optimal positions of breakpoints through an iterative process by minimizing the distance among all the data points included between two consecutive breakpoints and their centroid (calculated average center). Eventually, the symbol can be assigned for each window (τ), creating a word of length W for the given sub-sequence (T_i). The original numerical time series y(t) is then transformed into an alphabetic string ($y(\alpha)$) of length W∗N.

Figure 2 shows an example of time series temporal abstraction conducted with the aSAX process. An electrical load time series ($y(t) = \{y_1, \ldots y_{192}\}$ (black line)) with a 15 min sampling frequency, is divided into two sub-sequences T_i and T_{i+1} of 24 h each. In this example, five-time windows (W = 5) of unequal length are identified for each sub-sequence, and the alphabet size is set to five ($\alpha = 5$), meaning that four breakpoints $\beta = \{\beta_1, \beta_2, \beta_3, \beta_4\}$ are identified. The time series is then approximated through PAA (red segments), and for each segment, the corresponding symbol is assigned. The PAA values distribution is shown on the right side of the figure in red and the breakpoints, evaluated through the aSAX, in dashed blue lines. The original time series for the time window (T_{i+1}) is converted from a numerical vector into an alphabetic string "a-b-d-c-a", reducing it from a 96-dimensional object to a 4-dimensional one.

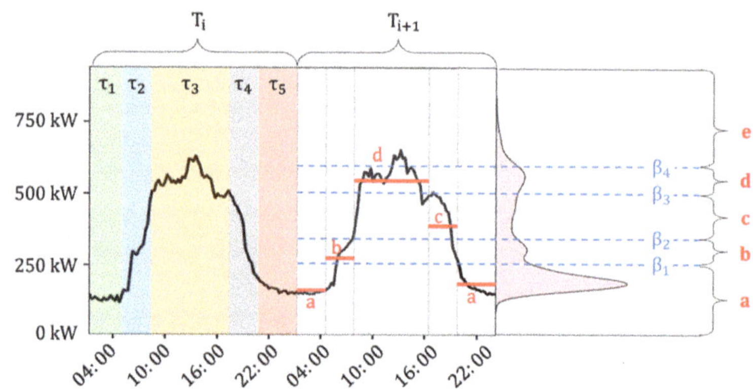

Figure 2. Example of an adaptive symbolic aggregate approximation (aSAX) process applied to an electrical load time series (T = 24 h, W = 5, $\alpha = 5$).

2.2. *Recursive Partitioning and Globally Optimal Evolutionary Tree*

Classification is the task of assigning a class label to unlabelled data instances through a classifier model, providing prediction or description of a given dataset [43]. The classification model is created through an inductive learning algorithm using a training set, which is a data frame with attributes and labeled instances. Once the model has been created, its performance is evaluated on a test set through the comparison between the predicted and real labels. The decision tree is the most commonly used model for classification, thanks to its understandable graphical representation. Depending on the type of target attribute, discrete categorical or continuous numerical, a decision tree is called, either a classification tree or regression tree, respectively. The tree consists of a root, internal nodes, and leaves, all connected by branches. The construction of a tree classifier can be performed through different algorithms; in this framework, recursive partitioning and globally optimal evolutionary tree are considered.

The most commonly used recursive partitioning method is the classification and regression tree (CART), which is a binary decision tree based on the splitting of the instances in purer subsets (i.e., nodes) through decision rules [44]. It proceeds in a forward step-wise approach by maximizing homogeneity in each child node, yielding to a local optimal tree.

Conversely, the so-called evolutionary decision tree is based on a stochastic algorithm that aims to construct a globally optimum classification model [37]. This process randomly initializes the root node split, then at each iteration, variation operators (i.e., split, prune, major split rule mutation, minor split rule mutation, crossover) are applied. The survivor is selected, and the process is repeated until the stopping criterion is satisfied. The evolutionary tree algorithm used in this paper is implemented in the R package "evtree" [37].

One of the most important hyper-parameter that can be set for this algorithm is the variation operator probability, which refers to the probability that a given variation operator is chosen at a generic iteration. The default operator probability considered is c20m40sp40, meaning that the algorithm has a 20% probability of selecting the crossover operator, a 40% probability for selecting one of the mutation operators (20% for minor split rule mutation and 20% for major split rule mutation) and a 40% probability for selecting one of the split (with 20% probability) or the prune operators (with 20% probability).

The advantage of an evolutionary tree algorithm is that it tends to offer higher accuracy in prediction than recursive partitioning algorithms [37] while maintaining the same interpretable tree structure.

2.3. Association Rules Mining (ARM)

ARM is a widely used technique that allows extracting static causal relationships and correlations between attributes in a dataset. The objective is to find a group of variables (items) that frequently occur together in a database. This technique can only handle categorical variables, and it is usually computationally costly. One of the most used ARM algorithms is the iterative Apriori algorithm based on a frequent itemset that allows the extraction of static rules from a categorical transactional dataset [45]. Association rules are defined between a set of items (or itemset) in the form $A \Rightarrow B$, where A is the itemset called antecedent (LHS = left-hand side of the rule) and B consequent (RHS = right-hand side of the rule) and $A \cap B = \emptyset$. Rule extraction is usually restricted to only an item in the consequent.

Some user-defined parameters (confidence, support, and lift) need to be set in order to evaluate the significance of the obtained rules and filter out the less important. A domain expert sets those parameters according to each specific case. The support is calculated as the probability of the intersection between the antecedent A and consequent B ($supp(A \Rightarrow B) = P(A \cap B)$), expressing the co-occurrence of the two events. The confidence ($conf(A \Rightarrow B) = P(B \mid A)$), defined as the conditional probability between A and B, allows assessment of the reliability of a rule. It gives the probability of the consequent event in all transactions containing the antecedent. The lift is the ratio between the confidence and support of consequent B ($lift(A \Rightarrow B) = P(B \mid A) \ / \ P(B)$). When the lift is higher than 1, it means that B is positively correlated with A, while if the lift is lower than 1, it suggests a negative correlation; otherwise, if the lift is equal to 1, there is no correlation at all. This parameter is particularly important since it allows one to select the most interesting rules [31]. In this paper, the ARM Apriori algorithm was used to extract interesting associations between the total building load and its sub-loads, especially during events detected as anomalous. This data analytics method was perfectly integrated with the outcome of the aSAX and classification processes, of which the results consist of categorical values.

3. Case Study

The case study analyzed refers to the energy consumption of a MV/LV transformer cabin identified as "substation C", that serves a part of the main campus of Politecnico di Torino (PoliTo), an Italian university located in Turin. Data related to the total electrical load and to some sub-loads are available with 15 min timesteps from 1 January 2015 to 31

December 2019. The hierarchical structure of the available data is shown in Figure 3: the first level refers to the total electrical load of substation C, while the second level shows the available sub-loads. In addition, the load breakdown in terms of average annual energy consumption was provided.

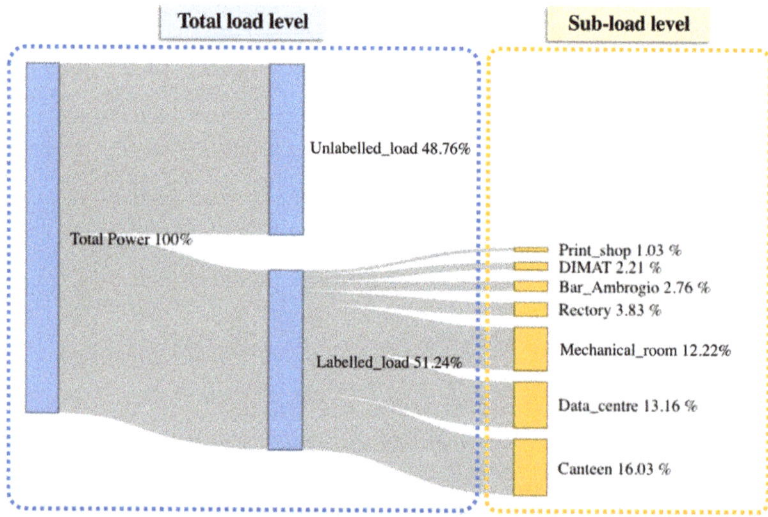

Figure 3. Hierarchical structure of the electrical load database under study.

In particular, a bar and a canteen were at the disposal of students and campus staff and accounted for 2.75% and 16.03%, respectively, of the total electrical energy consumption of substation C. The university data center accounted for 13.16% of the total energy consumption. The administration offices (rectory) corresponded to 3.83% of energy consumption and the mathematics department (DIMAT) for 2.21%. A large share of energy consumption (12.22%) was related to the mechanical room. The equipment located in this room included hot and chilled water circuits and auxiliaries such as recirculation pumps. The chilled water was provided by two chillers of nominal electrical power of 220 kW and a rated cooling capacity of 1120 kW, and a reversible water-water heat pump, with nominal a power and cooling capacity of 165 kW and 590 kW, respectively.

The remaining energy consumption was aggregated under a unique instance tagged as "Unlabelled_load" as showed in Figure 3. It accounted for 48.76% of the total energy consumption, and since it was not directly measured, cannot be assigned to a specific sub-load.

4. Methodological Framework

In this section the conceived ADD methodology is presented and described. The proposed methodology aims to develop a two-level ADD analysis capable of making in a first step a high-level detection on total electrical load time series (at meter level) and in a second step performing the anomaly diagnosis on sub-loads (at sub-meter level). The methodology follows the flow chart structure shown in Figure 4.

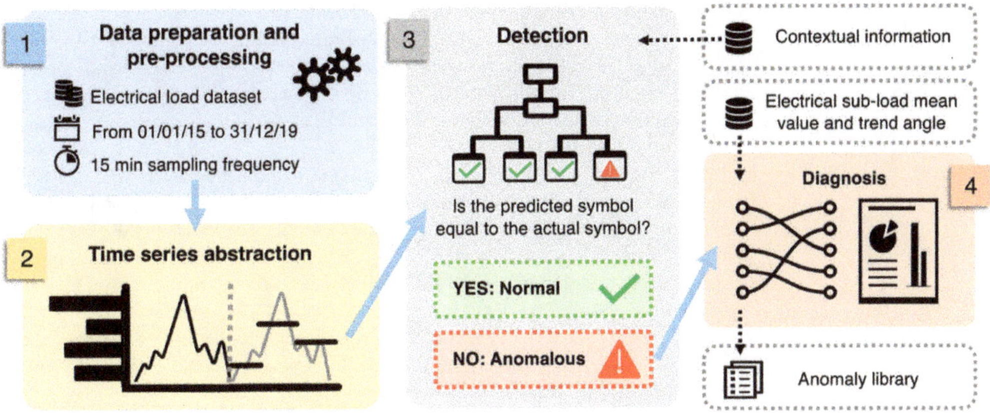

Figure 4. Flow chart explaining the adopted methodology.

In particular, four steps of analysis are considered.

- **Pre-processing:** The first step consists of pre-processing data that was aimed at removing punctual anomalies and inconsistencies from the datasets. The dataset used in this study included electrical load data (collected from substation C) from 1 January 2015 to 31 December 2019 with a 15 min sampling frequency. Negative measurements were removed a priori. Nearly-zero values of electrical load related to continuously operating systems (refrigerators, emergency lighting) were considered inconsistent and removed. Statistical outliers (e.g., data affected by transmission problems) were also identified and removed by means of boxplot analysis. Then all statistical inconsistencies and missing values are replaced through a linear interpolation;
- **Temporal abstraction of the time series**: In the second step of the analysis, the temporal abstraction of the electrical load time series was performed according to the procedure introduced in [17]. Temporal abstraction consists of the reduction and transformation of the time series in a sequence of alphabetic symbols. In particular, a recursive partitioning regression tree (RT) was used to identify sub-daily time windows with an unequal length for dimensionality reduction, considering the total electrical load from 2015 to 2019 as a numerical target and the hours of the day as a predictive attribute, as performed in [17]. Once time windows were evaluated, the PAA approximation is performed. The breakpoint identification was carried out through the aSAX method procedure by choosing the appropriate alphabet size through a k-means clustering process. The identification of the optimal number of clusters (i.e., alphabet size) was implemented through the R package "NbClust" [46];
- **Anomaly detection at total electrical load level**: Anomaly detection was performed on the encoded total electrical load time series of substation C. In each sub-daily time window, the total electrical load symbol obtained through aSAX was predicted through a globally optimal evolutionary tree [37], using as explanatory attributes contextual information such as calendar variables (day type and holiday) and energy variables (electrical demand of sub-loads). The model was developed through a test-train-validation process and was able to predict the expected symbol in each time widow with high accuracy. However, when the model failed to correctly predict the symbol in a time window, the occurrence of a potential anomaly was assumed. Referring to Figure 5, the predicted symbol is the one with the higher occurrence in a given leaf node (green bar). All other symbols were infrequent and then potentially anomalous (yellow and red bars). Given the interest in detecting higher electrical load than normal, only the tree leaves nodes that showed infrequent symbols corresponding to a high electrical load (red bars in Figure 5) were considered and investigated in the following diagnostic phase;

- **Diagnosis at sub-load level:** Once the classification models were developed, a post-mining phase was performed. The post-mining phase was aimed at searching historical relationships between misclassified total electrical load symbols and specific trends of sub-loads occurred in same time window. The process is described in Figure 6. The anomalous symbols identified in the training phase of the models were extracted and stored in a categorical data frame (Step-1 in Figure 6). From time series of sub-loads, the mean value and the trend angle were extracted. They were categorised through the aSAX process and then added to the categorical data frame (Step-2 in Figure 6). This data frame was then transformed into a transactional database on which ARM was applied (Step-3 in Figure 6). The LHS is composed of the additional categorical variables related to sub-loads, while RHS contains only the total electrical load anomalous symbol. ARM automatically extracts a set of rules which connects the historical infrequent behaviour of the total electrical load with the sub-load conditions. This process was implemented through the R package "arules" [47]. Resulting rules were then sorted and filtered setting appropriate interest measures parameters such as support, confidence and lift (Step-4 in Figure 6). Filtered rules were then stored within an anomaly library where they were ranked to show which sub-load condition (for example high electrical load or significantly uptrend) was responsible for the anomalous total electrical load behaviour. The tool gives a critical insight of the historical energy behaviour and, when implemented in real time load analysis, can provide useful feedback on which energy management actions are needed.

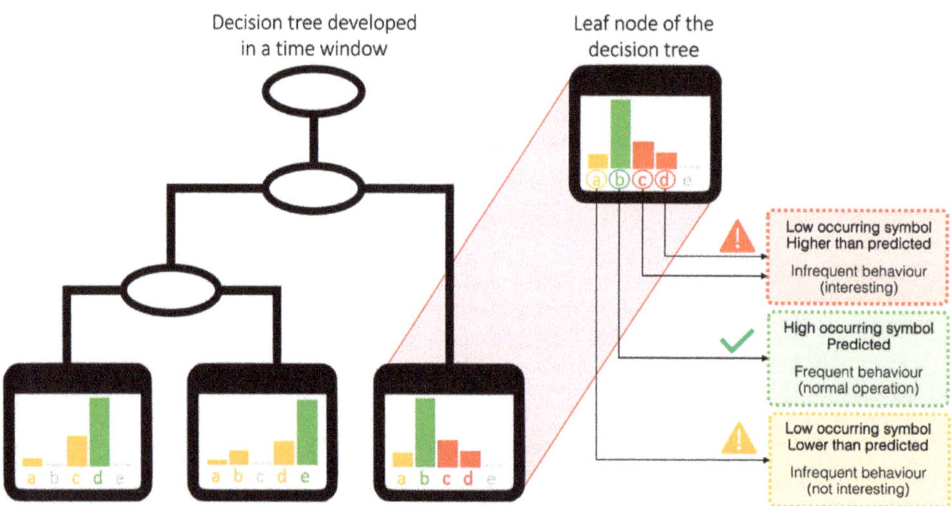

Figure 5. Interpretation of anomaly detection results.

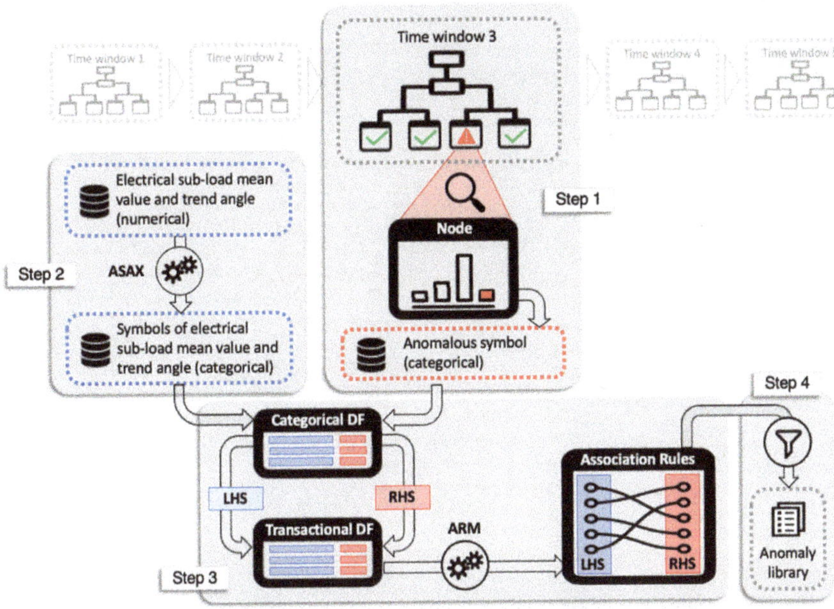

Figure 6. Sub-meter level diagnosis methodology description.

5. Results

The previously described methodology was applied to the case study presented in Section 3. The quantitative analysis of data was performed through the statistical software R [48], and results related to each stage are reported in the following sections.

5.1. Pre-Processing

The pre-processing phase allowed to handle missing values and to remove outliers. The procedure was applied to the total electrical load and sub-loads dataset.

In particular, punctual outliers due to data transmission problems were detected, removed, and replaced through linear interpolation. The carpet plots of the total electrical load of substation C are reported in Figure 7a (one for each year considered). It can be seen that the building energy systems were usually turned on at 6:00 and turned off at 19:00. The electrical load increased from the night baseload until 8:00 when teaching activities and office activities began and started decreasing after 16:00. This pattern was visible for every working day (from Monday to Friday) with an average electrical load (from 8:00 to 16:00) of more than 300 kW. During the weekend, on the other hand, there was a significant decrease in the average electrical load to 100 kW, mainly due to the weekly university break and the absence of teaching and office activities. The same carpet plot representation is reported in Figure 7 for some representative sub-loads. Figure 7b shows the electrical load of the mechanical room in the years from 2015 to 2019. Because of the intensive use of the chillers in summer, the highest monthly average electrical load was reached in July with a value of about 100 kW. During the winter months, the electrical load was not zero because of the electrical demand of the recirculation pumps. Figure 7c shows the electrical load of the campus canteen. Also, this load is strongly dependent on the weekly university occupancy schedule. In fact, a significant decrease in the average electrical load was visible during weekends, when the campus was unoccupied, and no teaching or office activity took place.

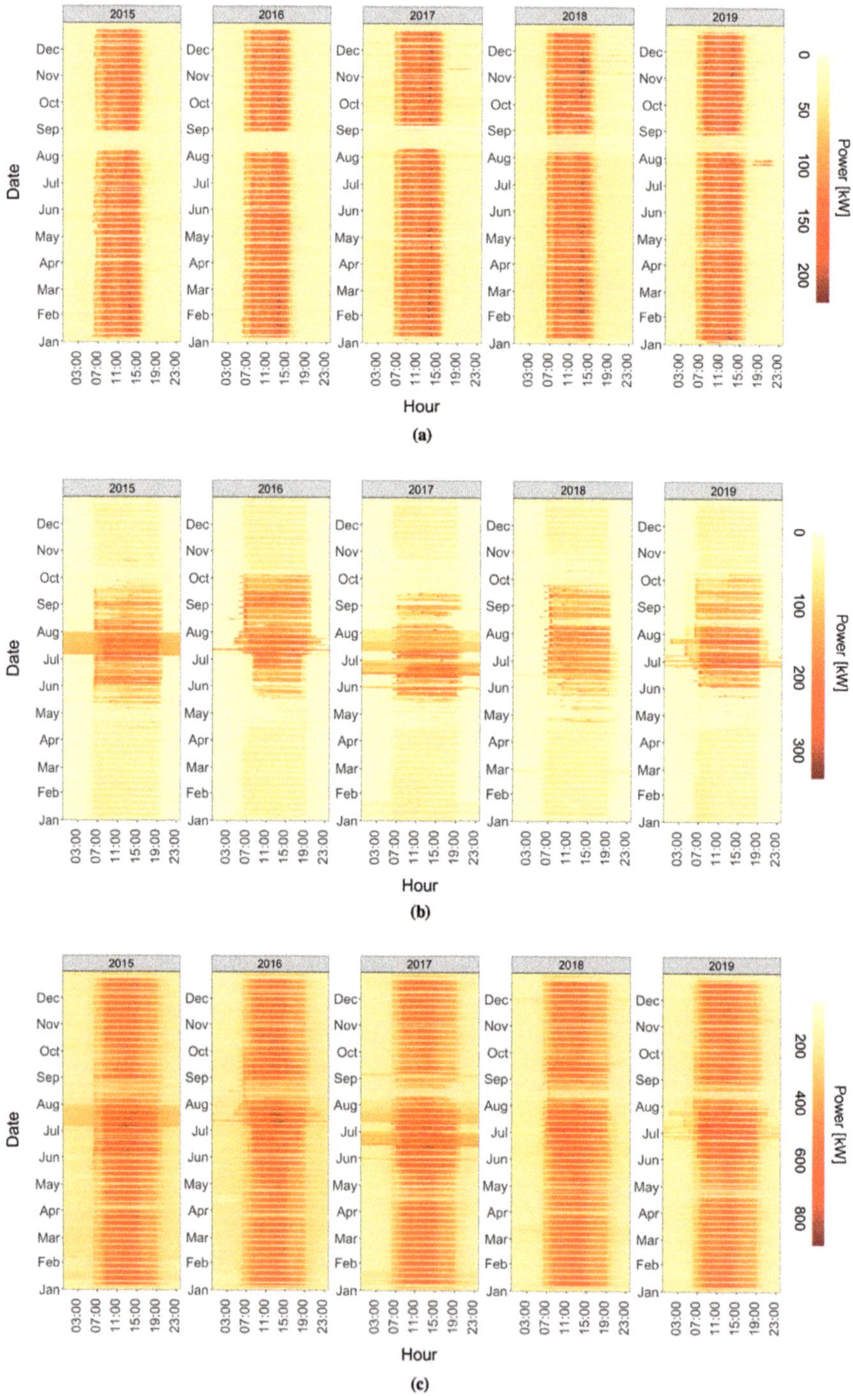

Figure 7. Carpet plot of the (**a**) total electrical load (substation C) (**b**) mechanical room (**c**) canteen.

5.2. Time Series Abstraction

In order to perform the data transformation and dimensionality reduction, the original time series of the electrical load was split into 24 h intervals since a daily periodical pattern was observed.

The time windows of daily load profiles were evaluated through a RT, considering the total electrical load as a numerical target and the hours of the day as a predictive attribute. The total electrical load from 2015 to 2019 was analyzed.

Holidays and weekends were excluded from the analysis since they usually present profiles that are flat or with low variance, and include those days in the model would have reduced the accuracy of the results. The splitting criterion adopted was based on the variance reduction around the numerical target's mean in each leaf node. In this way, the daily pattern was split into homogeneous consumption time windows. As a stopping criterion, a minimum number of objects in the child nodes at each split was set in order to have a time window length of at least 2 h.

The RT automatically identified the optimal number of windows thanks to a cost complexity pruning process. This procedure allowed us to choose the best tree by generating a fully expanded tree and then prune it iteratively. According to [17], this procedure enables the identification of an optimal trade-off between misclassification error and model complexity. The selection of the optimal tree size was performed according to the one standard error rule (i.e. 1-SE rule) [49].

The resulting tree had five leaves, which corresponded to five sub-daily time windows, which are summarised in Table 1. It can be seen that the first and fifth-time windows corresponded to the night hours during which the university was closed and not occupied. On the opposite, the remaining time windows correspond to occupied hours of the campus.

Table 1. Sub-daily time windows for total electrical load.

ID	Time Window	Duration
1	00:00–06:29	6 h 30 min
2	06:30–08:59	2 h 30 min
3	09:00–15:44	6 h 45 min
4	15:45–19:14	3 h 30 min
5	19:15–23:59	4 h 45 min

Once the time windows were identified, the PAA was performed in order to prepare the dataset for the encoding through the aSAX process.

A fundamental parameter to be set in the aSAX process is the alphabet size (α), which determines how many symbols are going to be used for the encoding, and as a consequence, also the number of breakpoints to search. While in the literature, the alphabet size is usually selected according to domain expertise [17,20,31], in this framework, an unsupervised technique consisting of k-means partitive clustering was used. In particular, the reduced data of the time series (through PAA) were clustered in order to find homogeneous groups and determine the optimal number of breakpoints. For this purpose, during the clustering process, several cluster quality indices, embedded in the R package NbClust [46], were calculated in order to assess the optimal number of clusters (k) according to a majority rule approach, setting a search space between k = 3 and k = 8. The results obtained suggested the partition with k = 6 as the optimal one, then determining the setting of the alphabet size value also equal to 6.

In detail, the positions of breakpoints, calculated under equally probability assumption, were used as initialization of the aSAX iterative algorithm [38]. As shown in Figure 8, those breakpoints (dotted lines) were not able to divide the distributions effectively, producing narrow intervals at low values and wider intervals for higher values of the reduced PAA time series. The final adaptive breakpoints (solid lines) were evaluated once a tolerance of 10^{-10} on the representation error was reached (after about 60 iterations).

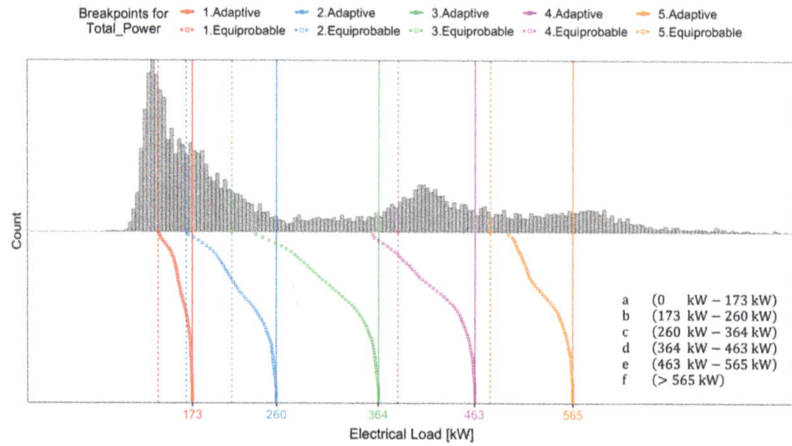

Figure 8. Step by step identification of adaptive breakpoints through the aSAX algorithm applied to the total electrical load.

Figure 9 shows the carpet plot and histograms, referring to the encoded total electrical load time series. In particular, Figure 9a shows that in the first and fifth-time window, the most frequent symbols were "a" and "b", which corresponded to a low electrical load during night hours. In the second and fourth-time windows, corresponding to early morning and late afternoon, there was a prevalence of medium electrical load identified with the symbol "d" describing the switch-on/off of the systems. In the third time window, the symbols "e" and "f" were the most frequent since the electrical load in the middle of the day is the highest.

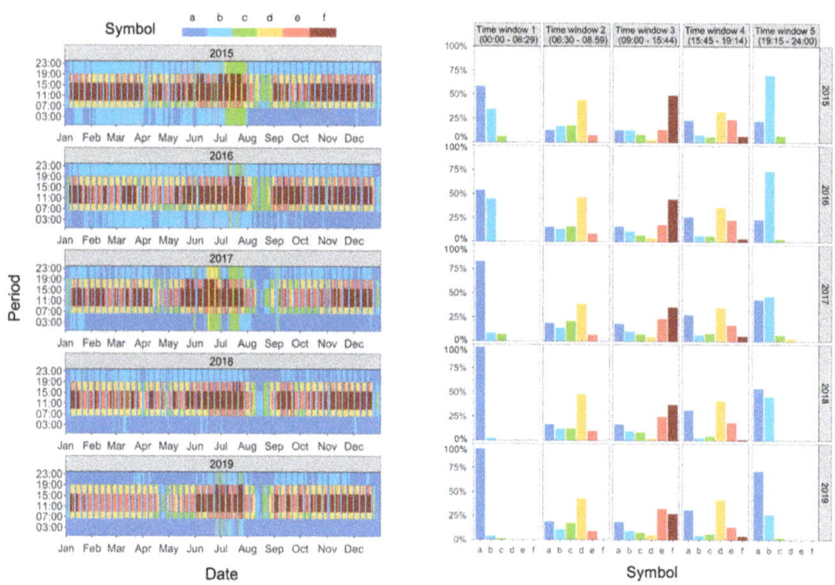

(a) (b)

Figure 9. aSAX representation of the total electrical load: (**a**) carpet plots (**b**) histogram distributions of symbols along the time windows and years.

Figure 9b shows the histograms of electrical load symbols divided by time windows and years. From this representation was evident how the load patterns had changed during the years from 2015 to 2019. In particular, in the first and fifth-time windows, a change of pattern from the symbol "b" to the symbol "a" was visible due to a lower baseload during night hours, when the campus was unoccupied. This behavior could be related to the refurbishment of buildings and/or systems served by substation C. The same trend was seen in the third time window where a change of pattern from the symbol "f" to the symbol "e" was visible, resulting in a lower electrical load during peak hours. This behavior suggests that the energy performance of the campus was improving over time. Further considerations about changes in the load patterns of the campus have been made in the following when the selection of a proper training period for the classification models is discussed.

5.3. Anomaly Detection at Total Electrical Load Level

For each time window, a globally optimal evolutionary tree was developed in order to further investigate the dependency of the total electrical load (i.e., target variable) from the boundary conditions (i.e., predictive variables).

To create a model that automatically learns new patterns as the building energy consumption changes, a training period that is consistent with the recent past was selected. In fact, as previously discussed, older patterns of energy consumption strongly differed from more recent ones, and including them in the learning training set could have compromised the capabilities of the models in terms of accuracy on the validation set. Therefore, the classification models were trained and tested on 2018 data and for simulating an online deployment of the process were validated on the first month of 2019. In particular, the 2018 dataset was split, with 80% placed into the train set and 20% into the test set, through a random sampling process.

The attributes considered in the evolutionary classification trees are listed in the following:

- Day type: input ordinal categorical variable representative of each day of the week with values from 1 (Monday) to 7 (Sunday);
- Holiday: input binary categorical variable YES/NO capable of distinguishing working from non-working days;
- Total Power pre: mean total electrical load (in kW) calculated in the previous time window to the one considered for the classification (numerical input variable);
- Canteen: mean electrical load (in kW) of the canteen calculated in the time window considered for the classification (numerical input variable);
- Mechanical room: mean electrical load (in kW) of the mechanical room calculated in the time window considered for the classification (numerical input variable);
- Symbol: target categorical variable representative of the encoded symbol of the total electrical load in a time window.

The choice to use as predictive values some sub-loads and not others was driven by a sensitivity analysis and by their percentage weight on the total electrical load. The canteen and the mechanical room weights were 12.22% and 16.03%, respectively, on the total electrical load (Figure 3). Moreover, among the labeled sub-loads, they showed the highest variance in 2018, as well as significant variations during the day. It is clear how they could be extremely useful in characterizing the relationships that existed between the normal operation of the substation C and their electrical demand.

For all the time windows, the maximum depth of the classification tree was set to 6, the minimum number of observations in each node was set to 20, and the default setting c20m40sp40 for variation operators was assumed (20% crossover, 40% mutation, and 40% split/prune).

Since the evolutionary algorithm and the splitting process were randomly initialized, the seed for the random number generator was set in the code in order to replicate the analysis easily.

Figure 10 shows the tree resulting from the training phase for the second time window. It shows that it effectively classified in each leaf node the most frequent symbol from the others while maintaining a readable and understandable structure. The developed set of evolutionary trees (one for each time window) was aimed at extracting very accurate decision rules so that in the leaf node, a high occurring symbol can be found. If this condition is satisfied, the low occurring symbols can be considered as potential anomalies for the considered time window. Those potential anomalies could then be subject to further investigation in order to understand which sub-load can be assumed as the cause for that infrequent behavior (anomaly diagnosis).

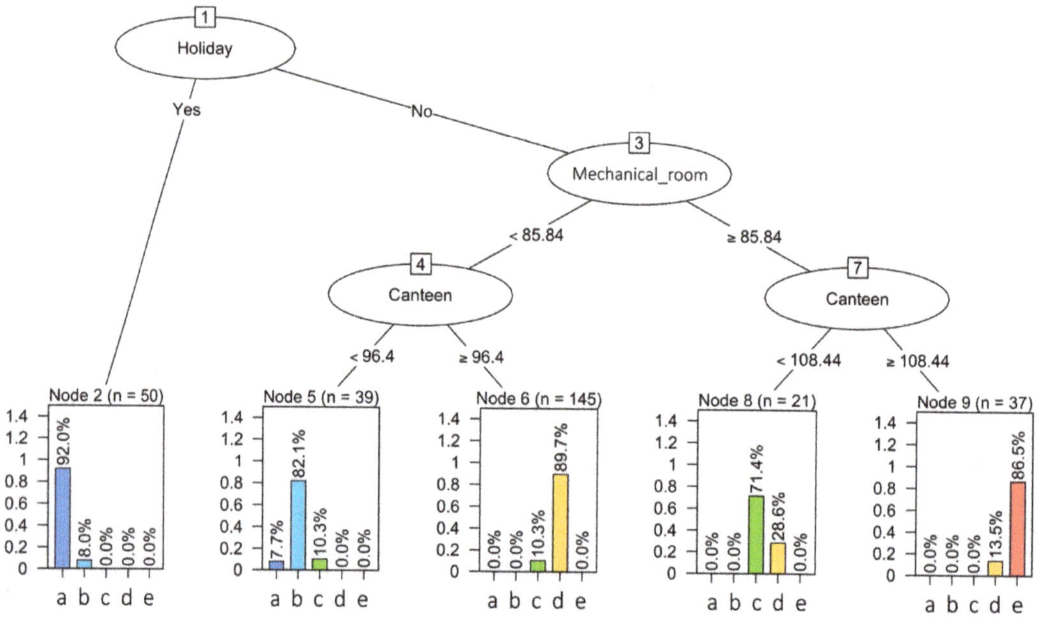

Figure 10. Globally optimum classification tree for the second time window (06:30–08:59).

Decision rules extracted from each tree (one for each time window) are reported in Table 2. It can be observed that the input variables used for the classification tree were able to explain the occurrence of each symbol with strong accuracy. Furthermore, it can be noticed that time window one was not associated with any decision rule. This window was found to be characterized by a very high occurrence (over 97% over the training period) of a single symbol. In this case, the available input variables were not able to further characterize the occurrence of other symbols.

The model performance for each time window is shown in Table 3. The table also reports that the overall accuracy in training testing and validation was 88.91%, 86.22%, and 89.03%, respectively. The results obtained suggest high generalizability for the classification models and the absence of overfitting issues.

Table 2. Decision rules extracted from globally optimal trees created in each time window on the training period.

Time Window	Node	Decision Rule	Symbol	Accuracy
00:00–06:29	1	-	⇒ a	97.3%
06:30–08:59	2	IF Holiday = Yes	⇒ a	92.0%
	5	IF Holiday = No AND Mechanical room < 85.84 kW AND Canteen < 96.4 kW	⇒ b	82.1%
	6	IF Holiday = No AND Mechanical room < 85.84 kW AND Canteen ≥ 96.4 kW	⇒ d	89.7%
	8	IF Holiday = No AND Mechanical room ≥ 85.84 kW AND Canteen < 108.4 kW	⇒ c	71.4%
	9	IF Holiday = No AND Mechanical room ≥ 85.84 kW AND Canteen ≥ 108.4 kW	⇒ e	86.5%
09:00–15:44	3	IF Canteen < 54.4 kW AND Holiday = Yes	⇒ a	96.0%
	5	IF Canteen < 54.4 kW AND Holiday = No AND Total Power pre < 257.1 kW	⇒ b	76.5%
	6	IF Canteen < 54.4 kW AND Holiday = No AND Total Power pre ≥ 257.1 kW	⇒ c	85.0%
	8	IF Canteen ≥ 54.4 kW AND Canteen < 143.5 kW	⇒ e	73.9%
	10	IF Canteen ≥ 143.5 kW AND Mechanical room < 38 kW	⇒ e	86.0%
	11	IF Canteen ≥ 143.5 kW AND Mechanical room ≥ 38 kW	⇒ f	81.1%
15:45–19:14	2	IF Total Power pre < 388.8 kW	⇒ a	87.4%
	4	IF Total Power pre ≥ 388.8 kW AND Total Power pre < 614 kW	⇒ d	86.5%
	5	IF Total Power pre ≥ 388.8 kW AND Total Power pre ≥ 614 kW	⇒ d	85.4%
19:15–23:59	2	IF Holiday = Yes	⇒ a	96.0%
	4	IF Holiday = No AND Day Type = {6,7}	⇒ a	97.2%
	6	IF Holiday = No AND Day Type = {1,2,3,4,5} AND Canteen < 16.5 kW	⇒ a	85.5%
	7	IF Holiday = No AND Day Type = {1,2,3,4,5} AND Canteen ≥ 16.5 kW	⇒ b	87.6%

Table 3. Accuracy results from a comparison between test and validation.

Time Window	Training (80% 2018)	Test (20% 2018)	Validation (Jan. 2019)
00:00–06:29	97.30%	96.89%	100%
06:30–08:59	87.33%	82.19%	93.55%
09:00–15:44	83.56%	79.45%	58.06%
15:45–19:14	86.64%	86.30%	96.77%
19:15–24:00	89.72%	86.30%	96.77%
Mean	88.91%	86.22%	89.03%

5.4. Diagnosis at Sub-Load Level

Once the classification models were created, the subset of anomalous symbols (higher than expected symbols) included in each node was transformed into a transactional database that contains the categorical target variable (total electrical load symbol) and some additional explanatory variables related to the sub-loads.

To extract those additional categorical variables, the sub-loads were subjected to the same time series abstraction process described for the total electrical load in Section 5.2. Using the same time window discretization as the total electrical load and the same alphabet size ($\alpha = 6$), each time series of the available sub-loads was encoded through the aSAX process.

In order to further enrich information about sub-loads, the trend angle was also extracted and encoded (see Figure 11).

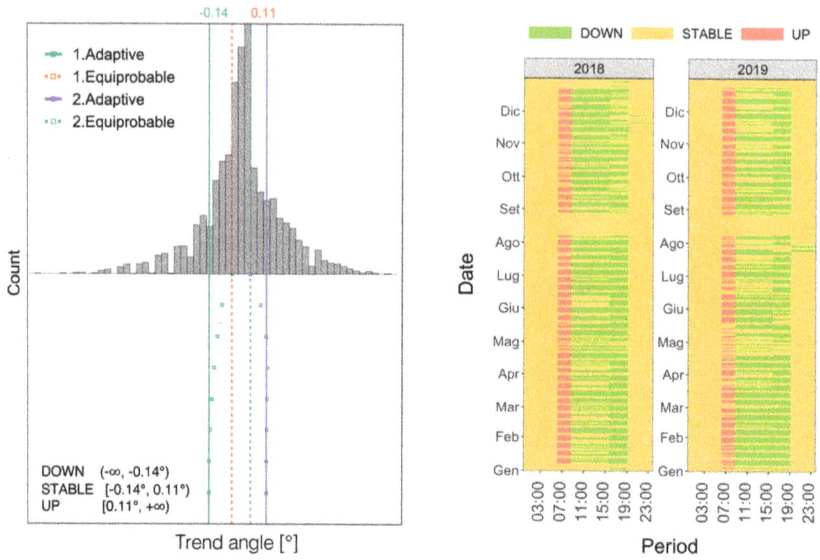

(a) (b)

Figure 11. Results of trend angle aSAX encoding applied to the rectory sub-load: (**a**) identification of adaptive breakpoints through the aSAX algorithm, (**b**) encoded trend angle carpet plot for 2018 and 2019.

This feature allows tracking of the trend of the time series in each time window, making it possible to know if the load is increasing, decreasing, or it is stable. In this case, the alphabet size was set to three ($\alpha = 3$) in order to reflect those three possible trends (respectively encoded as Up, Down, and Stable). The initial breakpoints, calculated under equally probability assumption, were used as initialization of aSAX iterative algorithm, and the final adaptive breakpoints were evaluated once a tolerance of 10^{-10} on the representation error was reached. Then the Apriori ARM algorithm was applied to the transactional database structured, as depicted in Figure 12.

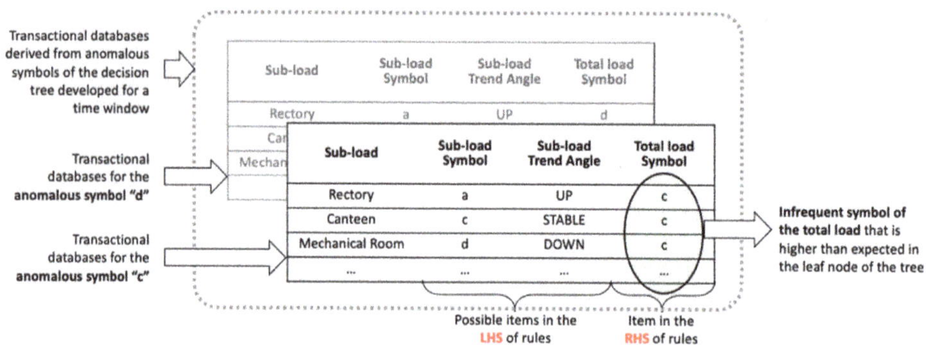

Figure 12. Representation of the transactional databases used for the extraction of association rules. LHS: left-hand side of the rule; RHS: right-hand side of the rule.

In particular, the RHS was the anomalous total electrical load symbol extracted from the leaf node of the classification tree for a specific time window, while the LHS was composed of all possible combinations of electrical load symbols and trend angles symbols of sub-loads. The minimum and the maximum number of items in a transaction was set in order to obtain rules with one or maximum of two items in the LHS. The minimum support to mine rules was set to 0.005, and the minimum confidence to 0.005. Redundant rules, equally or less predictive of a more general rule with the same confidence [50], were removed, and the remaining ones were represented in a scatter plot (Figure 13). The scatter plot helps the analyst to understand how interesting rules were filtered out by setting $lift(A \Rightarrow B) > 1$ and $conf(A \Rightarrow B) > 0.5$. Those rules were then stored in the anomaly library, where they were ranked according to the lift value. LHS of those rules represents the sub-load conditions that were found to be significantly influencing the abnormal total electrical load.

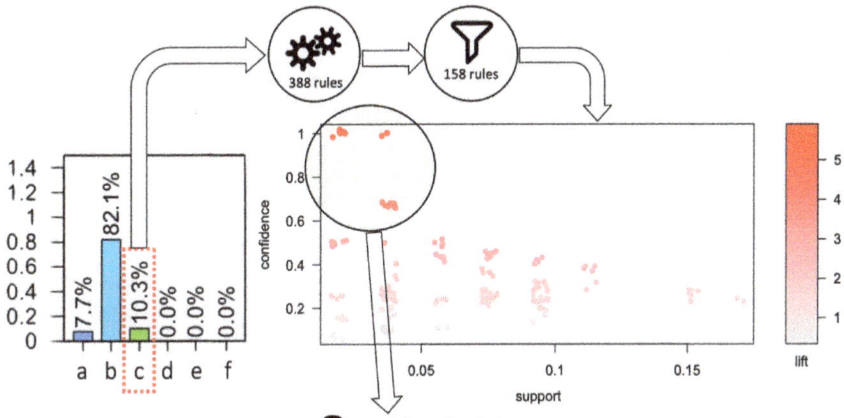

ID	LHS	RHS	Support	Confidence	Lift
1	{sym_Data_centre = e}	{sym_Total_Power = c}	0.01886792	1.00000000	5.300001
2	{sym_Rectory = d, sym_Data_centre_trend = DOWN}	{sym_Total_Power = c}	0.01886792	1.00000000	5.300001
3	{sym_Mechanical_room = d, sym_Canteen_trend = DOWN}	{sym_Total_Power = c}	0.01886792	1.00000000	5.300001
4	{sym_Mechanical_room = d, sym_Canteen = c}	{sym_Total_Power = c}	0.01886792	1.00000000	5.300001

Figure 13. Diagnosis procedure of extracting, filtering, and selecting only relevant association rules from node five of the second time window.

An example of the procedure is shown in Figure 13 for node five of the second time window. In this node, the most frequent symbol was "b", and the only infrequent interesting symbol (higher electrical load) was "c". The transactional database was then constructed: the LHS was composed of the additional categorical variables related to sub-loads (electrical load symbol and trend angle symbol), while RHS contained only the total electrical load anomalous symbol (symbol "c"). ARM automatically extracts 338 rules, of which 180 resulted redundantly, and 158 rules were significant. After filtering, only 19 rules were stored in the anomaly library. In this particular case, the most frequent items in the anomaly library were: mechanical room symbol "d"; canteen symbol "c"; rectory symbol "d". For example, among the 19 rules considered, rule four (IF sym_Mechanical_room = "d" AND sym_Canteen = "c" ⇒ sym_Total_Power = "c") had a lift value of about five and confidence of 100%. It means that if during the operation of the ADD process this rule was matched, then the diagnosis was extremely robust, given that the anomaly detected was already present in the analyzed historical database.

5.5. Deployment of the ADD Tool

The methodology was conceived to be implemented in a real-time data acquisition tool connected to a smart metering infrastructure. The metering infrastructure continuously collects data, and once a time window ends, the symbol of the total electrical load was calculated through aSAX and compared to the one predicted by the globally optimal tree. Three possible cases could then occur:

- The actual symbol was the same as the predicted one. This means that given the boundaries conditions, the total electrical load of that time window is behaving as expected, then no further diagnosis is requested;
- The actual symbol was different from the predicted symbol and indicated a lower electrical load than expected. This means that even though the total electrical load of that time window is not behaving as expected, no further diagnosis is required. This is due to the focus of the methodology for which an anomaly is related only to higher consumption than expected;
- The actual symbol was different from the predicted symbol and indicated a higher electrical load than expected. This means that given the boundaries conditions, the total electrical load of that time window is higher than expected, and then a further investigation is needed.

In the latter case, the diagnosis analysis is enabled. Given the boundary conditions, the corresponding leaf node of the evolutionary tree is identified, and the tool automatically retrieves the library of association rules extracted on the historical dataset for that specific anomaly condition (i.e., a specific symbol of the total electrical load). The following step was then to extract the additional features from sub-loads and encoding them in symbols/categorical values. Once all the potential LHS items had been computed, a scan of the rules included in the anomaly library was performed to detect any perfect match. If a perfect match of a rule exists, it means that a full diagnosis of the anomaly could be performed considering that the same anomaly condition (i.e., the relation between anomalous total load and sub-loads) was present in the historical dataset. Otherwise, if a perfect match does not exist, a partial match with the single item was searched. In the case of a partial match, the diagnostic capability is not as strong as for the perfect rule match. However, useful insight can be obtained about new possible configurations of sub-loads that could be included in the anomaly library during future updates. In order to make the whole ADD process flexible in learning new patterns, a full retraining of the classification models and anomaly library is supposed to be performed every month, considering a historical dataset of one year.

The deployment of the methodology was performed on the validation set that consisted of the data referred to in January 2019. The process of detection through the evolutionary tree was performed on all-time windows. Only for reference, it was considered the classification performance achieved in the second time window for the whole month. The confusion matrix related to the classifier is reported in Figure 14. In particular, it can be seen that the classification tree achieved an accuracy of 93.55%, and only one time the actual symbol was different from the predicted one revealing a higher electrical load than expected, respectively "c" instead of "b". In particular, the anomaly occurred on 4 January 2019.

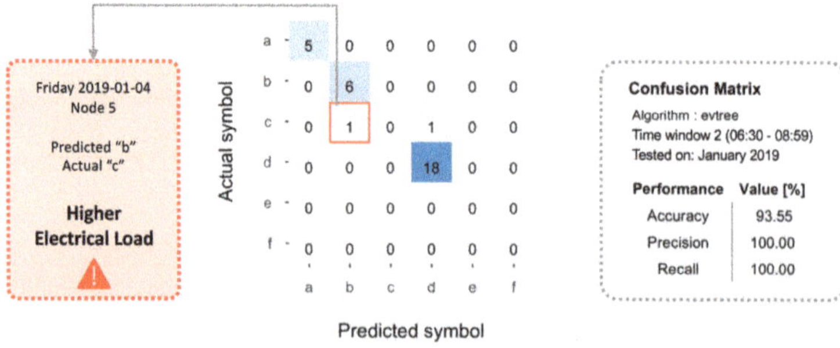

Figure 14. Confusion matrix for the globally optimal classification tree predicting January 2019 total electrical load symbol in the 2nd time window. Red square: the detected anomalous behavior.

Once identified the day and the time window of the anomaly, the corresponding tree's leaf node was identified as well. In the example, the anomalous symbol of the total electrical load of 4 January 2019 was detected in the tree leaf node five. The diagnosis process was then enabled, and the sub load conditions were compared with the anomaly reference library. In the considered example, there was not a perfect rule match but a partial one on the following items:

- Printshop electrical load symbol "c".
- Mechanical room electrical load symbol "c".
- Canteen electrical load symbol "c".
- Canteen trend angle symbol "UP".

As previously discussed, a partial match is not as strong as a perfect rule match but provides useful suggestions to be considered for conducting the anomaly diagnosis. This aspect was demonstrated through further graphical analysis, reported in Figure 15. The figure shows a comparison between the anomalous and normal pattern of the total electrical load and the loads related to the mechanical room, printshop, and canteen. Only the second time window is reported in the plot. In particular, in red, the anomalous data related to the 4 January 2019 are reported, while in green are shown the frequent "normal" patterns of the given loads extracted from the training period (part of 2018). Along with the actual electrical loads (solid lines) were reported the relative PAA segments (dashed lines) and, for the "normal" pattern, the standard deviation (grey areas).

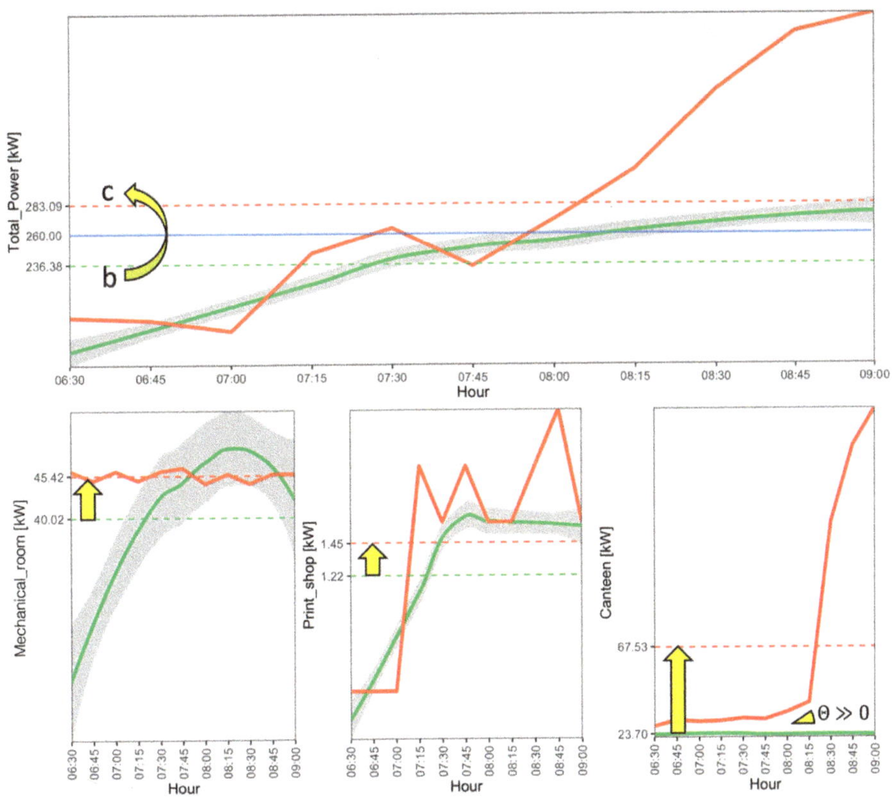

Figure 15. Comparison between the actual (red lines) and expected (green lines) electrical load with the relative standard deviation (grey areas) on 4 January 2019. The dashed green and red lines represent the PAA segments for the actual and expected load respectively. The blue horizontal solid line on the top graph represents the aSAX breakpoint related to the total power.

The combined effect of the three sub-loads (i.e., mechanical room, printshop, canteen) led to an overall electrical load higher than expected. The mean total electrical load rose from 236 kW (symbol "b") to 283 kW (symbol "c"), and it was easy to verify that the identified sub-loads contributed almost 90% to the power shift upward of the total electrical load. It is worth noting that although the printshop presented an anomalous electrical load pattern, the observed profile (red line) did not significantly deviate from the normal one (green line).

6. Discussion of the Results

This paper focused on the development of ADD methodology able to analyze meter-level electrical load data in order to detect anomalous patterns and perform a diagnosis process on sub-loads. This methodological framework was conceived to be highly scalable and reliable in order to be implemented in energy data monitoring infrastructure for supporting a prompt detection of anomalies avoiding energy wastes over time.

The time window size and alphabet size for the aSAX encoding are key parameters. In [20] is reported an interesting sensitivity analysis based on these two parameters, showing that a trade-off between window numbers and alphabet size has to be found in order to minimize the variance between patterns and resolution needed. In this paper, the time window number was chosen by using an RT and the alphabet size by a k-means clustering evaluation. Once those parameters are set, the aSAX encoding procedure can be considered completely automatic. Moreover, the conducted analysis showed that considering the trend angle as an additional feature, a robust sub-loads characterization could be performed without adding computational burden.

Moreover, the selection of the predictive variables for the globally optimal classification tree needs particular attention. The overall energy consumption of a building is strongly related to the occupancy schedule, environmental conditions, thermo-physical features of the building, and the behavior of users. For this reason, those variables should be all included in the classification model and could help in describing infrequent but non-anomalous patterns. On the other hand, trustworthy values are difficult to retrieve or measure with continuity. Surely, the inclusion of those variables could qualitatively increase the model predictions.

A further interesting aspect of being considered is related to the data that should be used for training and how often training is needed. It is well known that building electrical load varies over the years due to the electrification of end-uses and the seek of the higher performance of appliances and facilities. For this reason, a good trade-off between retraining rate and computational effort should be performed. In our study, we validated the model in the first month of 2019 in order to assess its accuracy.

In addition, in order to prove the effectiveness of monthly retraining of the tool, a comparison was performed between two different deployment approaches. The first deployment considered was static, with the hypothesis of using the same classification models trained in 2018 for six months in 2019. The second deployment was dynamic, considering monthly retraining of the classification models with a one-year moving window training set. Results showed that the average classification accuracy was 82.85% for the dynamic deployment and was 78.77% for the static one. Therefore, with a dynamic deployment, the anomaly detection capabilities improved, given that the classifiers are able to learn new patterns that change over time. Following the same reasoning, the authors propose to implement a monthly update also for the association rules included in the anomaly library.

7. Conclusions and Future Work

This paper proposed a multiple-step ADD methodology to automatically detect at whole-building meter level anomalous energy consumption and then perform a diagnosis on the sub-loads responsible for that anomalous pattern. Frequent and infrequent electrical load patterns, properly transformed through an adaptive symbolic aggregate approximation process, were discovered by means of globally optimum evolutionary classification trees. Association rule mining was employed to discover the main sub-loads, which mostly affected the anomaly detected at the whole-building level.

In the future, the ADD process presented in this paper is expected to be implemented online within the energy information system of Politecnico di Torino and supplied through an energy data analytics dashboard developed with the R packages "shiny" [51] and "shinydashboard" [52]. Figure 16 reports a demo of the dashboard that is currently under construction and under offline testing. Moreover, the authors aim to integrate this ADD process together with other complementary tools able to perform electrical load forecasting and energy performance tracking (i.e., benchmarking).

Figure 16. Energy data analytics dashboard developed by the building automation and energy data analytics (BAEDA) Lab, which implements the anomaly detection and diagnosis (ADD) procedure presented in this paper.

Further research will also be focused on the testing of alternative configurations of algorithms (i.e., data clustering, forecasting) with respect to the one considered in this study. In fact, the proposed algorithms cannot always be assumed as the best solution for performing such kind of analysis on energy consumption time series. As a reference, the aSAX transformation, the development of classification trees, and the extraction of association rules perfectly match with the need to provide a fully interpretable tool to the final user. However, this constraint, in some cases, can also determine an information loss and accuracy decrease. For this reason, a future analysis may well consider the use of more sophisticated algorithms (e.g., deep learning algorithms) that are characterized by their non-interpretable nature but makes it possible to achieve higher performance in detecting and diagnosing energy anomalies. This option still remains valuable if an explanation layer is included in the analytical process. Nowadays, such a task corresponds to the main goal of the machine learning field of the so-called explainable artificial intelligence (XAI), which offers new opportunities for effectively embedding advanced algorithms in AI-based energy management solutions where explanations of the black-box model predictions are often compulsory.

Author Contributions: Conceptualisation, M.S.P. and A.C.; Data curation, M.S.P.; Formal analysis, R.C.; Investigation, M.S.P.; Methodology, M.S.P., R.C. and A.C.; Project administration, A.C.; Software, R.C.; Supervision, A.C.; Validation, M.S.P. and A.C.; Writing—original draft, R.C.; Writing—review and editing, M.S.P. and A.C. All authors have read and agreed to the published version of the manuscript.

Funding: This research received no external funding.

Institutional Review Board Statement: Not applicable.

Informed Consent Statement: Not applicable.

Data Availability Statement: No new data were created or analyzed in this study. Data sharing is not applicable to this article.

Acknowledgments: The authors express their gratitude to the Living Lab of Politecnico di Torino for providing data and to Giovanni Carioni for the support in data preparation and collection.

Conflicts of Interest: The authors declare no conflict of interest.

References

1. IEA. Buildings A Source of Enormous Untapped Efficiency Potential. Available online: https://www.iea.org/topics/buildings (accessed on 7 September 2020).
2. Fan, C.; Yan, D.; Xiao, F.; Li, A.; An, J.; Kang, X. Advanced data analytics for enhancing building performances: From data-driven to big data-driven approaches. *Build. Simul.* **2020**. [CrossRef]
3. Capozzoli, A.; Mechri, H.E.; Corrado, V. Impacts of architectural design choices on building energy performance applications of uncertainty and sensitivity techniques. In Proceedings of the IBPSA 2009 International Building Performance Simulation Association, Glasgow, Scotland, 27–30 July 2009; pp. 1000–1007.
4. Capozzoli, A.; Cerquitelli, T.; Piscitelli, M.S. Enhancing Energy Efficiency in Buildings Through Innovative Data Analytics Technologies. In *Pervasive Computing*; Elsevier: Amsterdam, The Netherlands, 2016; ISBN 9780128037027.
5. Miller, C.; Meggers, F. The Building Data Genome Project: An open, public data set from non-residential building electrical meters. *Energy Proc.* **2017**, *122*, 439–444. [CrossRef]
6. Miller, C.; Kathirgamanathan, A.; Picchetti, B.; Arjunan, P.; Park, J.Y.; Nagy, Z.; Raftery, P.; Hobson, B.W.; Shi, Z.; Meggers, F. The Building Data Genome Project 2, energy meter data from the ASHRAE Great Energy Predictor III competition. *Sci. Data* **2020**, *7*, 1–13. [CrossRef] [PubMed]
7. Attanasio, A.; Piscitelli, M.S.; Chiusano, S.; Capozzoli, A.; Cerquitelli, T. Towards an automated, fast and interpretable estimation model of heating energy demand: A data-driven approach exploiting building energy certificates. *Energies* **2019**, *12*, 1273. [CrossRef]
8. Manfren, M.; Nastasi, B.; Groppi, D.; Astiaso Garcia, D. Open data and energy analytics—An analysis of essential information for energy system planning, design and operation. *Energy* **2020**, *213*, 118803. [CrossRef]
9. Kramer, H.; Lin, G.; Granderson, J.; Curtin, C.; Crowe, E. *Synthesis of Year One Outcomes in the Smart Energy Analytics Campaign Building Technology and Urban Systems Division*; Lawrence Berkeley National Laboratory: Berkeley, CA, USA, 2017.
10. Zhang, C.; Zhao, Y.; Li, T.; Zhang, X. A post mining method for extracting value from massive amounts of building operation data. *Energy Build.* **2020**, *223*. [CrossRef]
11. Fan, C.; Sun, Y.; Shan, K.; Xiao, F.; Wang, J. Discovering gradual patterns in building operations for improving building energy efficiency. *Appl. Energy* **2018**, *224*, 116–123. [CrossRef]
12. Himeur, Y.; Ghanem, K.; Alsalemi, A.; Bensaali, F.; Amira, A. Anomaly detection of energy consumption in buildings: A review, current trends and new perspectives. *arXiv* **2020**, arXiv:2010.04560.
13. Esling, P.; Agon, C. Time-series data mining. *ACM Comput. Surv.* **2012**, *45*. [CrossRef]
14. Chou, J.S.; Telaga, A.S. Real-time detection of anomalous power consumption. *Renew. Sustain. Energy Rev.* **2014**, *33*, 400–411. [CrossRef]
15. Fan, C.; Xiao, F.; Zhao, Y.; Wang, J. Analytical investigation of autoencoder-based methods for unsupervised anomaly detection in building energy data. *Appl. Energy* **2018**, *211*, 1123–1135. [CrossRef]
16. Pereira, J.; Silveira, M. Unsupervised Anomaly Detection in Energy Time Series Data Using Variational Recurrent Autoencoders with Attention. In Proceedings of the 17th IEEE International Conference on Machine Learning Applications ICMLA, Orlando, FL, USA, 17–20 December 2018; pp. 1275–1282. [CrossRef]
17. Capozzoli, A.; Piscitelli, M.S.; Brandi, S.; Grassi, D.; Chicco, G. Automated load pattern learning and anomaly detection for enhancing energy management in smart buildings. *Energy* **2018**, *157*, 336–352. [CrossRef]
18. Capozzoli, A.; Piscitelli, M.S.; Brandi, S. Mining typical load profiles in buildings to support energy management in the smart city context. *Energy Proc.* **2017**, *134*, 865–874. [CrossRef]
19. Zhao, Y.; Zhang, C.; Zhang, Y.; Wang, Z.; Li, J. A review of data mining technologies in building energy systems: Load prediction, pattern identification, fault detection and diagnosis. *Energy Built Environ.* **2020**, *1*, 149–164. [CrossRef]
20. Miller, C.; Nagy, Z.; Schlueter, A. Automated daily pattern filtering of measured building performance data. *Autom. Constr.* **2015**, *49*, 1–17. [CrossRef]
21. Li, K.; Yang, R.J.; Robinson, D.; Ma, J.; Ma, Z. An agglomerative hierarchical clustering-based strategy using Shared Nearest Neighbours and multiple dissimilarity measures to identify typical daily electricity usage profiles of university library buildings. *Energy* **2019**, *174*, 735–748. [CrossRef]

22. Piscitelli, M.S.; Mazzarelli, D.M.; Capozzoli, A. Enhancing operational performance of AHUs through an advanced fault detection and diagnosis process based on temporal association and decision rules. *Energy Build.* **2020**, *226*, 110369. [CrossRef]
23. David, M.C.; Zareipour, H. Data association mining for identifying lighting energy waste patterns in educational institutes. *Energy Build.* **2013**, *62*, 210–216. [CrossRef]
24. Rossi, B.; Chren, S.; Buhnova, B.; Pitner, T. Anomaly Detection in Smart Grid Data: An Experience Report. In Proceedings of the 2016 IEEE International Conference on Systems, Man, and Cybernetics (SMC), Budapest, Hungary, 9–12 October 2016; pp. 2313–2318.
25. Xiao, F.; Fan, C. Data mining in building automation system for improving building operational performance. *Energy Build.* **2014**, *75*, 109–118. [CrossRef]
26. Piscitelli, M.S.; Brandi, S.; Capozzoli, A.; Xiao, F. A data analytics-based tool for the detection and diagnosis of anomalous daily energy patterns in buildings. *Build. Simul.* **2020**, 1–17. [CrossRef]
27. Imayakumar, A.A.; Dubey, A.; Bose, A. Anomaly Detection for Primary Distribution System Measurements using Principal Component Analysis. In Proceedings of the 2020 IEEE Texas Power and Energy Conference (TPEC), College Station, TX, USA, 6–7 February 2020; pp. 1–6.
28. Zhang, L.; Wan, L.; Xiao, Y.; Li, S.; Zhu, C. Anomaly Detection method of Smart Meters data based on GMM-LDA clustering feature Learning and PSO Support Vector Machine. In Proceedings of the 2019 IEEE Sustainable Power and Energy Conference (iSPEC), Beijing, China, 20–24 November 2019; pp. 2407–2412.
29. Khoshrou, A.; Pauwels, E.J. Data-driven pattern identification and outlier detection in time series. *Adv. Intell. Syst. Comput.* **2019**, *858*, 471–484. [CrossRef]
30. Lin, J.; Keogh, E.; Wei, L.; Lonardi, S. Experiencing SAX: A Novel Symbolic Representation of Time Series. *Data Min. Knowl. Discov.* **2007**, *15*, 107–144.
31. Fan, C.; Xiao, F.; Madsen, H.; Wang, D. Temporal knowledge discovery in big BAS data for building energy management. *Energy Build.* **2015**, *109*, 75–89. [CrossRef]
32. Yan, R.; Ma, Z.; Zhao, Y.; Kokogiannakis, G. A decision tree based data-driven diagnostic strategy for air handling units. *Energy Build.* **2016**, *133*, 37–45. [CrossRef]
33. Liu, J.; Shi, D.; Li, G.; Xie, Y.; Li, K.; Liu, B.; Ru, Z. Data-driven and association rule mining-based fault diagnosis and action mechanism analysis for building chillers. *Energy Build.* **2020**, *216*, 109957. [CrossRef]
34. Tightiz, L.; Nasab, M.A.; Yang, H.; Addeh, A. An intelligent system based on optimized ANFIS and association rules for power transformer fault diagnosis. *ISA Trans.* **2020**, *103*, 63–74. [CrossRef] [PubMed]
35. Huang, R.; Liu, J.; Chen, H.; Li, Z.; Liu, J.; Li, G.; Guo, Y.; Wang, J. An effective fault diagnosis method for centrifugal chillers using associative classification. *Appl. Therm. Eng.* **2018**, *136*, 633–642. [CrossRef]
36. Zhang, T.; Lu, J.; Zhang, G.; Ding, Q. Fault diagnosis of transformer using association rule mining and knowledge base. In Proceedings of the 2010 10th International Conference on Intelligent Systems Design and Applications, Cairo, Egypt, 29 November–1 December 2010; pp. 737–742.
37. Grubinger, T.; Zeileis, A.; Pfeiffer, K.P. Evtree: Evolutionary learning of globally optimal classification and regression trees in R. *J. Stat. Softw.* **2014**, *61*, 1–29. [CrossRef]
38. Pham, N.D.; Le, Q.L.; Dang, T.K. HOT aSAX: A Novel Adaptive Symbolic Representation for Time Series Discords Discovery. In *Lecture Notes in Computer Science*; (including subseries Lecture Notes in Artificial Intelligence and Lecture Notes in Bioinformatics); Springer: Berlin, Germany, 2010; Volume 5990, pp. 113–121. ISBN 3642121446.
39. Keogh, E.; Chakrabarti, K.; Pazzani, M.; Mehrotra, S. Locally adaptive dimensionality reduction for indexing large time series databases. In Proceedings of the 2001 ACM SIGMOD International Conference on Management of Data, Santa Barbara, CA, USA, 21–24 May 2001; pp. 151–162.
40. Keogh, E.; Chakrabarti, K.; Pazzani, M.; Mehrotra, S. Dimensionality Reduction for Fast Similarity Search in Large Time Series Databases. *Knowl. Inf. Syst.* **2001**, *3*, 263–286. [CrossRef]
41. Zhang, Y.; Duan, L.; Duan, M. A new feature extraction approach using improved symbolic aggregate approximation for machinery intelligent diagnosis. *Meas. J. Int. Meas. Confed.* **2019**, *133*, 468–478. [CrossRef]
42. Yu, Y.; Zhu, Y.; Wan, D.; Liu, H.; Zhao, Q. A novel symbolic aggregate approximation for time series. *Adv. Intell. Syst. Comput.* **2019**, *935*, 805–822. [CrossRef]
43. Tan, P.-N.; Steinbach, M.; Karpatne, A.; Kumar, V. Cluster Analysis: Basic Concepts, and Algorithms. In *Introduction to Data Mining*; Pearson: London, UK, 2019; p. 526.
44. Piscitelli, M.S.; Brandi, S.; Capozzoli, A. Recognition and classification of typical load profiles in buildings with non-intrusive learning approach. *Appl. Energy* **2019**, *255*, 113727. [CrossRef]
45. Aggarwal, C.C. *Data Data Mining: The Textbook*; Springer: Berlin, Germany, 2012.
46. Charrad, M.; Ghazzali, N.; Boiteau, V.; Niknafs, A. NbClust: An R Package for Determining the. *J. Stat. Softw.* **2014**, *61*, 1–36. [CrossRef]
47. Michael, H.; Buchta, C.; Gruen, B.; Hornik, K.; Johnson, I.; Borgelt, C. *Package ' arules ': Mining Association Rules and Frequent Itemsets Description*; R Foundation for Statistical Computing: Vienna, Austria, 2020; pp. 1–109.
48. R Core Team. *R: A Language and Environment for Statistical Computing*; R Foundation for Statistical Computing: Vienna, Austria, 2017.

49. Atkinson, E.J.; Therneau, T.M. An Introduction to Recursive Partitioning Using the RPART Routines. *Mayo Clin. Sect. Biostat. Tech. Rep.* **2000**, *61*, 33.
50. Hahsler, M.; Chelluboina, S. Visualizing Association Rules: Introduction to the R-extension Package arulesViz. In *R Project Module*; R Foundation for Statistical Computing: Vienna, Austria, 2011; pp. 1–24.
51. Chang, W.; Cheng, J.; Allaire, J.; Xie, Y.; McPherson, J. *Package ' shiny ': Web Application Framework for R*; R Foundation for Statistical Computing: Vienna, Austria, 2020; p. 238.
52. Chang, W.; Ribeiro, B.B. Package "ShinyDashboard": Create Dashboards with "Shiny". *J. Stat. Softw.* **2018**, *14*, 1–27.

energies

Article

The Use of Energy Models in Local Heating Transition Decision Making: Insights from Ten Municipalities in The Netherlands

Birgit A. Henrich [1,2], Thomas Hoppe [1,*], Devin Diran [2] and Zofia Lukszo [1]

1 Faculty of Technology, Policy and Management (TPM), Delft University of Technology, Jaffalaan 5, 2628 BX Delft, The Netherlands; birgit.henrich@tno.nl (B.A.H.); z.lukszo@tudelft.nl (Z.L.)
2 Department of Strategy and Policy, Netherlands Organisation for Applied Scientific Research TNO, Anna van Buerenplein 1, 2595 DA The Hague, The Netherlands; devin.diran@tno.nl
* Correspondence: T.Hoppe@tudelft.nl; Tel.: +31-15-27-82783

Abstract: In 2018, the Dutch national government announced its decision to end natural gas extraction. This decision posed a challenge for local governments (municipalities); they have to organise a heat supply that is natural gas-free. Energy models can decrease the complexity of this challenge, but some challenges hinder their effective use in decision-making. The main research question of this paper is: What are the perceived advantages and limitations of energy models used by municipalities within their data-driven decision-making process concerning the natural-gas free heating transition? To answer this question, literature on energy models, data-driven policy design and modelling practices were reviewed, and based on this, nine propositions were formulated. The propositions were tested by reflecting on data from case studies of ten municipalities, including 21 experts interviews. Results show that all municipalities investigated, use or are planning to use modelling studies to develop planning documents of their own, and that more than half of the municipalities use modelling studies at some point in their local heating projects. Perceived advantages of using energy models were that the modelling process provides perspective for action, financial and socio-economic insights, transparency and legitimacy and means to start useful discussions. Perceived limitations include that models and modelling results were considered too abstract for analysis of local circumstances, not user-friendly and highly complex. All municipalities using modelling studies were found to hire external expertise, indicating that the knowledge and skill level that municipal officials have is insufficient to model independently.

Keywords: energy modelling; heating transition; modelling practices; data-driven policy design; local policy; municipality; multi-model ecologies

check for updates

Citation: Henrich, B.A.; Hoppe, T.; Diran, D.; Lukszo, Z. The Use of Energy Models in Local Heating Transition Decision Making: Insights from Ten Municipalities in The Netherlands. *Energies* **2021**, *14*, 423. https://doi.org/10.3390/en14020423

Received: 14 December 2020
Accepted: 12 January 2021
Published: 14 January 2021

1. Introduction

1.1. The Dutch Heating Transition

In 2016, the heating and cooling sector accounted for half of the EU's energy consumption [1]. In The Netherlands, 53% of the national heat supply is provided by natural gas [2]. In March 2018, the Dutch national government announced its decision to end natural gas extraction from the Groningen gas field by 2030 [1] to help reach the climate goals of the Paris Agreement and to reduce the negative impact of natural gas extraction in the province of Groningen [2]. This is also referred to as the so-called 'heating transition' in The Netherlands and was later defined by the RVO (The Netherlands Enterprise Agency) as removing natural gas from industry, the built environment and the agricultural sector [2], and replacing it by (sustainable) heating alternatives. According to the Climate Agreement, the main climate policy program in The Netherlands, a sufficient level of sustainable heating must be made available to replace the natural gas supply and to meet the climate change mitigation target of reducing CO_2 emissions by 3.4 megatons in the built environment. To reach this goal, 1.5 million existing residential homes have to be supplied with sustainable heating by 2030 [3].

However, this is challenging because decision-making and policymaking in this transition are far from simple, as actors, technology and institutions interact in a complex manner [1]. The heating transition requires a change of the supply of renewable energy, the infrastructure, residential heating systems and of thermal insulation in residential houses, which all raise questions about the division of costs and the freedom of choice [4]. Next to these dependencies, the heating transition poses significant financial challenges. Natural gas is currently cheaper than sustainable alternatives and residents do not always have sufficient funds available to provide the needed investments or to deal with increased living expenses [5].

To organise this complex transition, every municipality is expected to formulate a "Transition Vision Heat" (See Table A1, Appendix A, Glossary) and an implementation plan in their local government plans, to show how they will organise a heat supply that is natural gas-free and affordable, according to the Environment and Planning Act. This means that municipalities are expected (by the national government) to take a leading role in the heating transition. This is new for municipalities and requires them to collect new knowledge, expertise and competences. To this end, the national government has set up Test Beds for Natural Gas-Free districts (i.e., pilot projects) and a knowledge and learning programme to learn and experiment [3] within the National Programme for Natural Gas-Free Districts. The latter has a 120 million euro budget.

1.2. The Use of Energy Models in Data-Driven Policymaking

To enable the heating transition, municipalities need to answer questions such as, which heating source would lead to low end-user costs, low societal costs and low CO_2 emissions? To evaluate the effect and impact of potential policy measure or decisions on, for example, a preferred technology for natural gas-free heating in city districts, evidence-based policymaking entails the derivation of fact-based knowledge to support the decision making by policymakers. One way to approach evidence-based policymaking is with data-driven policies. A data-driven policy uses data and tools for processing and analysing data to design policies and to facilitate collaboration with citizens to co-create [6]. Currently, municipalities make limited use of data and data processing and analysis tools for decision-making support. This is partly due to a lack of guidelines. New guidelines are to be developed that can make use of new data sources and tools [6]. Historically, the first decision-making support tool developed for environmental planning was the multi-criteria decision aid (MCDA). The MCDA is considered a qualitative decision support tool [7]. One drawback of MCDA tools is that they do not allow for analysis to compare whether doing an action is better than doing nothing [8]. In the last years, the number of quantitative tools to support decision-makers has been growing, which include energy models. The advantage of energy models, compared to more qualitative tools such as MCDA, include a higher degree of traceability, easier implementation in computing environments and better opportunities for ex-ante analysis [8]. Dutch municipalities are increasingly trying to include energy models when designing policy for the heating transition are energy models. In the present study, an energy model is defined as a computer model of an energy system that introduces a structured way of thinking about the implications of changing parts of the system [9]. Energy models may help analysts and policymakers to better understand the increasingly complex energy sector. However, clear guidelines on how to use these models while designing policies are still lacking.

Next to a lack of guidelines on how to integrate energy models, practitioners, such as policymakers, also experience challenges with energy models themselves. This hinders the use of energy models for policy design and decision-making [10]. When interpreting modelling results, caution is needed, because when modelling, it is unavoidable to make use of assumptions and estimates, which may not be valid under all circumstances [11,12]. According to a recently published research report, in The Netherlands [10] no less than six different models focusing on the heating transition sometimes provide different results for the same research question, due to differences in approach, assumptions and input data. This makes it difficult for policymakers to interpret, understand and trust modelling results.

Another significant challenge of current energy models is that they fail to take into account social aspects. This is problematic since the heating transition is highly dependent on humans and their intentions. Social aspects, such as behaviour and attitude of the public, affect proposed or implemented policies and should, therefore, not be ignored [13]. At present, building owners (either citizens/homeowners, institutional investors, private landlords or housing associations) have the right and responsibility to make investment decisions about the heating supply of their buildings [14]. In other words, they need to be incentivised to change their current gas-based heat supply. For this reason, building owners and local communities form an essential part of the heating system and their contribution to the heating transition, by deciding to adopt sustainable heating technologies and/or thermal insulation for their homes, is key in making the transition happen.

1.3. Research Focus

The present study focuses on the use of energy models in local heating transition projects to assess to what extent energy models are used in the decision-making process, how, and which advantages and limitations this has. The present paper aims to provide insight into the practice of energy modelling and insight into the needs and challenges of practitioners when using energy models in the heating transition. Thus far, no academic studies have addressed these issues. Insights therein can provide a starting point for more structured guidelines of effective energy modelling. The research question of this study was, therefore, as follows: What are the perceived advantages and limitations of using energy models for municipalities within their data-driven decision-making process concerning the natural-gas free heating transition? To answer the research question, a review of the literature and multiple embedded case studies was conducted in which different heating transition projects in ten Dutch municipalities were investigated. The scope was limited to energy models used by practitioners in the Dutch heating transition, as further explained in Section 3.

The paper is structured as follows. In Section 2, a literature review is presented on the use of energy models in heating transition projects, as well as on data-driven policy design and good modelling practices. Section 2 concludes with a set of theoretical propositions. In Section 3, research design and methodology are presented. In Section 4, the results of the analysis are presented. This includes testing of key propositions regarding the use of energy modelling. In Section 5, the results are discussed, and the academic merit of the present study is presented. The paper ends with a conclusion, the limitations of the study and suggestions for future research.

2. Literature Review

2.1. Data-Driven Policymaking

To plan for a transition to sustainable heating in the built environment, municipalities need data and evidence to support their decision-making processes [10]. One way to approach this is by formulating data-driven policies. Multiple studies agree that using a data-driven approach using new data sources and tools, such as energy models, can improve policymaking practices [6,13,15–18], but a systematic approach to do so is still missing [6,13]. Moreover, various studies express concerns about the capabilities of policy-makers and stakeholders to deal with new data sources and technologies [16,18]. Thus far, no academic studies have been conducted addressing how the use of energy models affects practitioners within the heating transition, indicating a research gap. In addition, multiple studies call for more clear guidelines for the use of new data and tools by governmental institutions [6,13,19]. Argyrous [20] offers some guidelines on ensuring transparency and accountability, but only Koussouris et al. [17] offer concrete suggestions for practitioners besides ensuring the governmental organisation has the right expertise.

2.2. Challenges of Using Energy Models in Heating Transition Policymaking

Considering the academic literature regarding the use of energy models to support policymaking in the heating transition one thing becomes clear: there is a large variety of models and tools being used to support decision making within the energy transition, and few comparisons are being made between these models and tools. An overview of the literature found describing different modelling methods used for a sustainable heating transition is shown in Table A2, Appendix B (relevant findings for the present study). Reviewing this sample [1,21–41] shows that although modelling approaches have the potential to reduce the uncertainty of complex social issues, there is currently no systematic approach on how to apply models to make policy decisions and how to consider not only objective facts but also social and socio-economic factors. As the complexity of heating transition projects is partly due to the dependency on social factors such as human behaviour, models which consider not only objective techno-economic factors but also social and socio-economic factors, could increase the value of modelling approaches in heating transition projects [13,22,35,38,39].

Furthermore, the literature shows a large variety of models that are currently used, based on different theories and mathematical principles. A few common challenges can be recognised among this variety. First, the correctness and sensitivity of assumptions. Second, the transparency and usability for practitioners. Third, the need to integrate both economic, environmental and social factors. Another interesting aspect concerns the lack of energy modelling research, particularly in the heating transition of The Netherlands. Thus far, in this country, only one academic study was conducted addressing a model focused on the heating transition [1].

Although there is limited academic literature available, grey literature is abundant. A whitepaper by Nikolic et al. [19] offers general principles for good modelling practice and red flags that indicate inadequate modelling practices. It concludes that there is a need for modelling guidelines that are more practical and easier to communicate, and that there is a need for more interaction between academia and practitioners. Both Nikolic et al. [19] and De Ridder et al. [42] suggest that municipalities need to develop more internal knowledge to understand and make use of models. Diran et al. [43,44] claim that better access to data regarding buildings, infrastructure and energy production is needed to utilise current energy models, especially within the utility sector. Figure 1 presents an overview of the energy models and tools regarding the heating transition as used in The Netherlands.

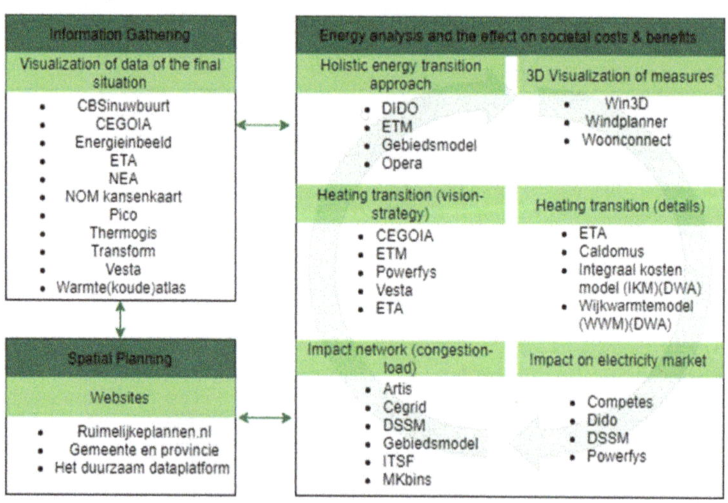

Figure 1. Overview of Dutch energy models used for decision-making support in the heating transition. Image translated and adapted from [45].

A study by Brouwer et al. [10] compares six models that are often used by municipalities, i.e., the Vesta MAIS (Multi Actor Impact Simulation) model, the CEGOIA model, the Energy Transition Model (ETM), a DWA model and the Caldomus model. The characteristics of these models are discussed in Table 1. The study [10] reveals that these models provide significantly different results for the same research question due to differences in assumptions and modelling approach. Differences identified [10] include differences in building types and geographical borders, differences in renovations to improve surpassing energy label 'B'; differences in costs of all-electric networks; differences in the order of steps within the approach; different assumptions regarding the scarcity of heat sources and different assumptions regarding learning curves; different heating technologies included; and differences in optimisation research questions.

Table 1. Overview of the six energy models often used by municipalities for heating transition policymaking.

Model [a]	Developer [b]	Type of Model [c]	Format	Availability	Geographical Scope
Vesta MAIS	PBL (English: Dutch Planning Bureau for the living environment)	Techno-economic optimisation	C++ (GeoDMS software)	Open access	National, regional, city, neighbourhood
CEGOIA	CE Delft	Techno-economic optimisation	Excel model	Model owned by CE Delft	National, regional, city, neighbourhood
Energietransitie model (ETM) (English: Energy Transition Model)	Quintel	Techno-economic simulation	Website	Open access	International, national, regional, city
Warmtetransitie model (WTM) (English: Heating Transition Model)	Over Morgen	Techno-economic optimisation	Unknown	Model owned by Over Morgen	Unknown
Integraal kostenmodel (IKM) (English: Integral cost model)	DWA (A Dutch engineering consultancy)	Techno-economic optimisation	Excel model	Model owned by DWA	Regional, city
Wijkwarmtemodel (WWM) (English: District heating model)	DWA	Techno-economic optimisation	Excel model	Model owned by DWA	Neighbourhood
Caldomus	Innoforte	Techno-economic optimisation	Excel model	Model owned by Innoforte	Regional, city, neighbourhood

[a] English translation provided by the authors. [b] English translation provided by the authors. [c] Optimisation models find the optimal solution for a chosen criterion and constraints, whereas simulation models merely allow the end-user to explore how a system responds to different inputs.

2.3. *Propositions on the Use of Energy Models by Municipalities*

Based on the literature it can be deduced that clear guidelines for the use of energy models are missing and that there are serious concerns about the lack of expertise regarding energy models and data management at public organisations. Among energy models used for energy policy design, there are challenges regarding the correctness and sensitivity of assumptions, regarding the transparency and usability for practitioners (such as policymakers) and regarding the need to integrate more social factors. Moreover, although there is grey literature available, there is a lack of academic research about the use of energy models by municipalities. Based on the literature reviewed, propositions were formulated regarding current practices, advantages and limitations of municipalities using energy models in the heating transition. Table 2 presents these propositions with argumentative justifications provided for each of them. Note that some of these propositions were formulated in an if-then structure to improve readability. However, this structure only has conversational implication and is not in line with formal logical implication, i.e., if X then Y" is only false in case "X" is true and "Y" is false.

Table 2. Overview of the theoretical propositions and their respective justifications.

Proposition	Justification
1. Different municipalities use different energy models (if any) with different aims.	Due to the large share of energy models available that use different approaches and assumptions and that have a different focus [10], it is expected that different municipalities will use different energy models with different aims.
2. If energy models are complex to use, then practitioners will make limited use of them while planning for the heating transition.	Current energy models are not usable for non-experts, such as practitioners [21,28]. It is therefore expected that practitioners make limited use of energy models due to the complexity of energy models.
3. If energy models do not integrate social or socio-economic factors, then practitioners will make limited use of them while planning for the heating transition.	The complexity of heating projects is partly due to the dependency on social factors such as human behaviour and that models which consider not only-objective but also social and socio-economic factors could increase the value of modelling approaches in heating transition projects [13,22,31–35,38,39]. Therefore, it is expected that practitioners currently make limited use of energy models because energy models currently used do not include social factors.
4. If assumptions within energy models are uncertain, than this will decrease the trust within energy models for practitioners.	The correctness and sensitivity of assumptions influence trust and willingness of practitioners to use energy models in their heating transition projects [10].
5. If data is uncertain or unavailable, then this will decrease the trust within energy models for decision making among practitioners.	More data is needed about buildings, infrastructure and energy production to utilise current energy models [44].
6. Practitioners seek the help of external parties to use and interpret energy models.	Current energy models are often not usable for non-experts, such as practitioners [21,28]. Therefore, it is expected that practitioners seek external expertise when using an energy model.
7. External parties have commercial reasons to not be transparent about their energy model design.	According to the data-driven approach and good modelling practices discussed in the literature review, models and modelling studies require a high degree of transparency [19,20]. Since many energy model developers are commercial parties, it is expected that external parties sometimes have commercial reasons to not be fully transparent.
8. Practitioners need new (in-house) expertise to effectively use energy models	Municipalities need to develop more internal knowledge and expertise to understand and make use of models [10,18–20,42].
9. Interactive visualisation and different interfaces for different stakeholders improve the usability of energy models.	Interactive visualisation can help in making models and their results more understandable for non-experts [17].

3. Research Design

3.1. Embedded Case Study Research Design

To answer the research question, multiple embedded case studies were conducted. Based on the embedded case study design of Yin [46], the nine propositions formulated based on the literature review, guided design, data collection and analysis will be reflected upon [46]. In the present study, multiple cases represented a variety of heating transition projects. Key actors involved included heating transition practitioners and energy model developers. Practitioners, such as policymakers and project managers, are closely involved in the heating transition project of the municipality and/or in the development of the local heating vision document. Energy model developers are involved in the developing models that are used by municipalities.

3.2. Case Selection

The first generation of pilot projects from the National Programme for Natural Gas-Free Districts (see Table A1, Appendix A), consisting of 27 municipalities, served as an initial source of case study selection. It was predicted that these cases would produce similar results or contrasting results for anticipatable reasons. All of these projects started at a similar time in 2018, received government funding and had a similar manner of publicly documenting their progress. Differences in results between these projects are expected to be based on the size of the municipality, based on specific neighbourhood characteristics of the pilot projects and on different energy models that are used. In total, ten municipalities participated in the present study. This entailed a sample of three large municipalities (>100,000 residents), five medium-sized municipalities (>30,000 residents) and two small municipalities (<30,000 residents), across ten provinces (out of twelve provinces in the country; showing high geographical variation), with ten different approaches to natural gas alternatives analysis, and a variety of different selected heating alternatives. Table 3 presents an overview of the ten municipalities that participated and the potential alternatives for natural gas for their respective pilot projects, based on the information that was published in the project implementation reports of 2018.

3.3. Pattern Matching

To enable reflection from the empirical study to the theoretical propositions, the "pattern matching" technique was used. According to Yin [46], pattern matching is one of the most desirable techniques used in case study analysis. Pattern matching entails comparing empirically-based patterns with the predicted patterns made before collecting data, e.g., the theoretical propositions. The ATLAS.ti 8 [47] software was used to support the process of pattern matching. As there is a risk of collecting too little data with this approach [46] data were also collected on emerging themes that were present in the academic and grey literature but that were not captured in the propositions. After finalising the empirical study, each of the nine propositions will be reviewed separately and will be either confirmed or rejected based on confirmatory evidence that follows from the empirical analysis, as described in Section 3.4.

Table 3. Overview of the ten case studies, presenting the size and the proposed alternative heating technology options of each of the municipalities analysed.

#	Municipality	Number of Residents (2019)	The Technological Heating Alternative Proposed for the Local Project
1	Loppersum	9614 [48]	Heating network, heat pumps and thermal energy storage [49]
2	Tytsjerksteradiel	31,780 [48]	Individual heat pumps [50]
3	Assen	67,963 [48]	Unknown [51]
4	Noordoostpolder	46,849 [48]	Heat network [52]
5	Katwijk	65,302 [48]	Aquathermic solution, Medium-Temperature Heat Network [53]
6	Rotterdam	644,618 [48]	High-temperature Heat Network (possibly later Medium-Temperature) [54]
7	Utrecht	352,866 [48]	High-Temperature heat network and heat pumps [55]
8	Eindhoven	231,642 [48]	Heat Network [56]
9	Brunssum	28,103 [48]	Low-Temperature Heat network [57]
10	Middelburg	48,544 [48]	High-Temperature Heat network [58]

3.4. Data Collection, Treatment and Analysis

The types of data per case study that were used concerned: (1) governmental reports (for example heating transition implementation plans); (2) in-depth interviews with practitioners from municipalities and; (3) in-depth interviews with model developers. The information of these three sources was converged in a triangulating fashion. The documents (such as project implementation plans and model guidelines) provided secondary data that were used to structure the interviews. Only publicly available documents were used. Twenty-one in-depth, (expert) interviews provided primary data of the case studies.

All twenty-one interviewees were provided with informed consent forms and all interviewees provided, among others, permission for the use of their statements for the present study. An anonymised overview of respondents is shown in Tables A3 and A4, Appendix C. All interviews were conducted via video call or telephone, and audio was recorded. Interviews with both practitioners (14) and model developers (7) were fully transcribed. Transcripts were provided to the interviewees after the interviews and interviewees were given ample opportunity to read and alter the transcript. All interviews were conducted between the first of May and the first of September of 2020. The average duration of individual interviews was 55 min.

The interviews were semi-structured with open-ended questions to allow for in-depth analysis. Although a set of pre-defined questions was used, interviewees were also given the opportunity to explore questions in greater depth and to introduce new topics. This type of in-depth interviews, according to Roller [45], increases the credibility of the data by reducing response bias (distortion due to the tendency of interviewees to provide answers that are considered socially accessible) and by reducing satisficing (providing an easy 'I do not know' answer). The data collection process, including the informed consent forms, was approved by the Ethical Committee of the Technology, Policy and Management faculty at Delft University of Technology.

Analysis of the interview transcripts was completed by thematic coding. Atlas.ti 8 [47] (computer-aided qualitative data analysis software) was used to perform the coding process and to create coding reports. A semantic analysis was conducted, meaning that data was coded at face value, i.e., at the explicit meaning. Thematic coding is viewed as a relatively simple qualitative method that offers a high level of flexibility. Quotations were created based on the theoretical propositions and the research questions, and a code was assigned to each quotation. As proposed in standards for theoretical thematic analysis [59], an initial set of codes was set-up to guide analysis of the transcripts. The coding frame, as expected, did not fully cover all aspects related to the topic and was adapted and supplemented where needed with codes such as 'motivation residents' and 'not familiar with energy models'. These adaptations were made rather inductively, meaning that the 'open coding' function of Atlas.ti 8 [47] was used to add codes during the first round of coding. After this first round of coding, all codes and their frequency were assessed to see whether splitting or merging of codes was necessary. To transform the raw data into meaningful information, all quotes were given an English title; code groups were created to show the relation between several codes and so-called network figures were created to show the focus of different quotes within one code. Moreover, code-occurrence tables (see Tables A5 and A6 in Appendices D and E) were made to quantify the findings, which reduced the subjectivity of result interpretation.

4. Results

The interviews conducted with practitioners yielded 820 quotes divided over 36 thematic codes. Seven interviews were conducted with model developers. These interviews yielded 561 quotes divided over 53 thematic codes (See for an overview of codes and code-occurrence Appendices B and C). The results of the case studies were used to either validate or reject the propositions (Section 2.3; Table 2). The findings regarding the testing of the propositions are presented in Table 4. The findings will be discussed in more detail in Sections 4.1–4.8 below.

Table 4. An overview of the findings that confirm or reject the propositions made.

Proposition	Confirmed/Rejected
1. Different municipalities use different energy models (if any) with different aims.	confirmed
2. If energy models are complex to use, then practitioners will make limited use of them while planning for the heating transition.	confirmed
3. If energy models do not integrate social or socio-economic factors, then practitioners will make limited use of them while planning for the heating transition.	rejected
4. If assumptions within energy models are uncertain, then this will decrease the trust within energy models for practitioners.	unclear
5. If data is uncertain or unavailable, then this will decrease the trust within energy models for heating transition decision making of practitioners.	unclear
6. Practitioners seek the help of external parties to use and interpret energy models.	confirmed
7. External parties have commercial reasons to not be transparent about their energy model design.	unclear
8. Practitioners need new (in-house) expertise to effectively use energy models.	confirmed
9. Interactive visualisation and different interfaces for different stakeholders improve the usability of energy models.	confirmed

4.1. Different Municipalities Use Different Energy Models with Different Aims

The proposition 'Different municipalities use different energy models (if any) with different aims' was confirmed based on the case studies. Six different energy models were used by the ten municipalities studied to support decision-making for heating transition pilot projects or the design of the Transition Vision Heat: the Vesta MAIS model, the CEGOIA model, the Caldomus model, DWA models (the IKM and the WWM), the ETM and the WTM). This is in line with [10] which mentioned these six models as the most used models for the Dutch heating transition. Moreover, two national modelling studies based on one or more of these energy models were used, the 'Startanalyse' (Start Analysis in English; translation by the authors) and the 'Openingsbod' (Opening Offer in English; translation by the authors) (see Table A1, Appendix A). In the case studies, these models were only seldom used by practitioners, with the only exception in this sample pertaining the municipality of Utrecht, where a modelling team was deployed to use the Vesta MAIS model to develop heat scenarios. More in general, municipalities were found to use models and modelling studies to support the decision-making process, to provide more legitimacy towards residents or as a basis for more detailed heating transition business cases. No socio-technical energy transition modelling methodologies or agent-based modelling methodologies were found though, indicating that these were not considered important in the current planning and implementation of the heating transition at the local level. All models, except for the ETM, were optimisation models. The ETM did not offer an automated optimisation function. All models, except for the ETM, aimed to find the heating alternative with the lowest societal costs.

4.2. Complexity and User-Friendliness of Energy Models

The propositions 'If energy models are complex to use, then practitioners will make limited use of them while planning for the heating transition', and 'Practitioners seek the help of external parties to use and interpret energy models' were confirmed based on the case studies analysed. The results showed that the use of energy models was not necessarily limited, seven out of the ten heating transition pilot projects investigated used an energy model in their decision-making process and seven out of seven Transition Vision Heat projects used or were planning on using an energy model. However, six interviewees mentioned there were issues regarding the complexity and user-friendliness of energy models that hindered effective usage in heating transition projects. Four out of seven model developers claimed that practitioners often did not have the right background or the time to master these complex tools independently. According to the same four interviewees, large-sized municipalities usually had more time and resources to learn how to use a model than their small-sized peers. If a third party conducted the modelling process, large-sized municipalities were therefore generally better able to critically reflect on the results. All seven municipalities from this sample that used energy models in their heating transition projects used third parties at some point during their heating transition projects to conduct modelling studies. Third-party expertise was used at all scope levels, Regional Energy Strategy development (see Table A1, Appendix A), Transition Vision Heat development and pilot projects. Municipalities were found to hire external parties to provide modelling calculations, home inspections, modelling result interpretation or to provide studies, for example into available heat sources. These findings confirm that there are indeed challenges with the complexity and user-friendliness of energy models and that these are usually overcome by seeking help from external parties.

4.3. Integration of Social or Socio-Economic Factors into Energy Models

The proposition 'If energy models do not integrate social or socio-economic factors, then practitioners will make limited use of them while planning for the heating transition' was rejected based on the case study analysis. All fourteen practitioners interviewees agreed that social and socio-economic factors are important and influence the success of heating transition projects. Three municipalities were found to use social or socio-economic data or information or were planning to use this to identify coupling opportunities (opportunities to combine activities for the heating transition with other improvement opportunities in a neighbourhood, such as sewer system updates, building renovations or traffic alterations), and two municipalities used or were planning to use social or socio-economic information to determine the prioritisation of neighbourhoods for heating transition activities. On the other hand, none of the practitioners or model developers interviewed claimed that social or socio-economic factors influenced the choice of heating alternatives, which is the focus of the six energy models municipalities of the sample used. The choice of heating alternative was based on the lowest societal costs in all municipal heating transition projects within the present study. All seven energy model developers agreed that that social, political and psychological aspects influence heating transition projects. However, all claimed that these factors should not and/or could not be included in their respective models and that it would be better to consider these factors alongside the techno-economic modelling results in energy modelling studies.

4.4. Unavailable Data and Uncertain Assumptions

The proposition 'If assumptions within energy models are uncertain than this will decrease the trust within energy models for practitioners' could neither be confirmed nor rejected based on the empirical results. Energy model developers were found to use different assumptions, and two energy model developers claimed that these are usually the reason why results between different energy models differ. Practitioners offered critiques of assumptions of models or modelling studies, in particular about assumptions regarding energy labels and the use of renewable gas. However, the impact this had on trust in energy models did not become clear in the interviews. The interviews showed that if practitioners did not agree with assumptions used in models or modelling studies that they requested model developers to change said assumptions or that they opted for a different model that used different assumptions. All seven model developers stated that they tried to be transparent about the assumptions they used and that, in collaboration with the practitioners, assumptions can be altered during the modelling process.

The proposition 'If data is uncertain or unavailable, then this will decrease the trust within energy models for heating transition decision making of practitioners' could not be confirmed nor be rejected. Data played an important role for municipalities and model developers in developing heating transition plans, and even though data was sometimes unavailable, this study offered no proof that this decreased the trust of practitioners in energy models. If municipalities decided to use a model, this energy model proved to be more useful if it was fed with local data. Unavailable data that could be useful according to practitioners and model developers is data about energy use per connection, data about the willingness to pay of residents and data about the potential impacts on the electricity grid. One energy model developer mentioned that the data collection process at public organisations was too time-consuming and two energy model developers mentioned that they ran into issues with the energy use data available from Statistics Netherlands ('CBS' in Dutch). These data were aggregated due to privacy laws and was often deemed too inaccurate to use for heating transition projects. Similarly, two energy model developers and one practitioner stated that the data from the Basic registration of addresses and buildings (BAG) (See Table A1, Appendix A) regarding energy labels provided too little insight into the level of thermal insulation present at residential houses. One of the most uncertain data sets used for heating transition projects was data about available heat sources. All model developers agreed that the datasets for heat source data were uncertain and that extra research was always needed to assess the local situation. However, whereas four energy models used the availability of heat sources as a determining factor for the choice of a natural gas alternative, two models did not use heat source availability as a determining factor.

4.5. The Use of Third Party Modelling Expertise

The proposition 'Practitioners need new (in-house) expertise to effectively use energy models' was confirmed based on the case studies. Only one municipality was yet capable of modelling scenarios individually. Others relied on the modelling expertise of third parties. Even if a municipality outsourced the modelling process, a minimum knowledge level was required to correctly interpret and critically reflect on results. According to energy model developers, practitioners, with only a few exceptions, did not meet this minimum condition. This also caused practitioners to propose incorrect or unsuitable research questions to model developers.

The proposition 'interactive visualisation and different interfaces for different stakeholders could improve the usability of energy models' was also confirmed based on the case studies. Three energy model developers had developed interactive models, maps or tools that, according to them, helped clients such as practitioners to better understand and interpret the modelling results. No statements from practitioners were gathered on the advantages of interactive models.

The proposition 'External parties have commercial reasons to not be transparent about their energy model design' could neither be confirmed nor be rejected. Two energy model developers stated that it was not always possible to gain access to underlying assumptions, data and parameters of models from other commercial agencies. However, all six models in this study were compared to each other in the benchmark study [10], indicating that model developers were at least willing to be transparent towards independent researchers. Moreover, one national modelling study compared the results and underlying assumptions, datasets and parameter sensitivities of multiple models (of which two were commercial). Besides, transparency was only mentioned as a limiting factor by one practitioner. Hence, one could state that even though transparency, especially at commercial model developers, could be improved, it did not seem to be a limiting factor for municipalities to use energy models.

4.6. Advantages and Limitations of Using Energy Models

According to the academic literature, energy modelling can aid in decision making and policymaking because it introduces a structured way of thinking about the implications of changing parts of the system [9]. The case studies provided more concrete benefits and limitations of using energy models for decision making in the Dutch heating transition. Practitioners stated that the use of energy models within heating transition projects provided perspective for action, financial insight, transparency and legitimacy, concrete propositions to residents and sparked useful discussions. Besides, one practitioner stated that nationally available modelling studies provided validation and robustness of (other) modelling results. Most of these advantages are related to creating public support for policy. Practitioners also mentioned limitations of using energy models. Interviewees argued that energy modelling results were considered too abstract, too general or too simplified for local analysis. In addition, models were considered not user-friendly and complex. Practitioners mentioned that modelling results provided no insight into available heat sources, limited insight into the impact of nearby heat networks and no or limited insight into end-user costs. Another challenge mentioned was that the Statistics Netherlands ('CBS' in Dutch) neighbourhood definitions do not provide a logical division of the city, which, among others, created the need to conduct a reality check after modelling to filter out odd results, especially for the utility sector.

4.7. Collaboration with Housing Associations, Network Operators and Citizen-Led Energy Cooperatives

Moreover, from the case studies, insights were gathered that suggest that collaboration with housing associations and network operators is important during heating transition projects to prepare implementation plans and to find coupling opportunities. Housing associations were considered important as they often have property within the municipality and because they have renovation plans that may or may not align with the municipal heating transition plans. Network operators were considered important because they are responsible for underground infrastructure and network reinforcements. Therefore they have to be made aware of the municipal heating transition plans, and they have to provide input about the current limitations of the infrastructure for specific heating options. Moreover, citizen-led energy cooperatives play an important role in heating transition pilot projects. In five out of thirteen interviews, it was mentioned that collaboration with citizen-led energy cooperatives is considered important. In one small and one medium-sized municipality, energy cooperatives even provided project leaders for heating transition pilot projects. For Transition Vision Heat development at larger municipalities citizen-led energy cooperatives were found to exercise less influence. Close collaboration with energy model developers happened only in municipalities that have established modelling teams that model energy systems independently; for this sample, those included the two largest municipalities (>300,000 residents).

4.8. The Use of Comparative Analysis and Multi-Model Ecologies

As mentioned, different models sometimes result in different outcomes, which can create confusion and uncertainty at practitioners. One practitioner interviewed explicitly mentioned experiencing such confusion. Three model developers of this sample actively used comparative analysis to reduce this issue, and one national energy modelling study, the 'Openingsbod', also offered comparative analysis. In such an analysis, differences in methodology, assumptions, data and results of different energy models or modelling studies are compared to one another. This indicated where result differences originate from and provided an overview of the robustness of results across models. One practitioner claimed that the latter helped in determining a priority of neighbourhoods to start with heating transition projects.

Finally, three practitioners mentioned the challenge of matching up heating transition plans at different levels of abstraction, which were found to influence each other and that were sometimes developed simultaneously and with different energy models. To decrease this challenge, one energy model developer tried to position his model in such a manner that he could assess how plans would fit together. This energy model developer envisioned a multi-model ecology in which their model provided a broad energy perspective and where other energy models would offer more detailed calculations on, for example, heating transition visions, heating transition business cases and the effects on power networks.

5. Discussion

5.1. Reflection vis-à-vis the Academic Literature

The present study has provided a more concrete image of the role of energy models in data-driven policymaking and decision-making in the heating transition. The literature review showed that modelling approaches have the potential to reduce the uncertainty and complexity of heating transition projects. The present study provided a concrete overview of the advantages of using energy models in heating transition decision making as experienced by practitioners and model developers. The advantages found seem to indicate that although energy models do not necessarily make a heating transition project less complex, they at least offer means to make legitimate choices. The advantages identified are in line with the advantages of data-driven policy design mentioned by Koussouris et al. [17] who stated that tools such as energy models, simplify decision-making processes, even under complicated conditions, by facilitating the opportunity to model complex processes and the opportunity to collaborate with different actors involved, and those mentioned by Adam et al. [15] who stated that providing evidence for the effectiveness of policy choices is one of the cornerstones of legitimate policymaking.

The results of the present study could provide a starting point for recommendations targeting policymakers and model developers to facilitate more effective use of energy models in heating transition decision-making. Such targeted recommendations were not found in the literature and could help towards designing a systematic approach for integrating energy models in data-driven policymaking, which is needed and currently lacking [6,13].

Moreover, the results of this study suggest that offering comparative model analysis would help practitioners to deal with the myriad of sometimes contrasting models, modelling studies and modelling results available and that setting up a multi-model ecology might decrease the challenges of aligning heating transition projects at different abstraction levels. This is in line with Manfren et al. [60] who state that multi-model ecologies could help in creating the integration between top-down and bottom-up modelling perspectives. Furthermore, it aligns with Nikolic et al. [61] who state that multi-model ecologies help get a more coherent and less biased understanding of the "right thing" to do in energy transition decision making as using multiple models allows multiple perspectives to be explored and be brought together.

Although this study confirmed certain advantages of using energy models it also shed light on the limitations of using energy models for decision-making. Designing modelling scenarios is considered a time-consuming and costly task. Modelling results are not absolute truths but rather results subject to calculation rules and assumptions, and if a model or its outcomes are incorrect, one might be worse off than when not using a model to begin with [19]. According to energy model developers interviewed in the present study, not all practitioners understood the limitations of energy models and interpreted modelling results as absolute truths.

Finally, the literature review suggested that it is problematic that current heating transition models do not include social and/or socio-economic factors, as the transition is highly dependent on humans and their behaviour [13]. However, the present study showed that practitioners were not always sure how social or socio-economic data should influence the choice of a heating alternative or the prioritisation of neighbourhoods. Moreover, accessing these data was sometimes difficult due to privacy restrictions. Model developers did not see added value in including social or socio-economic factors within their heating transition models, which all had a techno-economic focus. Their models were focused on finding the lowest societal and/or end-user costs for different heating alternatives and did not include social factors, as affordability for residents is seen as one of the main challenges of the Dutch heating transition [5]. The costs of a heating alternative are, as far as known, not only depending on social or socio-economic factors. Something that could be depending on such factors, for example, concerns the degree of participation and technology adoption rates.

In the present study, not one municipality was found using model methodologies focused on assessing social interactions, such as Agent-Based Modelling, System Dynamic Modelling or Socio-Technical Energy Transition Modelling. Instead, municipalities used models with a mere techno-economic focus and assessed social and socio-economic data alongside the results of these modelling efforts to identify coupling opportunities and/or to determine prioritisation of neighbourhoods.

5.2. The Influence of National Agreements and Municipality Size

All municipalities that provided information about their Transition Vision Heat planning design in the present study used or were planning to use models/modelling studies. This was expected as it was agreed in the national Climate Agreement of 2019 [3,62] that municipalities would use the 'Startanalyse' and its guidelines [63] to design their Transition Vision Heat. According to the Climate Agreement, this would provide all stakeholders with a "uniform frame of reference regarding the impact of the various natural gas alternatives in a district" [3]. This agreement might have incentivised municipalities to use energy models when designing their Transition Vision heat. However, three pilot projects did not use energy models to choose a natural gas alternative. The pilot projects analysed, all started before this statement was made in the climate agreement and before the 'Startanalyse' and its guidelines [63] were published. Therefore, practitioners in pilot projects might have been less familiar with available models and modelling studies, might have had less access to models and modelling studies and/or might have been less incentivised to use available models or modelling studies.

Secondly, pilot projects that did not use an energy model to choose a heating alternative had a few things in common. All three pilot projects were located in villages with less than 2000 residents. All of them had active citizen-led energy cooperatives, two pilots were organised by the local energy cooperative, two pilot project leaders were not familiar with energy models, and two pilot projects entailed only or mostly detached houses, from before 1940 with poor thermal insulation levels. Two practitioners claimed that they did not feel that they needed an energy model because the choice for a heating alternative could be made with common sense and information about the residential characteristics. This indicates that an energy model might not always be considered necessary or desirable for heating transition decision-making and that it is important to consider when the use of an energy model would be beneficial and when other sources of evidence might be sufficient to support decision-making.

6. Conclusions

6.1. Answering the Research Question

This study aimed to answer the research question 'What are the perceived advantages and limitations of using energy models for municipalities within their data-driven decision-making process concerning the natural-gas free heating transition?'. To answer this question, a literature review and embedded multiple case study research were conducted, which included different heating transition projects in ten Dutch municipalities.

Results inter alia show that energy models observed in the present study were mostly initiated and used by consultancy agencies to support Dutch municipalities in designing heating transition plans. Over half of the municipalities analysed were found to use models or modelling studies at some point during their respective heating transition pilot projects. All cases that provided information about local Transition Vision Heat development were using or planning to use models or modelling studies for the design of their vision document.

Models that were used pertained to the CEGOIA model, the Vesta MAIS model, DWA models, the ETM and the WTM. Modelling studies that were used concerned the 'Openingsbod' and the 'Startanalyse'. Municipalities that did not utilise models or modelling studies for their pilot projects belonged to the four smallest municipalities analysed in the present study, indicating a negative relation between municipality size and model usage.

All municipalities that used models or modelling studies requested external expertise at some point during the modelling process, indicating that the knowledge and skill level at municipalities was not sufficient to do this independently. This was confirmed by model developers who also stated that the knowledge level of practitioners is often insufficient to interpret results of modelling studies conducted by third parties.

Advantages of using models in heating transition projects mentioned in the interviews were that the modelling process and its results provided perspective for action, financial and socio-economic insights, transparency and legitimacy towards residents, concrete propositions for residents and means to start useful discussions. However, interviewees also mentioned several limitations. First, models and modelling results were found too abstract, too general or too simplified for local analysis, not user-friendly and were considered complex. Results were difficult to interpret for non-experts such as practitioners, and interactive models could provide practitioners with a better understanding of the answer and help with getting a feeling for parameter sensitivity. Second, modelling results provided too little insight into end-user costs and the effects on the electricity grid. Third, data sets regarding energy use, thermal insulation levels and heat sources proved to be insufficient for local analysis, and there was no consensus between model developers and practitioners about the different assumptions regarding green gas availability and energy labels used in different models.

This study also showed that model developers deemed it unpractical to integrate social and socio-economic factors in the energy models discussed, but agreed that this data should be incorporated in modelling studies/reports. Model developers usually did this by collecting social or socio-economic data and by presenting this data next to the modelling results to provide context for further decision-making.

Finally, the results suggest that offering comparative model analysis would help practitioners to deal with the myriad of sometimes contrasting models, modelling studies and modelling results available and that setting up a multi-model ecology might decrease the challenges of aligning heating transition projects at different abstraction levels.

6.2. Limitations

The external validity of the empirical results is limited by the context in which the present study was conducted, in selected municipalities in The Netherlands. This was a scoping choice motivated by the case study design and time constraints of the present study. The representativeness of these results to other geographical, political and cultural contexts might therefore be fairly limited. It is expected that representativeness will particularly be limited for countries where the heating transition is not organised in a decentral manner or where there are not multiple (national) energy models available to analyse the costs of this transition.

Limited access to background information on some commercial energy models limited the reflection on technical aspects of the models reviewed in the case studies. In the present study, the capabilities, limitations, underlying assumptions of models were only compared at the surface level, based on publicly available reports and the challenges and advantages mentioned by interviewees. This limited access to background information limited the potential for in-depth model comparison. On the other hand, the time constraints of this research and the focus on user experiences and the modelling process rather than the actual energy models also limited this potential. This choice was made because limited access to the background information of (commercial) models was foreseen and because there are already other studies, such as [10], that focus on in-depth model comparison.

The data collection tools chosen, interviews and thematic coding, also have their respective limitations. Interviews and thematic coding are research tools that require a high degree of interpretation from the researcher. During the coding process, quotes had to be translated and interpreted. The literal transcripts, the coding process and the coding reports ensured quotes were methodologically analysed and that it was possible to review the original quotes.

The present study used multiple sources of evidence in a triangulating fashion to decrease the subjectivity of the answers and to check their consistency over time. A remarkable observation was that within the pilot projects observed the views and plans of interviewees did not always align with the views as exhibited in the implementation plans of the pilot project, due to advancing insights.

6.3. Recommendations for Future Research

The present study did not provide an answer as to when heating transition projects should and when they should not use energy models to guide their heating transition decision-making process. The discussion offered some criteria that might indicate projects that do not need energy models such as municipality size, residential housing characteristics and the presence of an energy cooperative. It is therefore recommended to conduct more research into which criteria could indicate that projects would have an advantage of using an energy model. It is recommended to conduct more case studies, with different types of heating transition projects, to explore this topic. In addition, it is suggested to also include case studies that utilise other decision support tools, such as MCDA tools, in order to assess the relative advantages and limitations of energy models when compared to other tools.

Furthermore, it is recommended to further study the impact of social and socio-economic factors. The literature review revealed that that social and socio-economic factors are highly important for heating transition decision-making processes, but currently, the impact of social and socio-economic data within Dutch heating transition projects is limited and at best influences the prioritisation of neighbourhoods. More research into certain factors, for example, income or the presence of energy cooperatives, could provide insight into the correlation of these factors with heating transition project progress and into the potential value of models that include such factors. Such insights would not only benefit the Dutch heating sector but might also benefit a range of international energy transition projects. On the one hand, this might entail desk research into socio-technical transitions and models (such as Socio-Technical Energy Transition, System Dynamics or Agent-Based models). On the other hand, it might address practical case studies that test socio-technical transition theories and models within heat or energy transition projects. Ideally, such case studies are not restricted to The Netherlands but also include projects in countries with significantly different heating systems, energy markets, institutions, social and socio-economic values to compare and corroborate results.

Finally, it is recommended to conduct more research into the field of multi-model ecologies (e.g., systems of interacting models). The present study has shown the need for comparative analysis, for modelling at different abstraction levels and for assessing the impact of choices regarding the heating transition in other disciplines, such as electrical infrastructure and social welfare. More research into multi-model ecologies can benefit both the Dutch and the international academic modelling field as it offers the opportunity to add value to existing models, for example by making them more interactive with other national or international models. Nikolic et al. [61] and Manfren et al. [60] offer the first set of principles, challenges and guidelines that provide a conceptual basis for multi-model ecologies. Currently, the 'Mondaine Suite' project [64] is one of the first projects that is aiming to realise a multi-model ecology by developing a coupling mechanism for different (Dutch) energy models. However, this project does not yet couple Socio-Technical Energy Transition models, System Dynamics or Agent-Based models, which might offer an interesting opportunity for future research to include more social and behavioural components into multi-modal ecologies.

Author Contributions: Data curation, B.A.H.; Investigation, B.A.H.; Methodology, B.A.H., T.H.; Supervision, T.H., D.D. and Z.L.; Visualization, B.A.H.; Writing—original draft, B.A.H.; Writing—review and editing, T.H., D.D. and Z.L. All authors have read and agreed to the published version of the manuscript.

Funding: This research did not receive external funding.

Institutional Review Board Statement: The research design was approved by the Human Research Ethics Committee of the Technology, Policy and Management faculty of the Delft University of Technology.

Informed Consent Statement: Informed consent was obtained from all subjects involved in the study.

Data Availability Statement: Data used in this study entails anonymized interview transcripts and anonymized interview coding reports. Both can be found on the repository of the Delft University of Technology: https://repository.tudelft.nl/islandora/object/uuid%3Aae00908d-e89a-400e-819b-dd0d11cdba34.

Conflicts of Interest: The authors declare no conflict of interest.

Appendix A

Table A1. Glossary and definitions of Dutch (policy) concepts and abbreviations used in the present study.

Concept	Abbreviation	Definition Used
Startanalyse	SA	The 'Startanalyse' (Start Analysis in English) is a national modelling study conducted with the Vesta MAIS model by PBL. The Startanalyse is presented together with guidelines for local analysis (Handreiking) in a guidebook (Leidraad) for Dutch municipalities.
Openingsbod	OB	The 'Openingsbod' (opening offer in English) is a modelling study initiated by Stedin, a Dutch network operator. The study was developed as a tool to quicken decision making in the Dutch heating transition. The study compares the modelling approach and the results of three different energy models.
Programme for Natural Gas-Free Districts	PAW	A joint programme of the Ministry of the Interior and Kingdom Relations, the Ministry of Economic Affairs and Climate Policy, the Association of Netherlands Municipalities and the Association of Regional Water Authorities that, among others, provides subsidies and requirements for the Test Beds for Natural Gas-Free Districts (pilot projects) [3].
Regional Energy Strategy	RES	"Within the RES, public authorities work alongside social partners, network managers the business community and, where possible, residents to develop regionally supported choices. The RES aims to realise the generation of renewable electricity (35 TWh), to realise the heating transition in the built environment (from fossil to sustainable sources) and to realise the necessary storage and energy infrastructure" [3].
Transition Vision Heat	TVW	The TVW is a policy document in which a municipal council has to establish a realistic schedule within which to transition away from natural gas [3]. The focus of the first TVW is on the period until 2030 and every municipality has to show which building will become natural gas free or insulated, with which electrical infrastructure and when [65].
Neighbourhood Implementation Plans	WUP	A WUP is the follow up of the TVW and indicates how a municipality will make a specific neighbourhood natural gas-free by transitioning to sustainable heating and cooking systems.
Basic registration of addresses and buildings	BAG	The BAG-dataset is a national dataset. Municipalities are responsible for providing data for the BAG-dataset, the dataset is maintained by the Dutch Cadastre, Land Registry and Mapping Agency

Appendix B

Table A2. Overview of the literature found describing different modelling methods used for sustainable heating transition projects.

Model Type	Studies	Relevant Findings for This Study
Agent-based model	[1,22,23]	These studies emphasise the importance of trying to incorporate social factors within modelling.
TIMES (The Integrated MARKAL-EFOM System) energy model (linear optimisation)	[24–26]	The maximum surplus assumption used in the model is often challenged.
Simulation model (using Long-rangeEnergy Alternatives Planning (LEAP software)	[27]	All stakeholders involved could run and modify the model themselves and even modify it according to their needs. The model has a large sensitivity for a multitude of assumptions
METIS simulation model	[28]	A current understanding of quantitative tools by policymakers is often missing
HOMER optimisation	[29]	Most models and tools currently used do not provide both economic and environmental analysis of energy systems, which can lead to the design of sub-optimal systems
Housing Stock Energy Model (HSEM)	[30]	Many HSEMs are lacking in transparency and modularity and that they are often limited in scope and limited in their utility. Behavioural responses are blurred in HSEMs.
Optimisation model	[31–35]	The results of the model will always depend on the focus of the optimisation and that there are not many models yet that can incorporate economic, environmental and social factors at the same time.
Dynamic system modelling	[36]	Analyses behaviour over time by identifying elements within the system and their mutual correlations.
PRIMES (Price-Induced Market Equilibrium System) model	[37]	The study does not seem to involve social or behavioural aspects. Analysis of energy systems is based on the inputs from GIS mapping
Eco-district heat kit optimisation model	[21]	This study attempted to make the model usable for non-experts.
Socio-Technical Energy Transition (STET) models	[38,39]	(Optimisation) Models tend to simplify their depiction of societal and political factors. STET models try to integrate both quantitative modelling and conceptual socio-technical transitions
Area-based model	[40]	Emphasises the importance of modelling at the sub-city scale as this enables, among others, more accurate quantification of demand increases.
Econometric model	[41]	Suggests that we should combine spatial attributes with econometric models.

Appendix C

Table A3. An overview of the respondents from municipal heating transition projects.

Municipality	Interviewee	Interviewee Function
1	01	Part-time project leader pilot project
2	02	Project leader pilot project
2	03	Project manager TVW
3	04	Environment manager pilot project
4	05	Project leader pilot project
5	06	Project manager TVW
5	07	Project leader pilot project
6	08	Project manager pilot project
6	09	Project manager TVW
7	10	Project leader pilot project
8	11	Process director pilot project
9	12	Project leader pilot project
9	13	Project manager TVW
10	14	Project leader pilot project

Table A4. An overview of the respondents from mode development firms involved in municipal heating transition projects.

Model Development Firm	Interviewee	Interviewee Function
1	15	Partner and modeller
2	16	Consultant natural resources
3	17	Senior Consultant
4	18	Consultant and technical expert model
5	19	Researcher climate, air and energy
6	20	Consultant heating transition
7	21	Director and modeller

Appendix D

Table A5. The code-occurrence table shows an overview of the 37 thematic codes, the respective code frequencies and the number of transcripts that quotes were identified in.

Code	Code Frequency	# Transcripts
Approach	149	11
Third-party expertise	79	13
Coupling opportunities	70	13
Model/modelling study used	65	13
Motivation residents	52	12
Analysis tools used	51	12
Project progress	51	13
Data	43	9
Collaboration	42	13
Participation activities	42	12
Information Respondent	37	13
(Envisioned) natural gas alternative	35	10
Incentivizing Residents	30	5
Financial arrangement residents	27	9
Added value pilot project	25	1
Limitations approach:	22	9
Future approach	21	8
Limitations model: Vesta	19	5
Responsibility municipality	17	7
Added value model: CEGOIA	13	3
Limitations model: CEGOIA	11	4
Limitations models in general	11	5
Other reasons to opt for a heating alternative	8	3
Added value analysis tool: Resident questionnaire	7	1
Added value modelling study: Openingsbod	6	2
Limitations model: DWA	4	1
Added value model: Caldomus	4	2
Added value model: DWA	3	2
Added value modelling study: Startanalyse	3	3
Limitations model: Caldomus	3	1
Added value analysis tool: Greenvis	2	1
Not familiar with energy models	2	2
Added value analysis tool: Resident meetings	1	1
Limitations analysis tool: Resident game	1	1
Added value analysis tool: Susteen	1	1

Appendix E

Table A6. The code occurrence table shows an overview of the 53 thematic codes, their respective code frequency and the number of transcripts that quotes were identified in and the code group.

Code	Code Frequency	# Transcripts	Code Group
Modelling and Consultancy approach	100	7	Consultancy and modelling approach
Collaboration	68	7	Collaboration and Competition
data	46	7	
Challenges modelling approach	45	6	Consultancy and modelling approach
Inclusion of socio-economic factors	43	7	Consultancy and modelling approach
Information about assumptions	37	7	Consultancy and modelling approach
Information about parameter sensitivity	34	7	Consultancy and modelling approach
Information about input data	32	7	Consultancy and modelling approach
Feedback channels	30	6	Consultancy and modelling approach
Result interpretation	28	6	Consultancy and modelling approach
Coupling opportunities	22	5	Consultancy and modelling approach
Users and Uses of Vesta MAIS	19	2	Uses and Users models
Limitations Vesta MAIS model	18	5	Limitations model
Information respondent	17	5	
Advantages modelling approach	17	6	Consultancy and modelling approach
Information Caldomus	15	2	General information model/modelling study
Users and uses ETM	13	2	Uses and Users models
Information model ETM	13	1	General information model/modelling study
Limitations Startanalyse	13	4	Limitations modelling stud
Advantage Vesta MAIS model	13	2	Advantages model
Influence of municipality size	12	4	
Limitations CEGOIA model	12	4	Limitations model
Information DWA Model(s)	11	1	General information model/modelling study
Planned changes Caldomus model	11	1	Limitations model
Connection RES, TVW, WUP	12	3	Consultancy and modelling approach
Competition	9	2	Collaboration and competition
Information model: CEGOIA	9	2	General information model/modelling study
Information Startanalyse	7	3	General information model/modelling study
Limitations Caldomus model	6	1	Limitations model
Planned changes Startanalyse	6	2	Limitations modelling stud
Advantage ETM	6	1	Advantages model
Information WTM	6	1	General information model/modelling study
Advantage Startanalyse	6	2	Advantages modelling study
Information Vesta Mais	5	2	General information model/modelling study
Limitations ETM	5	1	Limitations model
Users and uses of CEGOIA	5	1	Uses and Users models
Uses and users Startanalyse	5	2	Uses and users modelling studies
Advantage Openingsbod	4	2	Advantages modelling study
Users and uses WTM	4	1	Uses and Users models
Advantage CEGOIA model	3	1	Advantages model
Planned changes ETM	3	1	Limitations model
Planned changes CEGOIA model	3	1	Limitations model
Users and uses of the Caldomus model	3	1	Uses and Users models
Advantage WTM	3	1	Advantages model
Users and Uses DWA model(s)	3	1	Uses and Users models
Information Openingsbod	2	1	General information model/modelling study
Limitations DWA model(s)	2	1	Limitations model
Advantage Caldomus model	2	1	Advantages model
Uses and Users Openingsbod	2	1	Uses and users modelling studies
Limitations WTM	1	1	Limitations model
Information modelling study: Openingsbod	1	1	General information model/modelling study
Advantage DWA model(s)	1	1	Advantages model

References

1. Nava Guerrero, G.; Korevaar, G.; Hansen, H.; Lukszo, Z. Agent-Based Modeling of a Thermal Energy Transition in the Built Environment. *Energies* **2019**, *12*, 856. [CrossRef]
2. RVO. *Samen Aan de Slag Met Aardgasvrij, Inspiratie Voor Gemeenten*; Rijksdienst voor Ondernemend Nederland: Utrecht, The Netherlands, 2017; p. 13.
3. Government of The Netherlands. *Climate Agreement*; Government of The Netherlands: The Hague, The Netherlands, 2019; p. 247.
4. Buttelaar, S.; Heeger, A. *Burgerparticipatie in de Warmtetransitie*; Platform 31: Den Haag, The Netherlands, 2018; p. 56.
5. Schellekens, J.; Oei, A.; Haffner, R. *De financiele Gevolgen van de Warmtetransitie*; Ecorys: Rotterdam, The Netherlands, 2019; p. 58.
6. van Veenstra, A.F.; Kotterink, B. Data-Driven Policy Making: The Policy Lab Approach. In *Electronic Participation*; Parycek, P., Charalabidis, Y., Chugunov, A.V., Panagiotopoulos, P., Pardo, T.A., Sæbø, Ø., Tambouris, E., Eds.; Lecture Notes in Computer Science; Springer International Publishing: Cham, Switzerland, 2017; Volume 10429, pp. 100–111. ISBN 978-3-319-64321-2.
7. Pereira, A.; Quintana, S. From Technocratic to Participatory Decision Support Systems: Responding to the New Governance Initiatives. *J. Geogr. Inf. Decis. Anal.* **2002**, *6*, 95–107.
8. Horschig, T.; Thrän, D. Are Decisions Well Supported for the Energy Transition? A Review on Modeling Approaches for Renewable Energy Policy Evaluation. *Energy Sustain. Soc.* **2017**, *7*, 5. [CrossRef]
9. Pfenninger, S.; Hawkes, A.; Keirstead, J. Energy Systems Modeling for Twenty-First Century Energy Challenges. *Renew. Sustain. Energy Rev.* **2014**, *33*, 74–86. [CrossRef]
10. Brouwer, M. *Het Ene Model Is Het Andere Niet, Zes Rekenmodellen Voor de Energietransitie in de Gebouwde Omgeving Onderzocht*; Provincie Zuid-Holland: Den Haag, The Netherlands, 2019; p. 44.
11. Herbst, A.; Toro, F.; Reitze, F.; Jochem, E. Introduction to Energy Systems Modelling. *Swiss J. Econ. Stat.* **2012**, *148*, 111–135. [CrossRef]
12. van Beeck, N.M.J.P. Classification of Energy Models. *FEW Res. Memo.* **1999**, *77*, 26.
13. Androutsopoulou, A.; Charalabidis, Y. A Framework for Evidence Based Policy Making Combining Big Data, Dynamic Modelling and Machine Intelligence. In Proceedings of the 11th International Conference on Theory and Practice of Electronic Governance—ICEGOV '18, Galway, Ireland, 4–6 April 2018; ACM Press: Galway, Ireland, 2018; pp. 575–583.
14. Warmtenetwerk Kennissessie Participatie in de Energietransitie. Available online: https://warmtenetwerk.nl/nieuws/item/verslag-kennissessie-participatie-in-de-energietransitie/ (accessed on 5 March 2017).
15. Adam, C.; Steinebach, Y.; Knill, C. Neglected Challenges to Evidence-Based Policy-Making: The Problem of Policy Accumulation. *Policy Sci.* **2018**, *51*, 269–290. [CrossRef]
16. Janssen, M.; Helbig, N. Innovating and Changing the Policy-Cycle: Policy-Makers Be Prepared! *Gov. Inf. Q.* **2018**, *35*, S99–S105. [CrossRef]
17. Koussouris, S.; Lampathaki, F.; Kokkinakos, P.; Askounis, D.; Misuraca, G. Accelerating Policy Making 2.0: Innovation Directions and Research Perspectives as Distilled from Four Standout Cases. *Gov. Inf. Q.* **2015**, *32*, 142–153. [CrossRef]
18. Poel, M.; Schroeder, R.; Treperman, J.; Rubinstein, M.; Meyer, E.; Magieu, B.; Scholten, C.; Svetachova, M. *Data for Policy: A Study of Big Data and Other Innovative Data-Driven Approaches for Evidence-Informed Policymaking, Report about the State of the Art*; Centre for European Policy Studies: Oxford, UK, 2015; p. 120.
19. Nikolic, I.; Lukszo, Z.; Chappin, E.; Warmier, M.; Kwakkel, J.; Bots, P.; Brazier, F. Guide for Good Modelling Practice in Policy Support. *White Pap.* **2019**, *27*. [CrossRef]
20. Argyrous, G. Evidence Based Policy: Principles of Transparency and Accountability: Evidence Based Policy. *Aust. J. Public Adm.* **2012**, *71*, 457–468. [CrossRef]
21. Erker, S.; Lichtenwoehrer, P.; Zach, F.; Stoeglehner, G. Interdisciplinary Decision Support Model for Grid-Bound Heat Supply Systems in Urban Areas. *Energy Sustain. Soc.* **2019**, *9*, 11. [CrossRef]
22. Busch, J.; Roelich, K.; Bale, C.S.E.; Knoeri, C. Scaling up Local Energy Infrastructure; An Agent-Based Model of the Emergence of District Heating Networks. *Energy Policy* **2017**, *100*, 170–180. [CrossRef]
23. Maya Sopha, B.; Klöckner, C.A.; Hertwich, E.G. Exploring Policy Options for a Transition to Sustainable Heating System Diffusion Using an Agent-Based Simulation. *Energy Policy* **2011**, *39*, 2722–2729. [CrossRef]
24. Kerimray, A.; Suleimenov, B.; De Miglio, R.; Rojas-Solórzano, L.; Amouei Torkmahalleh, M.; Gallachóir, B.P.Ó. Investigating the Energy Transition to a Coal Free Residential Sector in Kazakhstan Using a Regionally Disaggregated Energy Systems Model. *J. Clean. Prod.* **2018**, *196*, 1532–1548. [CrossRef]
25. Venturini, G.; Pizarro-Alonso, A.; Münster, M. How to Maximise the Value of Residual Biomass Resources: The Case of Straw in Denmark. *Appl. Energy* **2019**, *250*, 369–388. [CrossRef]
26. Sarbassov, Y.; Kerimray, A.; Tokmurzin, D.; Tosato, G.; De Miglio, R. Electricity and Heating System in Kazakhstan: Exploring Energy Efficiency Improvement Paths. *Energy Policy* **2013**, *60*, 431–444. [CrossRef]
27. Novikova, A.; Csoknyai, T.; Szalay, Z. Low Carbon Scenarios for Higher Thermal Comfort in the Residential Building Sector of South Eastern Europe. *Energy Effic.* **2018**, *11*, 845–875. [CrossRef]
28. Sakellaris, K.; Canton, J.; Zafeiratou, E.; Fournié, L. METIS—An Energy Modelling Tool to Support Transparent Policy Making. *Energy Strateg. Rev.* **2018**, *22*, 127–135. [CrossRef]
29. Siraganyan, K.; Perera, A.; Scartezzini, J.-L.; Mauree, D. Eco-Sim: A Parametric Tool to Evaluate the Environmental and Economic Feasibility of Decentralized Energy Systems. *Energies* **2019**, *12*, 776. [CrossRef]

30. Sousa, G.; Jones, B.M.; Mirzaei, P.A.; Robinson, D. A Review and Critique of UK Housing Stock Energy Models, Modelling Approaches and Data Sources. *Energy Build.* **2017**, *151*, 66–80. [CrossRef]
31. Nakata, T.; Kubo, K.; Lamont, A. Design for Renewable Energy Systems with Application to Rural Areas in Japan. *Energy Policy* **2005**, *33*, 209–219. [CrossRef]
32. Qadrdan, M.; Fazeli, R.; Jenkins, N.; Strbac, G.; Sansom, R. Gas and Electricity Supply Implications of Decarbonising Heat Sector in GB. *Energy* **2019**, *169*, 50–60. [CrossRef]
33. Nässén, J.; Holmberg, J. On the Potential Trade-Offs between Energy Supply and End-Use Technologies for Residential Heating. *Energy Policy* **2013**, *59*, 470–480. [CrossRef]
34. Åberg, M.; Henning, D. Optimisation of a Swedish District Heating System with Reduced Heat Demand Due to Energy Efficiency Measures in Residential Buildings. *Energy Policy* **2011**, *39*, 7839–7852. [CrossRef]
35. Zvingilaite, E.; Klinge Jacobsen, H. Heat Savings and Heat Generation Technologies: Modelling of Residential Investment Behaviour with Local Health Costs. *Energy Policy* **2015**, *77*, 31–45. [CrossRef]
36. Ziemele, J.; Gravelsins, A.; Blumberga, A.; Vigants, G.; Blumberga, D. System Dynamics Model Analysis of Pathway to 4th Generation District Heating in Latvia. *Energy* **2016**, *110*, 85–94. [CrossRef]
37. Connolly, D.; Lund, H.; Mathiesen, B.V.; Werner, S.; Möller, B.; Persson, U.; Boermans, T.; Trier, D.; Østergaard, P.A.; Nielsen, S. Heat Roadmap Europe: Combining District Heating with Heat Savings to Decarbonise the EU Energy System. *Energy Policy* **2014**, *65*, 475–489. [CrossRef]
38. Li, F.G.N.; Trutnevyte, E.; Strachan, N. A Review of Socio-Technical Energy Transition (STET) Models. *Technol. Forecast. Soc. Change* **2015**, *100*, 290–305. [CrossRef]
39. Li, F.G.N.; Strachan, N. Take Me to Your Leader: Using Socio-Technical Energy Transitions (STET) Modelling to Explore the Role of Actors in Decarbonisation Pathways. *Energy Res. Soc. Sci.* **2019**, *51*, 67–81. [CrossRef]
40. Calderón, C.; Underwood, C.; Yi, J.; Mcloughlin, A.; Williams, B. An Area-Based Modelling Approach for Planning Heating Electrification. *Energy Policy* **2019**, *131*, 262–280. [CrossRef]
41. Fu, M.; Kelly, J.A.; Clinch, J.P. Residential Solid Fuel Use: Modelling the Impacts and Policy Implications of Natural Resource Access, Temperature, Income, Gas Infrastructure and Government Regulation. *Appl. Geogr.* **2014**, *52*, 1–13. [CrossRef]
42. de Ridder, J.; Cordeiro, C.N.; van Rooijen, L.; Wildeman, I. *Verduurzaming Warmtevoorziening Met Warmtenetten, Bestuurlijk Rapport*; De Rekenkamer: Amsterdam, The Netherlands, 2019; p. 32.
43. Diran, D.; van Veenstra, A.F.; Brus, C.; Geerdink, T. *Data voor de Transitievisie Warmte en Wijkuitvoeringsplannen*; TNO: Den Haag, The Netherlands, 2020; p. 35.
44. Diran, D.; Hoppe, T.; Ubacht, J.; Slob, A.; Blok, K. A Data Ecosystem for Data-Driven Thermal Energy Transition: Reflection on Current Practice and Suggestions for Re-Design. *Energies* **2020**, *13*, 444. [CrossRef]
45. Expertgroep Energietransitie Rekenmodellen Gebruik van Rekenmodellen in Een Proces Voor Regionale Energiestrategie. Available online: https://www.netbeheernederland.nl/_upload/Files/Rekenmodellen_21_4f58f0d451.pdf (accessed on 12 March 2020).
46. Yin, R.K. *Case Study Research and Applications: Design and Methods*, 6th ed.; SAGE: Los Angeles, CA, USA, 2018; ISBN 978-1-5063-3616-9.
47. Atlas.ti, Atlas.ti 8 Qualititative Data Analysis. 2016. Available online: https://atlasti.com/ (accessed on 25 April 2020).
48. CBS, Bevolking Op 1 Januari En Gemiddeld; Geslagcht, Leeftijd En Regio, Statline. Available online: https://opendata.cbs.nl/statline/#/CBS/nl/dataset/03759ned/table?ts=1586943147971 (accessed on 15 April 2020).
49. Gemeente Loppersum, Projectplan Aanvraag proeftuin aardgasvrije wijken, Rijksoverheid. Available online: https://www.rijksoverheid.nl/binaries/rijksoverheid/documenten/convenanten/2019/02/26/loppersum---convenant-over-grootschalige-proeftuin-met-een-aardgasvrije-wijk/3Uitvoeringsplan.pdf (accessed on 30 March 2020).
50. Gemeente Tytsjerksteradiel, Aardgasvrij Garyp energietransitie naar gasvrij wonen, Rijksoverheid. Available online: https://www.rijksoverheid.nl/binaries/rijksoverheid/documenten/convenanten/2019/01/29/tytsjerksteradiel---convenant-over-grootschalige-proeftuin-met-een-aardgasvrije-wijk/2Uitvoeringsplan.pdf (accessed on 30 March 2020).
51. Gemeente Assen, Uitvoeringsplan Aardgasvrije Wijk Assen, Rijksoverheid. Available online: https://www.rijksoverheid.nl/binaries/rijksoverheid/documenten/convenanten/2019/01/29/assen---convenant-over-grootschalige-proeftuin-met-een-aardgasvrije-wijk/3Uitvoeringsplan.pdf (accessed on 30 March 2020).
52. Gemeente Noordoostpolder, Aanvraag Aardgasvrije Wijken "Nagele in Balans". Available online: https://www.rijksoverheid.nl/binaries/rijksoverheid/documenten/convenanten/2019/02/26/noordoostpolder---convenant-over-grootschalige-proeftuin-met-een-aardgasvrije-wijk/1Aanvraag.pdf (accessed on 30 March 2020).
53. Gemeente katwijk, Uitvoeringsplan Smartpolder/Wijk Hoornes Aardgasvrij. Available online: https://www.rijksoverheid.nl/binaries/rijksoverheid/documenten/convenanten/2019/03/21/katwijk---convenant-over-grootschalige-proeftuin-met-een-aardgasvrije-wijk-kopie-3/2Bijlage+1+Uitvoeringsplan.pdf (accessed on 30 March 2020).
54. Gemeente Rotterdam, Uitvoeringsplan Proeftuin Rotterdam-Zuid Aardgasvrij. Available online: https://www.rijksoverheid.nl/binaries/rijksoverheid/documenten/convenanten/2019/02/26/rotterdam---convenant-over-grootschalige-proeftuin-met-een-aardgasvrije-wijk/3Uitvoeringsplan.pdf (accessed on 30 March 2020).
55. Stedin. Infrastructrurele Footprint, Toelichting op de Methodiek en Rekentool. 2017, p. 14. Available online: https://www.stedin.net/-/media/project/online/files/zakelijk/infrastructurele-footprint/infrastructurele-footprint.pdf (accessed on 5 April 2020).

56. Gemeente Eindhoven, Uitvoeringsplan Aardgasvrije Wijk't Ven. Available online: https://www.rijksoverheid.nl/binaries/rijksoverheid/documenten/convenanten/2019/05/22/eindhoven---aanvraag-en-convenant-proeftuin-aardgasvrij-wijk/2Uitvoeringsplan.pdf (accessed on 30 March 2020).
57. Gemeente Brunssum, Aanvraag Proeftuin Aardgasvrije Wijk Gemeente Brunssum. Available online: https://www.rijksoverheid.nl/binaries/rijksoverheid/documenten/convenanten/2019/03/21/brunssum---convenant-over-grootschalige-proeftuin-met-een-aardgasvrije-wijk-kopie-2/4+Aanvraag+proeftuin+en+Samenvatting+uitvoeringsplan.pdf (accessed on 30 March 2020).
58. Gemeente Middelburg, Dauwendael: Proeftuin voor toepassing van duurzame restwarmte in Middelburg. Available online: https://www.rijksoverheid.nl/binaries/rijksoverheid/documenten/convenanten/2019/01/29/middelburg---convenant-over-grootschalige-proeftuin-met-een-aardgasvrije-wijk/1Aanvraag.pdf (accessed on 30 March 2020).
59. Friese, S.; Soratto, J.; Pires, D. Carrying out a Computer-Aided Thematic Content Analysis with ATLAS.Ti. Available online: www.mmg.mpg.de/workingpapers (accessed on 31 August 2020).
60. Manfren, M.; Nastasi, B.; Groppi, D.; Astiaso Garcia, D. Open Data and Energy Analytics—An Analysis of Essential Information for Energy System Planning, Design and Operation. *Energy* **2020**, *213*, 118803. [CrossRef]
61. Nikolic, I.; Warnier, M.E.; Kwakkel, J.H.; Chappin, E.J.L.; Lukszo, Z.; Brazier, F.M.; Verbraeck, A.; Cvetkovic, M.; Palensky, P. *Multi-Model Ecology and Interface for Transition Models*; Delft University of Technology: Delft, The Netherlands, 2019; p. 21.
62. Gemeentelijke Transitievisie Warmte. Available online: https://www.pbl.nl/publicaties/achtergrondrapport-bij-de-startanalyse-aardgasvrije-buurten (accessed on 16 August 2020).
63. ECW. *Handreiking Voor Lokale Analyse*; ECW: Den Haag, The Netherlands, 2019; p. 32.
64. Zwamborn, A. Mondaine Connecting the Future. Available online: https://www.mondaine-suite.nl/ (accessed on 28 May 2020).
65. PAW Wat is een Transitievisie Warmte. Available online: https://www.aardgasvrijewijken.nl/klp/ro/transitievisie+warmte2/wat+is+een+transitievisie+warmte/default.aspx (accessed on 28 March 2020).

energies

MDPI

Review

Energy Modelling and Analytics in the Built Environment—A Review of Their Role for Energy Transitions in the Construction Sector

Massimiliano Manfren [1], Maurizio Sibilla [2,*] and Lamberto Tronchin [3]

[1] Faculty of Engineering and Physical Sciences, University of Southampton, Boldrewood Innovation Campus, Burgess Rd, Southampton SO16 7QF, UK; M.Manfren@soton.ac.uk
[2] School of the Built Environment, Oxford Brookes University, Headington Campus, Oxford OX3 0BE, UK
[3] Department of Architecture (DA), University of Bologna, Viale Europa 596, 47521 Cesena, Italy; lamberto.tronchin@unibo.it
* Correspondence: msibilla@brookes.ac.uk

Abstract: Decarbonisation and efficiency goals set as a response to global warming issue require appropriate decision-making strategies to promote an effective and timely change in energy systems. Conceptualization of change is a relevant part of energy transitions research today, which aims at enabling radical shifts compatible with societal functions and market mechanisms. In this framework, construction sector can play a relevant role because of its energy and environmental impact. There is, however, the need to move from general instances to specific actions. Open data and open science, digitalization and building data interoperability, together with innovative business models could represent enabling factors to accelerate the process of change. For this reason, built environment research has to address the co-evolution of technologies and human behaviour and the analytical methods used for this purpose should be empirically grounded, transparent, scalable and consistent across different temporal/spatial scales of analysis. These features could potentially enable the emergence of "ecosystems" of applications that, in turn, could translate into projects, products and services for energy transitions in the built environment, proposing innovative business models that can stimulate market competitiveness. For these reasons, in this paper we organize our analysis according to three levels, from general concepts to specific issues. In the first level, we consider the role of building energy modelling at multiple scales. In the second level, we focus on harmonization of methods for energy performance analysis. Finally, in the third level, we consider emerging concepts such as energy flexibility and occupant-centric energy modelling, considering their relation to monitoring systems and automation. The goal of this research is to evaluate the current state of the art and identify key concepts that can encourage further research, addressing both human and technological factors that influence energy performance of buildings.

Keywords: energy transitions; energy modelling; energy analytics; data-driven methods; building performance analysis energy efficiency; energy flexibility; occupant-centric design; open energy data

Citation: Manfren, M.; Sibilla, M.; Tronchin, L. Energy Modelling and Analytics in the Built Environment—A Review of Their Role for Energy Transitions in the Construction Sector. *Energies* **2021**, *14*, 679. https://doi.org/10.3390/en14030679

Academic Editor: Luisa F. Cabeza
Received: 15 December 2020
Accepted: 25 January 2021
Published: 28 January 2021

1. Introduction

In recent years, a notable research effort has been devoted to the conceptualisation of sustainability transitions [1] and, more specifically for energy, to the identification of "complementarities" at multiple levels [2,3]. Transition processes embody the necessity of radical-shifts and they represent an opportunity for innovation and entrepreneurship [4], with a clear focus on issues such as global warming and decarbonisation of energy systems [5]. In these innovation processes, the role of intermediaries and strategic niches appears to be crucial. In fact, understanding how actors can control and accelerate the energy transition is a key issue for research today [6] and intermediaries can play a fundamental role in this direction [7]. Intermediaries (i.e., public, non-profit, and private third-parties [8]) are actors which facilitate relations between key actors and enable knowledge sharing and pooling [9].The opportunities for the construction industry in this sense are relevant, because of the impact of built environment in terms of raw resources, energy and carbon emissions [10], but also because of the potential to exploit innovative technologies within emerging paradigms such as circular economy [11]. There is, however, the need to move from general instances to specific actions. These actions have to enable radical shifts compatible with societal functions and market mechanisms; for this reason, in this research we focus on energy modelling and analytics that can provide critical insights in this sense. At present, it possible to identify multiple enabling factors for radical shifts and acceleration of the process of change. First, the evolution of practices focused on concepts such as open data, open innovation, open science [12–14] and, in particular, open energy modelling principles [15,16]. Second, advances in building data interoperability (technical, informational and organizational) [17] and data availability at multiple levels, using technologies such as the Internet of Things (IoT) [18–20] and cyber-physical systems [21], which can enable, in turn, innovation in end-user energy delivery [22], and in energy infrastructures [23]. Third, the increasing decentralization of energy systems where the co-evolution of built environment and energy infrastructures [24] plays a fundamental role, that can be investigated by means of "soft-linking" of energy modelling approaches, from planning to operation [25]. Finally, innovative business models proposing concepts such as prosumer [26] and prosumager [27], which are determining changes in the way energy market works and energy trading takes place, for example using Peer-to-Peer automated exchange mechanisms, exploiting Blockchain technologies [28].

In this rapidly evolving framework, research aimed at radical changes in energy systems and built environment needs to consider the enabling factors reported above and to acknowledge the limitations and bottlenecks in view of energy efficiency and carbon reduction goals. The aim of this paper is to discuss to what extent and in what ways energy modelling and analytics can support the process of change for energy transitions in the construction sector. In Section 2 we illustrate the background of the research, explaining the fundamental elements that motivate it.

2. Background and Motivation

Energy transitions involve the transformation of the network of players and organisations traditionally working in the energy sector (e.g., policy-makers, regulators, transmission and distribution authorities, etc.) as well as the change of the role of customers, from passive to active (i.e., prosumers [26] and prosumagers [27]).In fact, socio-technical innovations are critically dependent on the possibility to access new information, knowledge and resources, which are key enablers for the development of innovative products and services [29], within a market mechanism. Construction sector can be conceptualized, for example, by considering three fundamental domains [30]: project, product and service. All these domains are going to be deeply influenced by socio-technical changes in energy transitions, which will transform the way buildings are designed, built and managed. Sharing knowledge among actors is crucial when addressing building energy performance in a comprehensive way, considering both human and technical factors [31]. In fact, the impact of occupants has to be considered from multiple stand-points [32] and users' behaviour can determine both "re-bound" [33] and "pre-bound" effects [34,35], that can create a substantial difference between expected and measured performance, which can be inscribed in the general category of "performance gaps" [36–38]. A "performance gap" can be found in all the stages of building life cycle [39] and the use of standardized assumption in modelling, e.g., to create Energy Performance Certificates, has to be critically questioned when using them to estimate actual energy consumption and potential savings [40].

Additionally, the dynamic interaction between building and energy infrastructures [41,42] has to be considered as well for multiple reasons (e.g., operational constraints, limitations of the penetration of renewables, innovative business model for the electricity market, etc.) and in light of possible developments in terms of "soft-linking" of energy models [25].

Finally, considering building performance from a whole life cycle perspective (indeed critical for emerging paradigms such as circular economy [11]), embodied energy in materials, technologies and processes represents another potential "performance gap" to be considered [43,44]. In fact, all these potential gaps create risks and lack of credibility when investing in energy efficiency and sustainability measures. Therefore, monitoring, verifying and tracking performance (i.e., energy, emission and cost in particular) using robust, transparent and empirically grounded methods is essential to evaluate the effectiveness of measures and share knowledge regarding practices. This, in turn, can contribute to investment de-risking and stimulate the growth of business "ecosystems" in energy and sustainability transitions, particularly for the construction sector. Additionally, the co-benefits of energy efficiency measures (e.g., improved indoor environmental quality, health, productivity, pollution reduction, etc.) [45] have to be considered both by policy makers and investors, to weight properly cost and benefits. Following the general trend towards open science, briefly outlined in Section 1, the research community in the energy field has stressed in recent years the fundamental importance of open energy data and models [46,47] and we can envisage an evolution towards systems of model [48] designed to address key problems in energy transitions, eventually taking advantage of "soft-linking" approaches [25,49]. Rather than being designed for separate applications, models can be potentially conceived and work like "ecosystems" [48] of interconnected applications, based on open data and modelling standard [46] where the researchers are opening their modelling "black-boxes" [47]. Indeed, transparent and robust models can become part of innovative business strategies, leading to techno-economically feasible pathways in transitions(thereby enabling a radical change to happen in practice). In fact, this review is part of a more extensive research work focused on "Buildings-as-Energy-Service" concept, in which separate literature reviews were conducted to explore both social and physical science perspectives on this topic. The concepts emerging from the reviews represent the basic elements of a Cognitive Mapping [50] process. The aim of this process is to create an inter-disciplinary research environment (a cognitive framework) [51] that is essential for innovation processes, where creativity is stimulated by the participation of user in the process of knowledge creation and sharing [52]. In Section 3 we describe the research methodology used to identify the role of energy modelling and analytical techniques in relation to the issues mentioned above.

3. Research Methodology

Considering the issues briefly outlined in Sections 1 and 2, the objective of this review study is to identify and analyse the features of energy modelling and analytical techniques that could be enabling factors in energy transition processes. The two fundamental research questions posed in this study are the following. First, what are the modelling techniques that can meet the criteria that will be described later in this section? Second, what are the essential characteristics (of modelling approaches) that can contribute to reduce the level of fragmentation of knowledge? The modelling framework proposed as outcome of the research attempts to reduce the level of fragmentation of the highly diversified body of knowledge available and to help in the conceptualization of processes of change (energy transition) by identifying opportunities, together with limitations and bottlenecks.

In this research both qualitative and quantitative data are analysed and it is therefore a "mixed approach" [53]. For this reason, we used concepts from Grounded Theory [54] as a reference for our research, in which both qualitative and quantitative data are utilised ("all is data" [55]). In brief, Grounded Theory (GT) can be defined as a "a set of integrated conceptual hypotheses systematically generated to produce an inductive theory about a substantive area" [56] and as "theory that was derived from data, systematically gathered and analysed through the research process" [57]. The results of a GT study are "a set of concepts, related to each other in an interrelated whole" [58].

The limitations of such approach depend on the fact that the selection in literature sampling depend on the subjective judgment (point of view) of the researcher and cannot stand outside of it [58]. However, the process can become more transparent and reproducible by stating the steps and the criteria used in it. In this research, we followed seven steps:

(1) Definition of knowledge domains of interest;
(2) Stratified search using domain and keywords in Web of Science database (WoS);
(3) Initial selection of pertinent literature on WoS;
(4) Definition of additional criteria for inclusion/exclusion of literature;
(5) Initial verification of literature using title, keywords and abstract;
(6) Final selection of literature;
(7) Detailed analysis of literature.

The fundamental knowledge domain of interest is "Building Energy Performance" (step 1) and the keywords considered initially are "Building stock", "Uncertainty" and "Flexibility" (step 2), to address fundamental topics in research. "Building stock" is chosen to identify examples of building energy modelling at multiple scales (e.g., for planning and policy, utility scale studies, etc.). "Uncertainty" is chosen to identify studies that analyse the critical dimension of energy performance uncertainty, which may create risks and lack of credibility for efficiency practices, starting from fundamental principles in Measurement and Verification (M&V) and Monitoring & Targeting (M&T). "Flexibility" is chosen to identify research regarding the interaction between building and infrastructures, which is strictly related to their technological co-evolution. The results obtained in step 2 are summarized in Table 1.

Table 1. Knowledge domain, keywords and criteria for literature selection.

Domain of Interest	Domain and Keywords	Sources in WoS Database	Sources in Categories: Architecture Construction Planning	Motivation for Criteria Selection	Source in Final Selection
"Building Energy Performance"	"Building Energy Performance" AND "Building stock"	1335	870	Building energy modelling for energy planning and policy targets, utility scale analysis, parametric building performance studies.	52
	"Building Energy Performance" AND "Uncertainty"	1551	705	Methods based on M&V and M&T principles that can help tracking energy performance transparently (and reducing uncertainty) and that can be applied at multiple temporal and spatial scales.	123
	"Building Energy Performance" AND "Flexibility"	1027	237	Strategies to control buildings and enhance their energy flexibility strategies in relation to energy demand in end-uses and user behaviour.	68

In order to obtain the final literature selection, additional criteria have been introduced and re-sampling of literature has been conducted iteratively until "theoretical saturation" was reached. Theoretical saturation term indicates "the phase of qualitative data analysis in which the researcher has continued sampling and analysing data until no new data appear and all concepts of the theory are well-developed and their linkages to other concepts are clearly described" [59].The criteria used in re-sampling have been summarized and motivated in Table 2. They are derived from previous research in the area of energy modelling [24,60] and consider the general trends towards the use of open data for energy research [46] and the necessity to increase of transparency in energy modelling [47]. In other words, the criteria introduced represent, in our opinion, limiting factors and constraints for the creation of "ecosystems" of models [48], which are briefly outlined in Section 2.

Table 2. Additional criteria introduced for energy modelling literature selection.

Criteria	Description	Motivation for Criteria Selection
Empirical Grounding	Based on empirical data, and tested on a relevant number of cases.	Reducing risk of investment in energy transitions and ensure the credibility of policies by means of evidence.
Harmonization	Methodologies in which redundancies and overlapping features are removed, ideally based on protocols and standard.	Avoid redundancy, multiplication of efforts and unnecessary increase of complexity of procedures. Streamline the implementation of models and procedures.
Scalability	Capability of analysing problems at multiple temporal and spatial scales.	Ability to work coherently and consistently on multiple temporal and spatial scales.
Interpretability	Ability to detect relevant cause-effect relationship, ideally combing statistical analysis techniques with physical understanding of phenomena.	Physical interpretation can help extract insights that are fundamental for the continuous improvements of processes and technologies.
Re-configurability	Able to be used in multiple stages of the building life-cycle, for example for design and operation, sharing similar underlying principles.	Creating a certain degree of continuity in the data analysis workflow during the life-cycle of projects.

In Section 4 the results of the review process are presented, structuring them according to three levels of analysis (related to the domain and keyword chosen, as explained before in this section) that correspond to the development, by means of iterative sampling, of the key concepts reported in Table 1. The overall research process is synthesized graphically in Figure 1.

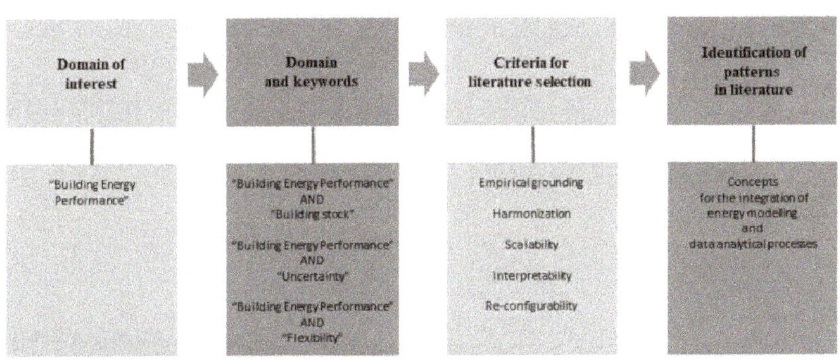

Figure 1. Diagram synthesizing the research process.

4. Results and Discussion

In this section we discuss how energy modelling and analytical tools could support energy transition processes for the construction industry, highlighting relevant insights for research across the three levels of analysis introduced in Section 3. The three levels proposed are indeed a strategy to perform a decomposition of the problem, going from general principles to specific issues that are emerging within the research framework. In Section 4.1 we analyse the topic of building energy performance analysis at multiple scales and its implications (e.g., in energy and planning policy, utility scale studies, etc.), which introduces the issues at general level (first level of analysis). In Section 4.2 we present harmonized methodologies (based on M&V principles and considering possible extensions) to analyse energy performance in buildings and we synthesize their characteristics (second level of analysis). Finally, in Section 4.3, we introduce innovative topics such as energy flexibility (infrastructures' interaction) and occupant-centric (users' interaction) energy modelling, which will contribute to redefine how buildings are actually designed and operated in the future (third level of analysis). Overall, throughout these three levels we show how many of the ongoing research developments are deeply related to the fundamental elements that motivate our research and are described in Section 2.

4.1. Building Energy Performance Analysis at Multiple Scales

Comprehensive reviews of building energy models have been published in recent years [61–63] and, while energy performance is particularly relevant, more comprehensive approaches to building performance analysis [64] are crucial for the evolution of the building sector. As anticipated, the analysis of building energy performance requires an understanding of both human and technical factors [31], and this confirms the inherent socio-technical dimension of energy modelling and analytics. It is therefore necessary to structure energy performance analysis with respect to both human and technical factors. In turn, this is important, for example, to address properly the gap between design and measured performance, i.e., the performance gap [36–38], introduced in Section 2. Further, the concept of statistical "Reference Buildings" [65] (RB) must be introduced to enable building performance benchmarking at multiple scales. RB models represent the common typologies, technologies and end-uses in the building stock, identified through statistical analysis and expert knowledge (e.g., on building technologies, types of end-uses, user behaviour, etc.) on a large-scale base. Building data are usually multi-level data, which makes it difficult to access the full information needed to describe in detail the performance of building stock. However, building energy modelling data can be organised in a hierarchical and standardized way; examples in this sense can be found at the EU level in the legislation on the definition of cost-optimal performance levels [66] and in EU Building Stock Observatory [67]. Further, in the US, technical standardisation has been tested with the definition of RB models [68,69], accounting also for the costs of various technological options [70].The role of energy modelling cycles and the importance of the level of detail (from conceptual to final design) are considered by the standard ASHRAE 209 [71]. Additionally, the use of hierarchical structures in datasets for building energy modelling can be found, for example, in performance gap studies [37], in the analysis of impact of automation systems [72], and in occupancy modelling [73].Further, with respect to building energy model calibration on measured data, we can find examples using multi-level data [74] and exploiting macro-parameters [75] (i.e., lumped quantities) to facilitate and guide the uncertainty and sensitivity analysis, together with the use of archetypes [76] (i.e., RB for a certain construction typology), and of additional information such as monitored internal temperature profiles [77].At the state of the art, multiple modelling options are available, depending on the scope of the analysis process, which range from physics based ("law driven") "white-box" models to statistics and machine leaning based ("data driven") "black-box" models. An analysis of the suitability of the different modelling strategies has been proposed by Koulamas et al. [78] and, more specifically for model calibration, by Manfren et al. [79]. Indeed, it is possible to use models to simulate performance (forward modelling) and to estimate model inputs from measured performance (inverse modelling) in multiple ways. Therefore, using forward and inverse modelling techniques [24] in a synergic way for calibration purposes is crucial. In this context, advanced techniques such as Bayesian analysis can help reconstructing built stock data under uncertainty [80–82], using probabilistic ranges for the model input parameters. The possibility to benchmark building performance on a large scale base [83,84] can increase the effectiveness of policies and can guarantee better decision-making processes, not only for policy makers but for multiple stakeholders (e.g., designers, energy managers, investors, etc.). In fact, the progressive convergence of bottom-up and top-down perspectives in energy modelling and planning for building stock [61] can contribute to the development of "soft-linking" approaches between various types of models [25] and, consequently, ensure consistency of actions in transition processes at multiple levels. Overall, a systematic statistical approach to building performance analysis [85] can be crucial to the evolution of design and operation paradigms for building stock. In recent years we assisted to an increasing commitment towards energy efficiency in buildings which led to the definition of paradigms such as Passive House [86], NZEB [87,88], and PEB [89], considering just the most relevant. Indeed, the possibility to deploy these paradigms at scale is subject to technical and economic constraints. In this sense, the use of statistical "Reference Buildings" can support techno-economic optimization studies [65,90], utility scale analysis of design [91] and operation of buildings [92] and energy planning at national scale [68–70], where innovative building paradigms are proposed and implemented. In terms of computation, the necessity of performing parametric (or probabilistic) simulation studies [93–95] is emerging and the algorithmic definition of simplified building models [96–98] can be exploited for building stock modelling at city scale [99–101] and regional scale [102]. In Table 3 we synthesize the outcomes of literature analysis regarding building energy performance analysis at multiple scales, highlight the main target of the different studies and their scale of analysis, namely national, regional, urban and stock. The latter indicates, in general, studies that are proposing building performance analysis on multiple typologies and end-uses.

Table 3. Building energy performance analysis—Target and spatial scale of analysis.

Source	Year	Target of Analysis			Spatial Scale of Analysis			
		Energy Planning and Policy	Utility Level Study	Parametric Building Analysis	National	Regional	Urban	Stock
Deru et al. [68]	2011	✔			✔			
Thornton et al. [70]	2011	✔			✔			
Goel et al. [69]	2011	✔			✔			
Ballarini et al. [102]	2017	✔				✔		
Delmastro et al. [99]	2016	✔					✔	
Ghiassi et al. [100]	2017	✔					✔	
Delmastro et al. [101]	2020	✔					✔	
Goel et al. [91]	2018		✔					✔
Meng et al. [92]	2017		✔					✔
Pernigotto et al. [96]	2014			✔				✔
Dogan et al. [97]	2016			✔				✔
Dogan et al. [98]	2016			✔				✔
Goel et al. [103]	2016			✔				✔
Badiei et al. [104]	2019			✔				✔

The examples reported before are clearly not exhaustive but they are used to illustrate the potential role of building energy performance analysis at large scale, using modelling methods that are transparent and reproducible, build upon (or compatible with) technical standardization. These topics are developed further in Section 4.2, consider two fundamental dimensions: the quantification of the impact of energy efficiency measures and the ability model dynamic behaviour (i.e., load profiles). Finally, at the beginning of this

Section we stressed the importance of a precise hierarchy for multi-level building energy modelling data. Another important aspect is that of "vertical integration" of information in energy modelling, from user up to infrastructures (e.g., user, individual spaces within the room, individual rooms, building zones, whole building, meter, energy infrastructure). Examples of research in this direction can be found in IEA Annexes on "Energy Flexibility in Buildings" [105] and "Occupant-Centric Building Design and Operation" [106]. These fundamental aspects of current research are discussed more in detail in Section 4.3.

4.2. Harmonizing Methodologies to Analyse Energy Performance

Appropriate spatial and temporal resolution of data is necessary to track building energy performance at multiple scales and energy metering data constitute, of course, the basic information layer. There is the need for harmonized methods that can ensure robust evidence (empirically grounded and validated) for efficiency measures (not only for research, but also for policy), by means of reliable statistics regarding the actual impact of efficient technologies [107,108] and especially by means of performance benchmarking of efficiency measures [109,110]. The term "harmonized" is used here to indicate, in general, methodologies in which redundancies and overlapping features are removed; harmonized methods can help documenting performance transparently, for example by tracking evidence of energy efficiency savings (and also related carbon and cost savings) in time and detecting the impact of influencing factors. Measurement and Verification (M&V) protocols [111,112] and methods represent the backbone in this sense and important research initiatives have been conducted in recent years to enhance and extend their applicability, such as the Uniform Methods Project (UMP) and other related projects [109,110,113]. The goal of these projects was harmonising the methods for the quantification of energy savings for different efficiency measures, both in residential and commercial buildings. Multiple measures (technologies) are included (HVAC, HP/chillers, CHP, lighting, envelope, variable-frequency drives, etc.). Another important project, focused on de-risking investment in energy efficiency, is the Investor Confidence Project (ICP) [114]. As already mentioned, the methods used in these projects represent an extension of the ones that can be found in M&V protocols [111,112] and technical standards [115–117], in which thresholds (expressed as statistical KPIs, representing the "goodness of fit") are given for the acceptability of models as calibrated [118] on measured data. Finally, open software is available [113,119,120] as a basis for further development that can potentially be enabled by open science principles (i.e., transparency and reproducibility of results, among others).

In general, these approaches are based on energy interval data (dependent variable) and weather data (independent variables) along with other independent variables (e.g., dummy variables for models of various occupancy and operational regimes) which can be derived from contextual knowledge and information. Instead of using energy data directly, it is possible to use the energy signature [115], which is the average power over the number of hours of operation in the interval considered. The most important independent variable for weather normalization of energy consumption is outdoor air temperature [121,122] and these methods are affine to variable-base degree days methods [92,123]. Temperature response methods are reviewed by Fazeli et al. [124]. Conceptual simplicity is one of their advantages (among others), compared to other meta-modelling techniques [125,126]. Automated model selection techniques [119,127] can be applied as well to compare the performance of multiple modelling options, using statistical KPIs representing their "goodness of fit". From an analytical perspective, it is important to be able to connect both the design and the operation phase analysis [128,129] in order to ensure consistency in the use of energy performance analysis techniques over the different phases of the life cycle [130]. In this way reliable limits for performance measured or estimated [131] can be produced and used against benchmarks, allowing a continuous improvement process (i.e., Plan Do Check Act is one of the key principles of Energy Management Systems [132]).

Far from being merely instruments for weather normalisation of energy use (i.e., outdoor temperature dependence), harmonised approaches can also help modelling dynamic loads (e.g., demand response) [109], ideally clustering operating conditions for typical profiles [133–135] to obtain specific insights on recurrent operating schedules (e.g., depending on the type of end-use).

In reality, understanding load dynamics at multiple scales is crucial for providing accurate estimates of the impact of flexibility measures that can inform policy [136] by creating a "soft link" between modelling approaches. Load modelling techniques can be used to complement "traditional" optimization approaches in cases where they are no longer sufficient and several operational configurations need to be studied [137]. Furthermore, the possibility of evaluating the thermal, electrical and fuel requirements with harmonised methods can extend further the principle of "soft-linking" of energy models in multi-commodity systems [138–142]. In this sense, harmonised methods should complement (in terms of general principles) open science-based approaches to energy research [16] because of their transparency. In addition, they may help to address related issues such as energy demand forecasts in future climate change scenarios [143–145] and definition of load profiles evolution due to efficiency measures and behavioural change, which are fundamental for optimizing decentralised energy systems in buildings [146] and communities [138,147,148].

In short, harmonised approaches can be used to discuss two main aspects of energy modelling research in a rigorous and transparent manner: the quantification of the effect of energy efficiency measures and the reconstruction of dynamic behaviour (i.e., time series modelling), such as load profiles analysis. Table 4 below provides a comparison of the main features of regression-based modelling methods that can meet the constraints set out in Section 3. We consider different types of end-uses, namely residential and non-residential, and different types of energy services, namely heating, cooling, domestic hot water (DHW), and appliances. First of all, the selected and reviewed literature reflects, in large part, empirically based studies in which the authors used operation phase data. The research is performed in all cases using regression-based (interpretable) methods that are significantly consistent with the harmonisation and standardisation principles outlined in this section. In terms of temporal scalability, the papers are categorised with respect to monthly, daily and hourly data. In certain cases, sub-hourly data are used, but we classify them as hourly data since this is the highest resolution considered by the model calibration thresholds proposed in the standards and protocols [118]; in any case, this resolution is adequate to capture the essence of building dynamic energy behaviour. In terms of spatial scalability, we consider building subsystems (building fabric and technological systems), building as a whole, building stock, and community and city scale. For the latter, the term design corresponds substantially to planning; the operational phase data are used as a basis for making accurate forecasts for the future. In addition, whole building energy balance is used in most situations, although in some cases (e.g., evaluation of building fabric characteristics) the energy balance at the zone or room level is used. Finally, with the term approximate physical approximation, we suggest the possibility of using regression coefficients to estimate physical quantities. Overall, the table illustrates how harmonized/standardized regression-based methods can cover several temporal and spatial scales of analysis and how they can theoretically combine design and operational phase performance analysis into the same analytical workflow (thereby satisfying re-configurability criteria, reported in Table 2). Finally, regression models can be used for both residential and non-residential end-uses to study energy services (heating, cooling, DHW, appliances) in multiple ways and can provide insights up to building system level when sub-metering data (e.g., thermal, electric) are available, while enabling, at the same time, the aggregation of results on a large scale base for building stock modelling.

Table 4. Harmonized regression-based modelling approaches for building performance analysis.

Source	Year	End-Use		Energy Services				Temporal Scale			Spatial Scale					Interpretation	Phase	
		Residential	Non-residential	Heating	Cooling	DHW	Appliances	Monthly	Daily	Hourly	Building fabric	Technical systems	Whole building	Building stock	Community	Physical approximated	Design	Operation
Lammers et al. [149]	2011		✔	✔		✔	✔	✔					✔					✔
Hallinan et al. [150]	2011		✔	✔		✔		✔					✔	✔	✔			✔
Hallinan et al. [151]	2011	✔		✔	✔	✔	✔	✔			✔	✔	✔	✔	✔	✔		✔
Danov et al. [152]	2011		✔	✔			✔		✔		✔		✔			✔		✔
Masuda and Claridge [153]	2012		✔	✔	✔		✔		✔		✔		✔			✔		✔
Bynum et al. [154]	2012		✔	✔	✔		✔		✔	✔		✔	✔			✔		✔
Masuda and Claridge [121]	2014		✔	✔	✔		✔		✔	✔		✔	✔	✔		✔		✔
Paulus et al. [127]	2015		✔	✔	✔	✔	✔		✔	✔		✔	✔					✔
Lin and Claridge [122]	2015		✔	✔	✔				✔			✔	✔			✔		✔
Hitchin and Knight [155]	2016		✔		✔				✔			✔	✔			✔		✔
Jalori and Reddy [156]	2015		✔	✔	✔	✔	✔	✔	✔	✔			✔					✔
Paulus [119]	2017	✔	✔	✔	✔	✔	✔	✔	✔	✔			✔					✔
Abushakra and Paulus [157]	2016		✔	✔	✔	✔	✔			✔			✔					✔
Bauwens and Roels [158]	2014	✔		✔					✔		✔					✔		✔
Erkoreka et al. [159]	2016		✔	✔			✔		✔	✔	✔					✔		✔
Giraldo-Soto et al. [160]	2018	✔	✔	✔			✔		✔	✔	✔					✔		✔
Uriarte et al. [161]	2019		✔	✔			✔		✔	✔	✔					✔		✔
Busato et al. [162]	2012	✔		✔				✔				✔	✔			✔	✔	✔
Busato et al. [163]	2013		✔	✔				✔		✔		✔	✔			✔	✔	✔
Krese et al. [164]	2018				✔		✔			✔			✔			✔		✔

Table 4. *Cont.*

Source	Year	End-Use		Energy Services				Temporal Scale			Spatial Scale					Interpretation	Phase	
		Residential	Non-residential	Heating	Cooling	DHW	Appliances	Monthly	Daily	Hourly	Building fabric	Technical systems	Whole building	Building stock	Community	Physical approximated	Design	Operation
Sjögren et al. [165]	2009	✔		✔	✔	✔	✔	✔	✔		✔	✔	✔			✔		✔
Vesterberg et al. [166]	2014	✔		✔	✔	✔	✔	✔	✔		✔	✔	✔			✔		✔
Meng and Mourshed [92]	2017		✔	✔					✔	✔			✔	✔		✔		✔
Meng et al. [167]	2020		✔	✔					✔	✔			✔	✔				✔
Oh et al. [168]	2020	✔		✔	✔	✔	✔			✔			✔	✔				✔
Westermann et al. [169]	2020	✔		✔		✔	✔			✔		✔	✔	✔				✔
Pasichnyi et al. [170]	2019	✔	✔	✔		✔				✔		✔	✔		✔	✔	✔	✔
Qomi et al. [171]	2016	✔		✔		✔		✔		✔	✔		✔		✔	✔	✔	✔
Afshari et al. [172]	2017	✔	✔		✔	✔	✔			✔			✔		✔		✔	✔
Afshari et al. [173]	2017	✔	✔		✔	✔	✔			✔	✔		✔		✔	✔	✔	✔
Allard et al. [129]	2018	✔			✔			✔	✔				✔			✔	✔	✔
Tronchin et al. [128]	2018	✔		✔	✔	✔	✔	✔		✔	✔	✔	✔				✔	✔
Manfren and Nastasi [131]	2020	✔		✔	✔	✔	✔	✔		✔	✔	✔	✔				✔	✔
Catalina et al. [174]	2008	✔		✔				✔		✔	✔		✔			✔	✔	
Hygh et al. [175]	2012		✔	✔	✔	✔	✔			✔	✔		✔				✔	
Asadi et al. [176]	2014		✔	✔	✔	✔	✔			✔	✔	✔	✔				✔	
Al Gharably et al. [177]	2016		✔	✔	✔	✔	✔			✔	✔		✔				✔	
Ipbüker et al. [178]	2016	✔		✔						✔	✔		✔				✔	
Goel et al. [103]	2016	✔	✔	✔	✔	✔	✔	✔		✔			✔	✔		✔	✔	

The possibility to employ advanced harmonized analytical techniques could, in principles, contribute to the development of innovative business models built upon Energy Performance Contracting (EPC) [179] principles, where dynamic operational conditions are clustered [134] and multiple regression models are combined together [156] to investigate performance, integrating data at multiple spatial and temporal resolutions, while retaining an approximated physical interpretation. Further, the graphical representation of regression-based methods can be combined with other visualization strategies used for energy (and exergy) flows at multiple scales, from building systems and sub-systems [180], to networks in multi-energy systems [181]. Physical-statistical (i.e., "grey-box") formulations [158,173,182–185], can extend the inherent capabilities of these modelling approaches even further and provide additional insights that may be particularly valuable in a continuous improvement logic, while retaining scalability [183,184].

Despite the variety of possible model formulations, we believe that data-driven approaches should use energy modelling definitions and quantities that are consistent with those proposed in the current technical standardization [186] to improve the comparability of results and consistency with policy objectives, for which standardisation plays a key role. For this reason, we report hereafter in Table 5 some experimental protocols (harmonized or standardized) with examples of applications at component level and building zone level. Indeed, the table highlights the potential continuity and integration of these experimental methods to estimate thermo-physical properties of building components and zones. Ideally, they could partially overlap with methods presented in Table 4, for example by alternating short-term measurement at higher frequency with long-term measurement at lower frequency [157] during building life cycle.

Table 5. Experimental protocols and applications.

		Type of Experimental Protocol				Application		Data Acquisition	
Source	**Year**	**ISO 9869**	**Co-heating**	**QUB**	**ISABELE**	**Component**	**Zone**	**Time Interval**	**Length of Data Acquisition**
Francis et al. [187]	2015	✔				✔		Subhourly	72 h
Rasooli and Itard [188]	2018	✔				✔		Subhourly	72 h
Erkoreka et al. [159]	2016	✔					✔	Subhourly	72 h, multiple periods
Uriarte et al. [161]	2019	✔					✔	Subhourly	72 h multiple periods
Bauwens et al. [158]	2014		✔				✔	Daily	2/3 weeks
Jack et al. [107]	2017		✔				✔	Daily	2/3 weeks
Alzetto et al. [189]	2018			✔		✔		Subhourly	1 night
Meulemans [190]	2018			✔		✔		Subhourly	1 night
Ahmad et al. [191]	2019			✔		✔		Subhourly	1 night
Rémi et al. [192]	2014				✔		✔	Subhourly	5–15 days
Thébault et al. [193]	2018				✔		✔	Subhourly	4 days

In QUB and ISABELE methods, the definitions used are in line with current technical standardisation; the physical parameters are represented by lumped quantities (thus reducing the number of parameters needed) and the model formulation greatly reduces the complexity compared to a physical "white-box" model, briefly recalled in Section 4.1. "White-box" models are detailed models based on physical laws used mainly for simulations during the design process and validated in accordance with energy simulation test standards [194,195].The potential contact point between "white-box" detailed modelling and "grey-box" (physical-statistical) lumped modelling parameters can be found in multi-level building energy model calibration [74] where "macro-parameters" (aggregated, lumped quantities) [75] are used to validate more detailed models, together with additional information such as internal temperature profiles [77] and other contextual information.

Indeed, the potential advantages of "grey-box" models are that they can be derived (and verified) from the basic concepts of energy analysis [196,197], built by using highly standardised rules [188], and they can employ efficient state-space [198] and analytical formulations [199]. Examples of validation of "grey-box" models using simulation test standards at the state of the art have been published by Lundström et al. [195] and Michalak [194,200]; a "grey-box" model for the detection of thermo-physical properties by inverse modelling has been implemented also in EnergyPlus, a detailed "white-box" modelling software [201]. Juricic et al. [202] considered the effect of natural weather variability in the identification of building envelope characteristics using these model types, showing how approximately two weeks of data are sufficient to achieve adequate accuracy. Finally, Baasch et al. [203] compared the performance of different "grey-box" methods in the derivation of thermo-physical properties from smart thermostat data acquisition (i.e., directly from temperature data instead of energy and temperature data), showing promising results.

"Grey-box" models can be also converted to "black-box" (i.e., statistical and machine learning models) for specific applications, for example control [204] or monitoring of internal conditions [205,206]."Black box" models are computationally efficient but they need to be trained on data before being deployed. As a result, "grey-box" models can be viewed as an intermediate stage between "white-box" and "black-box" models, and many examples of implementations have been found in recent years, ranging from experimental test facilities for building technologies [207] and construction components [208], to incorporation into the Building Information Modeling (BIM) workflow [209], and even to integrated room automation [210].

In addition, regression-based and "grey-box" model capabilities can be used in the Bayesian analysis framework. Bayesian analysis is suitable, for example, to 'reconstruct' building data (by estimating its characteristics) under uncertainty [80–82] or to evaluate the robustness of "grey-box" model estimates with respect to variable operating conditions [211] using Monte Carlo simulation methods [212], to reproduce realistically uncertain operating conditions.

What appears to be important for future research in this area is to increase the transparency of the modelling process by means of harmonised methodologies (using uniform rules and interpretable models as shown above) in order to verify and monitor output efficiently and to boost their level of automation without increasing complexity unnecessarily. Furthermore, the role of building automation [72,213] and monitoring systems [214,215] is crucial to understand the real dynamic behaviour of buildings by means of detailed data that can of course, complement energy metering, which represents the basic level of knowledge. Surrogate physical-statistical models (i.e., "grey-box" models) can be implemented also as "digital twins" (i.e., digital reproductions of the dynamic behaviour of their physical counterparts) at the level of construction technologies [216,217]. As a conclusion, in this Section we highlighted how harmonized methods for energy performance analysis are essential from multiple stand-points and how statistical and physical-statistical approaches are crucial for the evolution of energy research in buildings. Indeed, the methods reported and discussed in this Section can complement research on energy demand in end-uses based on epidemiology concept [218,219], providing however robust evidence on the performance of technologies and systems using empirically grounded methods, based on M&V principles.

4.3. Energy Flexibility and Occupant-Centric Energy Modelling

Energy flexibility in buildings [105] and occupant-centric energy modelling [106] for building design and operation are important research topics at present and they are directly addressing changes in fundamentals components of energy systems, such as users and energy infrastructures. Therefore, the topics discussed in this Section are complementing the ones in Section 4.1, focused on the potential of building performance analysis at scale, and Section 4.2, focused on harmonised methods for energy performance analysis (static and dynamic), showing how innovative concepts can contribute to reshape building design and operation strategies in the future. The analysis of the "mismatch" between building load profiles and on-site generation profiles (e.g., using PV power generation) has received a great deal of attention in recent years [41], due to the necessity of managing electric grid with increasing penetration of renewables. In this context, the concept of energy flexibility has been introduced to account for the dynamic interaction between end-user and electric infrastructures. Energy flexibility can be defined as the ability to control demand and supply according to consumer needs, grid conditions and climate [220]; an extensive review on this concept has been written by Reynders et al. [42]. There exist multiple options for increasing flexibility at the energy system level [136] and "soft-linking" of modelling approaches is increasingly important for energy planning and operation purpose [25,137]. More specifically, flexibility in buildings depends on the ability to use storage resources and to act on devices (including HVAC) after a trigger (e.g., time, power, energy price, etc.). Heating Ventilation and Air Conditioning (HVAC)systems are crucial because of their impact on the overall consumption of buildings and because of the potentially active role in energy infrastructure for demand response [221] and for absorbing surplus of energy from renewables [222]. From a technical perspective, energy flexibility in buildings can be exploited to shape building load profiles or to maximize the amount of energy that is self-consumed on-site [223,224], thereby increasing the matching between demand and on-site generation. The flexibility potential can be determined by the thermal inertia of building construction components (thermal mass) and by the presence of technical systems with storage (thermal and/or electric). Indeed, the exploitation of on-site renewables in buildings requires the adoption of technologies such as photovoltaics, heat pumps and energy storage [225]. Further, on the infrastructure side, flexibility requires an evolution of standardization of communication protocols to ensure efficient operation [226] and the results in this sense can determine a relevant change for the electric energy system as a whole [227], which may be combined with (and pushed forward by) consumer centric innovations in business models [228]. Specific KPIs [229] are required to describe flexibility potential and a large part of research at the state of the art concentrates on strategies to unlock it by means of control strategies [229,230], considering also related topics such as appropriate levels of modelling complexity and effort for their implementation [231]. In Table 6 we report an analysis of control strategies aimed at building flexibility for different end-uses and services using the same abbreviations as in Table 4. In Table 6 we consider the control objective in relation to flexibility, namely Load Shaping (LS) and On-site Renewable Maximization (ORM), following the arguments reported above. Additionally, the control types considered are Rule-Base Control (RBC), Optimal Control (OC) and Model Predictive Control (MPC). In Rule-Base Control rules are designed to fulfil a certain control objective but are not designed to achieve optimization of the overall system behaviour. In Optimal Control the control strategy is defined as an objective function to be optimized but doesn't include a prediction for the future. In Model Predictive Control the strategy is defined by means of an optimization performed with a certain control horizon (usually 24/48h); a comprehensive review on MPC has been written by Drgona et al. [232]. Further, we indicate the technical elements on which control strategies are focused. Also in this case, control strategies can be used for both residential and non-residential buildings a can exploit flexibility of heating, cooling and DHW demand by using the thermal storage capabilities of building fabric and technical system (e.g., water storage tanks). What appears to be fundamental, both in predictive and non-predictive cases, is the definition of dynamic operating schedules and set-points trajectories that are constrained by comfort requirements for heating and cooling services. However, the implementation of a detailed comfort model is challenging, due to the characteristics of control-oriented modelling approaches, and, for this reason, simplifications are generally considered when defining operational boundaries (i.e., the constraints for operation). Finally, the dynamic interaction with the grid is particularly important when dynamic tariffs are present and optimized control strategies have to consider the cost of imported and exported energy on a dynamic base.

Table 6. Control strategies aimed at building flexibility for different end-uses and services.

Source	Year	End-uses		Energy services				Control Objective	Control Type	Time Schedule	Set-points	Comfort Constraints	Load (Demand)	Production (On-Site)	Grid Connection (import/export)	Tariff
		Residential	Non-residential	Heating	Cooling	DHW	Appliances									
De Coninck et al. [233]	2014	✔				✔		LS, ORM	RBC	✔	✔				✔	
Klein et al. [234]	2015		✔	✔	✔			LS, ORM	RBC		✔		✔	✔		
Le Dréau and Heiselberg [235]	2016	✔		✔				LS	RBC		✔					✔
Dar et al. [236]	2014	✔		✔		✔		LS, ORM	RBC				✔	✔	✔	✔
Reynders et al. [237]	2015	✔		✔				LS	RBC		✔					
Turner et al. [238]	2015	✔			✔			LS	RBC	✔	✔	✔				
Esfehani et al. [239]	2016	✔		✔		✔		LS, ORM	RBC		✔		✔			
Alimohammadisagvand et al. [240]	2016	✔		✔		✔		LS	RBC		✔					✔
Salpakari and Lund [241]	2016	✔		✔		✔	✔	LS, ORM	RBC, OC		✔		✔	✔	✔	✔
Masy et al. [242]	2015	✔		✔		✔	✔	LS	RBC, OC	✔	✔					✔
Psimopoulos et al. [224]	2019	✔		✔		✔	✔	LS	RBC		✔	✔	✔	✔	✔	✔
Bee et al. [223]	2019	✔		✔	✔	✔		LS	RBC		✔		✔	✔	✔	✔
Oliveira Panão et al. [243]	2019	✔		✔				LS	RBC	✔	✔		✔			
Vivian et al. [244]	2020	✔		✔	✔			LS	RBC	✔	✔	✔	✔			
De Coninck and Helsen [245]	2016		✔	✔				LS	OC		✔	✔				✔
Halvgaard et al. [246]	2012	✔		✔				LS	MPC		✔	✔				✔
Maasoumy Haghighi [247]	2013		✔	✔	✔			LS	MPC		✔	✔	✔			
Corbin and Henze [248]	2017	✔		✔	✔	✔		LS	MPC		✔		✔	✔	✔	
Corbin and Henze [249]	2017	✔		✔	✔	✔		LS, ORM	MPC		✔		✔	✔	✔	
Lindelöf et al. [250]	2015	✔		✔				LS	MPC	✔	✔					
Garnier et al. [251]	2015		✔	✔	✔			LS	MPC	✔	✔	✔				
Kandler et al. [252]	2015	✔		✔				LS, ORM	MPC				✔	✔	✔	
Blum et al. [253]	2019	✔	✔	✔	✔			LS	MPC	✔	✔	✔				✔

It is worth noticing that there exists a potential methodological continuity between M&V practices at the state of the art, presented in Section 4.2, and innovative control strategies that represent an evolution of weather compensated control. This can be achieved, for example, using dynamic re-setting of heating and cooling curves [234] and machine learning algorithms whose performance can be tested and compared transparently in different weather conditions [250]. In general, by integrating regression modelling and clustering, it is possible to analyse variations of dynamic operational trajectories [134,156]. User behaviour has a huge impact on all the building services reported in Table 3 and, in recent years, an increasing research effort has been put on "Occupant-Centric Building Design and Operation" [106], as already mentioned before in the text.In particular, extensive reviews on this broad topic have been published recently [32], describing tools, methods and applications; more specific reviews have been dedicated to occupancy and behaviour modelling [254] and to occupant-centric control strategies [255]. The practical necessity to adapt modelling strategies in response to the purpose of the specific study (e.g., design, management, etc.) is indicated with the term "fit-for-purpose" [73]. Considering energy performance in a whole life cycle perspective, the variability of people behaviour and occupancy patterns has to be considered already at the early design stage, in particular in high efficiency and Nearly Zero Energy Buildings (NZEBs) [256]. After that, in the operation stage, occupancy can be measured in different ways [257] and data can be used to conduct realistic simulations [258]. In any case, as reported before, modelling occupancy patterns and user behaviour may require strategies that are customized (i.e., "fit-for-purpose") for the specific problem to be addressed: one possible solution is that of generating parametric or probabilistic occupancy profiles and modelling all the related variables (e.g., internal gains due to people and appliances, air change rates, etc.) in a transparent way [259,260]. This approach has been used, for example, to analyse building performance gap [261]. Realistic occupancy profiles are fundamental to address not only energy services but also to investigate related issues such as thermal comfort [262], Indoor Environmental Quality (IEQ) [263–265] and electric load profiles [266], among others.

As a conclusion, what appears to be important for future research in this area is increasing the transparency of the modelling process and linking it to harmonized methodologies (presented in Section 4.2) to verify and track performance efficiently without increasing unnecessarily the complexity of models themselves (i.e., maintaining an appropriate balance). Further, the role of building automation and monitoring systems is critical to understand the real dynamic behaviour of buildings. For example, data collected by monitoring systems [214,215] and/or automation systems [72,213] enable the performance characterization of envelope [160] and technical systems [267], together with occupancy patterns [257], already mentioned. Building performance monitoring and modelling can exploit also advances in IoT technologies [268] and open software [269], leading to innovative applications for energy and environmental management [270]. The possibility to rely on a combination of simulation methods and empirically grounded techniques for M&V can open interesting research opportunities in these areas.

4.4. *Summary of Research Findings*

In this section we describe the concepts emerging from studies that are in the intersections of the three levels of analysis presented in Sections 4.1–4.3, respectively. For this reason, we report in Table 7 the source, the level of analysis and the relevant concepts for the integration of energy modelling and data analytical processes. First, we can see how statistical reference buildings and parametric modelling represent the necessary basis for building energy modelling at multiple scales [80–82,103]. After that, "white-box" and "grey-box" modelling approaches can be integrated using a hierarchical multi-level approach [74] where "macro-parameters" [75] (aggregated, lumped quantities) are used as a mean to validated/calibrate more detailed model [80,118]. In turn, "grey-box" models based on regression and time series can guarantee empirically grounded "boundaries" for the estimation of building performance (providing harmonized methods) that may be used in multiple applications, while retaining a physical interpretation of the coefficients. The interpretability of models can provide multiple insights that can be exploited for the continuous improvement of technologies and practices (i.e., the PDCA approach [132]). Additionally, by combining regression, time series and clustering [134,156] it could possible to identify recurrent patterns in user behaviour [73,106] and in infrastructures' interaction [25,105,136], with a more precise quantification of the actual flexibility achievable. Both aspects (user behaviour and infrastructures' interaction) have to be considered in innovative business models for buildings where traditional Energy Performance Contracting is combined with innovative features [179] to ensure competitiveness and adequate level of services. Finally, data from automation and monitoring systems [72,160,213–215] are necessary to enable in depth analysis of performance, even though dynamic energy metering can be considered as the fundamental layer of information [214,215].

Table 7. Articles at the intersection of levels of analysis.

Source	Level	Year	Pattern Identified	Paper Title
Calleja Rodríguez et al. [75]	#1#2	2013	Reference building approach and parametric modelling	UK office buildings archetypal model as methodological approach in development of regression models for predicting building energy consumption from heating and cooling demands
Goel et al. [103]	#1#2	2016	Reference building approach and parametric modelling	Streamlining Building Efficiency Evaluation with DOE's Asset Score Preview
Zhao et al. [81]	#1#2	2016	Reference building approach and parametric modelling	Reconstructing building stock to replicate energy consumption data
Lim et al. [82]	#1#2	2017	Reference building approach and parametric modelling	Review on stochastic modeling methods for building stock energy prediction
Booth et al. [80]	#1#2	2013	Multi-level calibration	A hierarchical bayesian framework for calibrating micro-level models with macro-level data
Yang and Becerik-Gerber [74]	#1#2	2015	Multi-level calibration	A model calibration framework for simultaneous multi-level building energy simulation
Fabrizio et al. [118]	#2#3	2015	Multi-level calibration	Methodologies and advancements in the calibration of building energy models
Guyot et al. [77]	#1#2	2020	Multi-level calibration	Building energy model calibration: A detailed case study using sub-hourly measured data
Jalori et al. [134]	#2#3	2015	Regression-based approaches at multiple temporal and spatial scale of analysis	A new clustering method to identify outliers and diurnal schedules from building energy interval data
Jalori et al. [156]	#2#3	2015	Regression-based approaches at multiple temporal and spatial scale of analysis	A unified inverse modeling framework for whole-building energy interval data: Daily and hourly baseline modeling and short-term load forecasting
Ligier et al. [179]	#2#3	2017	Regression-based approaches at multiple temporal and spatial scale of analysis	Energy Performance Contracting Methodology Based upon Simulation and Measurement
Meng et al. [92]	#1#2	2017	Regression-based approaches at multiple temporal and spatial scale of analysis	Degree-day based non-domestic building energy analytics and modelling should use building and type specific base temperatures
Gaetani et al. [73]	#1#3	2016	User behavioural analysis	Occupant behavior in building energy simulation: Towards a fit-for-purpose modeling strategy
IEA-EBC [106]	#1#3	2017	User behavioural analysis	IEA EBC-Annex 79-Occupant-Centric Building Design and Operation
IEA-EBC [105]	#1#3	2014	Flexibility and dynamic interaction with infrastructures	EBC Annex 67 Energy Flexible Buildings
Lund et al. [136]	#1#3	2015	Flexibility and dynamic interaction with infrastructures	Review of energy system flexibility measures to enable high levels of variable renewable electricity
Dominkovic et al. [25]	#1#3	2020	Flexibility and dynamic interaction with infrastructures	Implementing flexibility into energy planning models: Soft-linking of a high-level energy planning model and a short-term operational model
Ahmad et al. [214]	#2#3	2016	Automation systems, measurements, sensors	Building energy metering and environmental monitoring—A state-of-the-art review and directions for future research
Aste et al. [72]	#1#2#3	2017	Automation systems, measurements, sensors	Building Automation and Control Systems and performance optimization: A framework for analysis
Carstens et al. [215]	#2#3	2018	Automation systems, measurements, sensors	Measurement uncertainty in energy monitoring: Present state of the art
Giraldo-Soto et al. [160]	#2#3	2018	Automation systems, measurements, sensors	Monitoring system analysis for evaluating a building's envelope energy performance through estimation of its heat loss coefficient
Serale et al. [213]	#2#3	2018	Automation systems, measurements, sensors	Model Predictive Control (MPC) for enhancing building and HVAC system energy efficiency: Problem formulation, applications and opportunities

As explained above, energy modelling and data analytical processes can be integrated in systems of models. Ideally, the creation of systems of standardized or harmonized "surrogate" physical-statistical models (i.e., "grey-box" models), which can be implemented in cyber-physical systems could represent a major breakthrough for energy modelling research. It can guarantee, for example, the possibility to act coherently at multiple levels in energy systems, using data analytics as a common background, and to create a certain degree continuity of performance analysis process during building life cycle, from design to operation phase. As discussed in Section 4.2, this result may be achieved by means of regression-based modelling approaches that combine conceptual simplicity and ease of implementation with adequate performance, in terms of analytics. In the next Section with indicate future research work that can be based on the outcomes of this research.

5. Further Work

Further research work could focus on knowledge mapping to enhance the integration and transparency of data within a modelling framework for energy in buildings, able to act at multiple levels. In Section 4.4 we described the points of contact between the multiple levels of analysis considered and we indicated how "surrogate" physical-statistical models (i.e., "grey-box" models that can be implemented in cyber-physical systems) could potentially work in "ecosystems" of applications. "Ecosystems" of models can address different types of end-uses (i.e., residential and non-residential), technological domains (i.e., heating, cooling, DHW, appliances) and applications (e.g., energy management, control, fault detection, environmental monitoring, etc.) while sharing a set of common underlying principles and rules. In this sense, surrogate models can act as "digital twins," that is to say digital reproductions of the dynamic behaviour of their physical counterparts (or systems). Harmonization and technical standardization play an essential role to avoid redundancy, multiplication of efforts and unnecessary increase of complexity of procedures. In fact, this could be the case of technical issues affecting multiple levels of information in the built environment, such as energy efficiency and flexibility or behavioural modelling and occupant-centric design and operation, described in Section 4.3. As mentioned in the introduction, building data interoperability [17] using common data exchange formats is necessary to increase the digitalisation and automation of buildings. The use of semantic web technologies [271] and standards based on IFC could support not only design but also operation (e.g., energy and environmental monitoring) [272], employing "surrogate" modelling strategies (physical/statistical, "grey-box") [209] compatible with the above mentioned principles. Finally, as introduced in Section 2, the research presented in this paper is part of a broader investigation, focused on the concept of "Buildings-as-Energy-Service": new forms of knowledge integration are needed to develop innovative services and products that can work as "ecosystems" and exploit this concept.

6. Conclusions

Energy transitions involve the transformation of the network of players and organisations that have traditionally worked in the energy sector along with new roles for customers. Radical innovation in the energy sector will have an impact on multiple domains in the construction sector (e.g., project, product and service). In this paper, we reviewed ongoing research on energy modelling and analytical tools that could support energy transition processes for the construction sector. In particular, we discussed how harmonised methods for analysing and tracking energy performance (Section 4.2) and innovative concepts such as flexibility and occupant-centric design and operation (Section 4.3) could contribute to a radical change in the built environment, using similar principles of analysis for actions that involve multiple scales (Section 4.1).The review process has been articulated according to three levels of analysis, introduced in Section 3 and reported in Section 4, ranging from general concepts to specific issues and we provided a summary of research findings as a set of interrelated concepts (Section 4.4). Overall, we identified criteria for energy modelling and analytical techniques (i.e., empirically grounding, scalability, harmonization, interpretability and re-configurability), that, in our opinion, constitute constraints to the creation of "ecosystems" of energy models aimed at supporting energy transition processes at multiple levels in the built environment. Regarding the first level of analysis (Section 4.1), systems of models can contribute to the creation of robust empirically grounded studies regarding efficiency for energy policy and utility scale actions. With respect to the second level (Section 4.2), they can be used to integrate data at multiple temporal and spatial scales, streamlining the analytical workflow (starting from consolidated M&V and M&T practices) and they can provide approximated physical interpretation of results, thereby increasing the transparency of modelling. Finally, in the third level (Section 4.3) they can help increasing energy flexibility in the interaction with infrastructures and improving the level of energy services in an occupant centric (design and operation) perspective. In all the levels considered in this review, we stressed the importance of studies that are empirically grounded and that can provide robust evidence for informing future research and policy.

As discussed in Section 5, these principles can constitute the basis for further research work, focused on developing specific applications built on top of them. In fact, the research proposed is part of a broader research activity focused on the "Buildings-as-Energy-Service" concept and the creation of a Tool Kit for knowledge integration regarding this topic, with the support of Cognitive Mapping technique. New forms of knowledge integration are needed to develop innovative services and products and this Tool Kit may be used to engage multiple users in the process of knowledge creation and sharing. Conceptualization is fundamental in innovation studies for energy and sustainability transitions but while general concepts can be clearly understood, what is still unclear is how these concepts can then translate into specific projects, products and services for energy transitions in the built environment, using innovative business models. Tools for knowledge integration can give a contribution in this sense.

Further, the problem of data accessibility has to be considered as well. The lack of detailed data or inadequate data reliability due to non-standardized collection procedures can be addressed using harmonized methodologies (described in Section 4.2). At present, this is causing a knowledge gap that undermines informed policy choices in the energy transition process (as well as in many other processes). Sensors, the Internet of Things (IoT), together with processes of automation and digitalisation described in this paper, could enable access to a greater amount of data for the building stock. In this context, it will be important to create open data repositories about technology, energy demand for end uses and weather data. Standardized and up-to-date data could enable transparent and consistent modelling processes at multiple scales of analysis, partially reducing the effort and stimulating the development of innovative energy technologies and services.

As a conclusion, in this paper we proposed a reflection on concepts that can help structuring future R&D activities and we highlighted a potential way to increase transparency, robustness and reproducibility in modelling by linking general principles emerging from the state of the art of research, to specific applications, employing harmonized methods as the core element. We believe that sharing information and making it more transparent and easily accessible can support multiple communities involved in R&Dfor energy transitions overcoming social and technical issues that may hinder the radical shifts that are necessary for long-term built environment sustainability.

Author Contributions: Conceptualization, M.M., M.S. and L.T.; Methodology, M.M. and M.S.; Investigation, M.M. and M.S.; Writing—original draft preparation, M.M.; Writing review and editing, M.M. and L.T. All authors have read and agreed to the published version of the manuscript.

Funding: This research was funded by UK Research and Innovation through the Industrial Strategy Challenge Fund, Transforming Construction Network Plus subcontract CID 3835572.

Institutional Review Board Statement: Not applicable.

Informed Consent Statement: Not applicable.

Data Availability Statement: Data is contained within the article.

Acknowledgments: The project "Developing a Tool Kit for Knowledge Integration: Envisioning Buildings-as-Energy-Service" is supported by The Transforming Construction Network Plus which is funded by UK Research and Innovation through the Industrial Strategy Challenge Fund. The N+ unites construction's academic and industrial communities to create a new research and knowledge base, dedicated to addressing the systemic problems holding back the sector. The N+ is a joint project between UCL, Imperial College London and WMG, University of Warwick.

Conflicts of Interest: The authors declare no conflict of interest. The funders had no role in the design of the study; in the collection, analyses, or interpretation of data; in the writing of the manuscript, or in the decision to publish the results.

References

1. Köhler, J.; Geels, F.W.; Kern, F.; Markard, J.; Onsongo, E.; Wieczorek, A.; Alkemade, F.; Avelino, F.; Bergek, A.; Boons, F.; et al. An agenda for sustainability transitions research: State of the art and future directions. *Environ. Innov. Soc. Transit.* **2019**, *31*, 1–32. [CrossRef]
2. Markard, J.; Hoffmann, V.H. Analysis of complementarities: Framework and examples from the energy transition. *Technol. Forecast. Soc. Chang.* **2016**, *111*, 63–75. [CrossRef]
3. Sibilla, M.; Kurul, E. Distributed Renewable and Interactive Energy Systems in Urban Environments. *TECHNE J. Technol. Archit. Environ.* **2018**, *1*, 33–39. [CrossRef]
4. Carayannis, E.G.; Campbell, D.F.J. *Mode 3 Knowledge Production in Quadruple Helix Innovation Systems: 21st-Century Democracy, Innovation, and Entrepreneurship for Development*; Springer: Berlin/Heidelberg, Germany, 2011; ISBN 9781461420620.
5. Carayannis, E.G.; Barth, T.D.; Campbell, D.F. The Quintuple Helix innovation model: Global warming as a challenge and driver for innovation. *J. Innov. Entrep.* **2012**, *1*, 2. [CrossRef]
6. Gliedt, T.; Hoicka, C.E.; Jackson, N. Innovation intermediaries accelerating environmental sustainability transitions. *J. Clean. Prod.* **2018**, *174*, 1247–1261. [CrossRef]
7. Smith, A.; Raven, R.R. What is protective space? Reconsidering niches in transitions to sustainability. *Res. Policy* **2012**, *41*, 1025–1036. [CrossRef]
8. Kolk, A.; Van Tulder, R.; Kostwinder, E. Business and partnerships for development. *Eur. Manag. J.* **2008**, *26*, 262–273. [CrossRef]
9. Bush, R.E.; Bale, C.S.E.; Powell, M.; Gouldson, A.; Taylor, P.G.; Gale, W.F. The role of intermediaries in low carbon transitions—Empowering innovations to unlock district heating in the UK. *J. Clean. Prod.* **2017**, *148*, 137–147. [CrossRef]
10. Berardi, U. A cross-country comparison of the building energy consumptions and their trends. *Resour. Conserv. Recycl.* **2017**, *123*, 230–241. [CrossRef]
11. Barrie, J.; Zawdie, G.; João, E. Leveraging triple helix and system intermediaries to enhance effectiveness of protected spaces and strategic niche management for transitioning to circular economy. *Int. J. Technol. Manag. Sustain. Dev.* **2017**, *16*, 25–47. [CrossRef]
12. Open Science Policy Platform (OSPP). Available online: https://ec.europa.eu/research/openscience/index.cfm?pg=open-science-policy-platform (accessed on 25 August 2020).
13. Mendez, E.; Lawrence, R.; MacCallum, C.J.; Moar, E.; Lossau, N.; Deketelaere, K.; Luyben, K.; Epure, M.; Bertero, M.; Garfinkel, M.; et al. Progress on Open Science: Towards a Shared Research Knowledge System. Final Report of the Open Science Policy Platform. 2020. Available online: https://ec.europa.eu/research/openscience/pdf/ec_rtd_ospp-final-report.pdf (accessed on 5 December 2020).
14. European Open Science Cloud (EOSC) of the European Commission. Available online: https://ec.europa.eu/research/openscience/index.cfm?pg=open-science-cloud (accessed on 25 August 2020).
15. Openmod Open Energy Modelling Initiative (Openmod)—Open Models. Available online: https://wiki.openmod-initiative.org/wiki/Open_Models (accessed on 25 August 2020).
16. Hilpert, S.; Kaldemeyer, C.; Krien, U.; Günther, S.; Wingenbach, C.; Plessmann, G. The Open Energy Modelling Framework (oemof)—A new approach to facilitate open science in energy system modelling. *Energy Strateg. Rev.* **2018**, *22*, 16–25. [CrossRef]
17. Hardin, D.; Stephan, E.G.; Wang, W.; Corbin, C.D.; Widergren, S.E. *Buildings Interoperability Landscape*; United State Department of Energy: Oak Ridge, TN, USA, 2015. [CrossRef]
18. Atzori, L.; Iera, A.; Morabito, G. The Internet of Things: A survey. *Comput. Netw.* **2010**, *54*, 2787–2805. [CrossRef]
19. Tan, L.; Wang, N. Future internet: The Internet of Things. In Proceedings of the 2010 3rd International Conference on Advanced Computer Theory and Engineering (ICACTE), Chengdu, China, 20–22 August 2010; Volume 5, pp. 5–376.
20. Breiner, S.; Subrahmanian, E.; Sriram, R.D. Modeling the Internet of Things: A Foundational Approach. In Proceedings of the Seventh International Workshop on the Web of Things, Stuttgart, Germany, 7 November 2016; ACM: New York, NY, USA, 2016; pp. 38–41.
21. Schmidt, M.; Åhlund, C. Smart buildings as Cyber-Physical Systems: Data-driven predictive control strategies for energy efficiency. *Renew. Sustain. Energy Rev.* **2018**, *90*, 742–756. [CrossRef]
22. Reka, S.S.; Dragicevic, T. Future effectual role of energy delivery: A comprehensive review of Internet of Things and smart grid. *Renew. Sustain. Energy Rev.* **2018**, *91*, 90–108. [CrossRef]

23. Arghandeh, R.; Von Meier, A.; Mehrmanesh, L.; Mili, L. On the definition of cyber-physical resilience in power systems. *Renew. Sustain. Energy Rev.* **2016**, *58*, 1060–1069. [CrossRef]
24. Tronchin, L.; Manfren, M.; Nastasi, B. Energy efficiency, demand side management and energy storage technologies—A critical analysis of possible paths of integration in the built environment. *Renew. Sustain. Energy Rev.* **2018**, *95*, 341–353. [CrossRef]
25. Dominković, D.F.; Junker, R.G.; Lindberg, K.; Madsen, H. Implementing flexibility into energy planning models: Soft-linking of a high-level energy planning model and a short-term operational model. *Appl. Energy* **2020**, *260*, 114292. [CrossRef]
26. Zafar, R.; Asif, M.; Razzaq, S.; Ali, W.; Naeem, U.; Shehzad, K. Prosumer based energy management and sharing in smart grid. *Renew. Sustain. Energy Rev.* **2018**, *82*, 1675–1684. [CrossRef]
27. Sioshansi, F.P. *Consumer, Prosumer, Prosumager: How Service Innovations Will Disrupt the Utility Business Model*; Academic Press: Cambridge, MA, USA, 2019.
28. Andoni, M.; Robu, V.; Flynn, D.; Abram, S.; Geach, D.; Jenkins, D.P.; McCallum, P.; Peacock, A. Blockchain technology in the energy sector: A systematic review of challenges and opportunities. *Renew. Sustain. Energy Rev.* **2019**, *100*, 143–174. [CrossRef]
29. Gui, E.M.; MacGill, I. Typology of future clean energy communities: An exploratory structure, opportunities, and challenges. *Energy Res. Soc. Sci.* **2018**, *35*, 94–107. [CrossRef]
30. Thuesen, C.; Koch-Ørvad, N.; Maslesa, E. Organising Sustainable Transition: Understanding the Product, Project and Service Domain of the Built Environment. In Proceedings of the 32nd Annual ARCOM Conference, Manchester, UK, 5–7 September 2016.
31. Yoshino, H.; Hong, T.; Nord, N. IEA EBC annex 53: Total energy use in buildings—Analysis and evaluation methods. *Energy Build.* **2017**, *152*, 124–136. [CrossRef]
32. Azar, E.; O'Brien, W.; Carlucci, S.; Hong, T.; Sonta, A.; Kim, J.; Andargie, M.S.; Abuimara, T.; El Asmar, M.; Jain, R.K.; et al. Simulation-aided occupant-centric building design: A critical review of tools, methods, and applications. *Energy Build.* **2020**, *224*, 110292. [CrossRef]
33. Herring, H.; Roy, R. Technological innovation, energy efficient design and the rebound effect. *Technovation* **2007**, *27*, 194–203. [CrossRef]
34. Sunikka-Blank, M.; Galvin, R. Introducing the prebound effect: The gap between performance and actual energy consumption. *Build. Res. Inf.* **2012**, *40*, 260–273. [CrossRef]
35. Rosenow, J.; Galvin, R. Evaluating the evaluations: Evidence from energy efficiency programmes in Germany and the UK. *Energy Build.* **2013**, *62*, 450–458. [CrossRef]
36. de Wilde, P. The gap between predicted and measured energy performance of buildings: A framework for investigation. *Autom. Constr.* **2014**, *41*, 40–49. [CrossRef]
37. Imam, S.; Coley, D.A.; Walker, I. The building performance gap: Are modellers literate? *Build. Serv. Eng. Res. Technol.* **2017**, *38*, 351–375. [CrossRef]
38. de Wilde, P. The building performance gap: Are modellers literate? *Build. Serv. Eng. Res. Technol.* **2017**, *38*, 757–759. [CrossRef]
39. Van Dronkelaar, C.; Dowson, M.; Spataru, C.; Mumovic, D. A Review of the Regulatory Energy Performance Gap and Its Underlying Causes in Non-domestic Buildings. *Front. Mech. Eng.* **2016**, *1*, 17. [CrossRef]
40. Cozza, S.; Chambers, J.; Deb, C.; Scartezzini, J.-L.; Schlüter, A.; Patel, M.K. Do energy performance certificates allow reliable predictions of actual energy consumption and savings? Learning from the Swiss national database. *Energy Build.* **2020**, *224*, 110235. [CrossRef]
41. Salom, J.; Marszal, A.J.; Widén, J.; Candanedo, J.; Lindberg, K.B. Analysis of load match and grid interaction indicators in net zero energy buildings with simulated and monitored data. *Appl. Energy* **2014**, *136*, 119–131. [CrossRef]
42. Reynders, G.; Lopes, R.A.; Marszal-Pomianowska, A.; Aelenei, D.; Martins, J.; Saelens, D. Energy flexible buildings: An evaluation of definitions and quantification methodologies applied to thermal storage. *Energy Build.* **2018**, *166*, 372–390. [CrossRef]
43. Pomponi, F.; Moncaster, A. Scrutinising embodied carbon in buildings: The next performance gap made manifest. *Renew. Sustain. Energy Rev.* **2018**, *81*, 2431–2442. [CrossRef]
44. De Wolf, C.; Pomponi, F.; Moncaster, A. Measuring embodied carbon dioxide equivalent of buildings: A review and critique of current industry practice. *Energy Build.* **2017**, *140*, 68–80. [CrossRef]
45. International Energy Agency. *Capturing the Multiple Benefits of Energy Efficiency*; IEA: Paris, France, 2014.
46. Pfenninger, S.; Decarolis, J.; Hirth, L.; Quoilin, S.; Staffell, I. The importance of open data and software: Is energy research lagging behind? *Energy Policy* **2017**, *101*, 211–215. [CrossRef]
47. Pfenninger, S.; Hirth, L.; Schlecht, I.; Schmid, E.; Wiese, F.; Brown, T.; Davis, C.; Gidden, M.J.; Heinrichs, H.; Heuberger, C.; et al. Opening the black box of energy modelling: Strategies and lessons learned. *Energy Strateg. Rev.* **2018**, *19*, 63–71. [CrossRef]
48. Bollinger, L.; Davis, C.; Evins, R.; Chappin, E.; Nikolic, I. Multi-model ecologies for shaping future energy systems: Design patterns and development paths. *Renew. Sustain. Energy Rev.* **2018**, *82*, 3441–3451. [CrossRef]
49. Deane, J.P.; Chiodi, A.; Gargiulo, M.; Gallachóir, B.P.Ó. Soft-linking of a power systems model to an energy systems model. *Energy* **2012**, *42*, 303–312. [CrossRef]
50. Novak, J.D. A Theory of Education: Meaningful Learning Underlies the Constructive Integration of Thinking, Feeling, and Acting Leading To Empowerment for Commitment and Responsibility. *Mean. Learn. Rev.* **2011**, *1*, 1–14.
51. Sibilla, M. A meaningful mapping approach for the complex design. *Int. J. Des. Sci. Technol.* **2017**, *23*. Available online: http://ijdst.europia.org/index.php/ijdst/article/view/2 (accessed on 10 December 2020).
52. Florida, R. Cities and the Creative Class. *City Community* **2003**, *2*, 3–19. [CrossRef]

53. Creswell, J.W.; Creswell, J.D. *Research Design: Qualitative, Quantitative, and Mixed Methods Approaches*; SAGE Publications: Thousand Oaks, CA, USA, 2017.
54. Birks, M.; Mills, J. *Grounded Theory: A Practical Guide*; SAGE Publications: Thousand Oaks, CA, USA, 2015.
55. Glaser, B.G. *Doing Grounded Theory: Issues and Discussions*; Sociology Press: Mill Valley, CA, USA, 1998.
56. Glaser, B.G.; Holton, J. Remodeling grounded theory. *Forum Qual. Soc. Res.* **2004**, *5*, 4.
57. Corbin, J.; Strauss, A. *Basics of Qualitative Research: Techniques and Procedures for Developing Grounded Theory*, 3rd ed.; SAGE Publications: Thousand Oaks, CA, USA, 2008; ISBN 9781412906432.
58. Charmaz, K. Constructionism and the grounded theory method. *Handb. Constr. Res.* **2008**, *1*, 397–412.
59. Lewis-Beck, M.; Bryman, A.; Liao, T.F. *The SAGE Encyclopedia of Social Science Research Methods*; College of Liberal Arts and Sciences: Chicago, IL, USA, 2004.
60. Manfren, M.; Nastasi, B.; Groppi, D.; Garcia, D.A. Open data and energy analytics—An analysis of essential information for energy system planning, design and operation. *Energy* **2020**, *213*, 118803. [CrossRef]
61. Kavgic, M.; Mavrogianni, A.; Mumovic, D.; Summerfield, A.; Stevanovic, Z.; Djurovic-Petrovic, M. A review of bottom-up building stock models for energy consumption in the residential sector. *Build. Environ.* **2010**, *45*, 1683–1697. [CrossRef]
62. Foucquier, A.; Robert, S.; Suard, F.; Stéphan, L.; Jay, A. State of the art in building modelling and energy performances prediction: A review. *Renew. Sustain. Energy Rev.* **2013**, *23*, 272–288. [CrossRef]
63. Fumo, N. A review on the basics of building energy estimation. *Renew. Sustain. Energy Rev.* **2014**, *31*, 53–60. [CrossRef]
64. de Wilde, P. *Building Performance Analysis*; Wiley & Sons: Hoboken, NJ, USA, 2018; ISBN 9781119341932.
65. Corgnati, S.P.; Fabrizio, E.; Filippi, M.; Monetti, V. Reference buildings for cost optimal analysis: Method of definition and application. *Appl. Energy* **2013**, *102*, 983–993. [CrossRef]
66. European Commission. *Commission Delegated Regulation*; (EU) No 244/2012; European Commision: Brussels, Belgium, 2012.
67. Arcipowska, A.; Rapf, O.; Faber, M.; Fabbri, M.; Tigchelaar, C.; Boermans, T.; Surmeli-Anac, N.; Pollier, K.; Dal, F.; Sebi, C.; et al. *Support for Setting up an Observatory of the Building Stock and Related Policies*; Buildings Performance Institute Europe (BPIE): Berlin, Germany, 2016.
68. Deru, M.; Field, K.; Studer, D.; Benne, K.; Griffith, B.; Torcellini, P.; Liu, B.; Halverson, M.; Winiarski, D.; Rosenberg, M.; et al. *U.S. Department of Energy Commercial Reference Building Models of the National Building Stock*; NREL: Denver, CO, USA, 2011.
69. Goel, S.; Athalye, R.A.; Wang, W. *Enhancements to ASHRAE Standard 90.1 Prototype Building Models*; No. PNNL-23269; Pacific Northwest National Laboratory (PNNL): Richland, WA, USA, 2014.
70. Thornton, B.A.; Rosenberg, M.I.; Richman, E.E.; Wang, W.; Xie, Y.; Zhang, J.; Cho, H.; Mendon, V.V.; Athalye, R.A.; Liu, B. *Achieving the 30% Goal: Energy and Cost Savings Analysis of ASHRAE Standard 90.1-2010*; No. PNNL-20405; Pacific Northwest National Laboratory (PNNL): Richland, WA, USA, 2011.
71. ASHRAE. *209-2018—Energy Simulation Aided Design for Buildings Except Low-Rise Residential Buildings*; (ANSI Approved); ASHRAE: Washington, DC, USA, 2018.
72. Aste, N.; Manfren, M.; Marenzi, G. Building Automation and Control Systems and performance optimization: A framework for analysis. *Renew. Sustain. Energy Rev.* **2017**, *75*, 313–330. [CrossRef]
73. Gaetani, I.I.; Hoes, P.P.-J.; Hensen, J.L. Occupant behavior in building energy simulation: Towards a fit-for-purpose modeling strategy. *Energy Build.* **2016**, *121*, 188–204. [CrossRef]
74. Yang, Z.; Becerik-Gerber, B. A model calibration framework for simultaneous multi-level building energy simulation. *Appl. Energy* **2015**, *149*, 415–431. [CrossRef]
75. Rodríguez, G.C.; Andrés, A.C.; Muñoz, F.D.; López, J.M.C.; Zhang, Y. Uncertainties and sensitivity analysis in building energy simulation using macroparameters. *Energy Build.* **2013**, *67*, 79–87. [CrossRef]
76. Korolija, I.; Marjanovic-Halburd, L.; Zhang, Y.; Hanby, V.I. UK office buildings archetypal model as methodological approach in development of regression models for predicting building energy consumption from heating and cooling demands. *Energy Build.* **2013**, *60*, 152–162. [CrossRef]
77. Guyot, D.; Giraud, F.; Simon, F.; Corgier, D.; Marvillet, C.; Tremeac, B. Building energy model calibration: A detailed case study using sub-hourly measured data. *Energy Build.* **2020**, *223*, 110189. [CrossRef]
78. Koulamas, C.; Kalogeras, A.; Pacheco-Torres, R.; Casillas, J.; Ferrarini, L. Suitability analysis of modeling and assessment approaches in energy efficiency in buildings. *Energy Build.* **2018**, *158*, 1662–1682. [CrossRef]
79. Manfren, M.; Aste, N.; Moshksar, R. Calibration and uncertainty analysis for computer models—A meta-model based approach for integrated building energy simulation. *Appl. Energy* **2013**, *103*, 627–641. [CrossRef]
80. Booth, A.; Choudhary, R.; Spiegelhalter, D. A hierarchical Bayesian framework for calibrating micro-level models with macro-level data. *J. Build. Perform. Simul.* **2013**, *6*, 293–318. [CrossRef]
81. Zhao, F.; Lee, S.H.; Augenbroe, G. Reconstructing building stock to replicate energy consumption data. *Energy Build.* **2016**, *117*, 301–312. [CrossRef]
82. Lim, H.; Zhai, Z.J. Review on stochastic modeling methods for building stock energy prediction. *Build. Simul.* **2017**, *10*, 607–624. [CrossRef]
83. Hong, S.M.; Paterson, G.; Burman, E.; Steadman, P.; Mumovic, D. A comparative study of benchmarking approaches for non-domestic buildings: Part 1—Top-down approach. *Int. J. Sustain. Built Environ.* **2013**, *2*, 119–130. [CrossRef]

84. Burman, E.; Hong, S.-M.; Paterson, G.; Kimpian, J.; Mumovic, D. A comparative study of benchmarking approaches for non-domestic buildings: Part 2—Bottom-up approach. *Int. J. Sustain. Built Environ.* **2014**, *3*, 247–261. [CrossRef]
85. Kneifel, J.; Webb, D. Predicting energy performance of a net-zero energy building: A statistical approach. *Appl. Energy* **2016**, *178*, 468–483. [CrossRef] [PubMed]
86. Wang, Y.; Kuckelkorn, J.; Zhao, F.-Y.; Spliethoff, H.; Lang, W. A state of art of review on interactions between energy performance and indoor environment quality in Passive House buildings. *Renew. Sustain. Energy Rev.* **2017**, *72*, 1303–1319. [CrossRef]
87. Deng, S.; Wang, R.; Dai, Y. How to evaluate performance of net zero energy building—A literature research. *Energy* **2014**, *71*, 1–16. [CrossRef]
88. Berardi, U. ZEB and nZEB (definitions, design methodologies, good practices, and case studies). In *Handbook of Energy Efficiency in Buildings: A Life Cycle Approach*; Butterworth-Heinemann: Oxford, UK, 2018.
89. Magrini, A.; Lentini, G.; Cuman, S.; Bodrato, A.; Marenco, L. From nearly zero energy buildings (NZEB) to positive energy buildings (PEB): The next challenge—The most recent European trends with some notes on the energy analysis of a forerunner PEB example. *Dev. Built Environ.* **2020**, *3*, 100019. [CrossRef]
90. Zangheri, P.; Armani, R.; Pietrobon, M.; Pagliano, L. Identification of cost-optimal and NZEB refurbishment levels for representative climates and building typologies across Europe. *Energy Effic.* **2018**, *11*, 337–369. [CrossRef]
91. Goel, S.; Baker, C.; Wolf, D.; Henderson, P.; Wang, N.; Rosenberg, M. A Simplified Energy Modeling Approach for Buildings (C009). In Proceedings of the 2018 Building Performance Analysis Conference and SimBuild, Chicago, IL, USA, 26–28 September 2018.
92. Meng, Q.; Mourshed, M. Degree-day based non-domestic building energy analytics and modelling should use building and type specific base temperatures. *Energy Build.* **2017**, *155*, 260–268. [CrossRef]
93. Kotireddy, R.; Hoes, P.-J.; Hensen, J.L. A methodology for performance robustness assessment of low-energy buildings using scenario analysis. *Appl. Energy* **2018**, *212*, 428–442. [CrossRef]
94. Schlueter, A.; Geyer, P. Linking BIM and Design of Experiments to balance architectural and technical design factors for energy performance. *Autom. Constr.* **2018**, *86*, 33–43. [CrossRef]
95. Jaffal, I.; Inard, C.; Ghiaus, C. Fast method to predict building heating demand based on the design of experiments. *Energy Build.* **2009**, *41*, 669–677. [CrossRef]
96. Pernigotto, G.; Prada, A.; Gasparella, A.; Hensen, J.L.M. Development of sets of simplified building models for building simulation. In Proceedings of the 3rd International High Performance Buildings Conference, Purdue, IL, USA, 14–17 July 2014.
97. Dogan, T.; Reinhart, C.; Michalatos, P. Autozoner: An algorithm for automatic thermal zoning of buildings with unknown interior space definitions. *J. Build. Perform. Simul.* **2016**, *9*, 176–189. [CrossRef]
98. Dogan, T.; Reinhart, C. Shoeboxer: An algorithm for abstracted rapid multi-zone urban building energy model generation and simulation. *Energy Build.* **2017**, *140*, 140–153. [CrossRef]
99. Delmastro, C.; Mutani, G.; Corgnati, S.P. A supporting method for selecting cost-optimal energy retrofit policies for residential buildings at the urban scale. *Energy Policy* **2016**, *99*, 42–56. [CrossRef]
100. Ghiassi, N.; Mahdavi, A. Reductive bottom-up urban energy computing supported by multivariate cluster analysis. *Energy Build.* **2017**, *144*, 372–386. [CrossRef]
101. Delmastro, C.; Gargiulo, M. Capturing the long-term interdependencies between building thermal energy supply and demand in urban planning strategies. *Appl. Energy* **2020**, *268*, 114774. [CrossRef]
102. Ballarini, I.; Corrado, V. A New Methodology for Assessing the Energy Consumption of Building Stocks. *Energies* **2017**, *10*, 1102. [CrossRef]
103. Goel, S.; Wang, N.; Gonzalez, J.; Horsey, H.; Long, N. Streamlining Building Efficiency Evaluation with DOE's Asset Score Preview. In Proceedings of the 2016 ACEEE Summer Study on Energy Efficiency in Buildings, Pacific Grove, CA, USA, 21–26 August 2016.
104. Badiei, A.; Allinson, D.; Lomas, K. Automated dynamic thermal simulation of houses and housing stocks using readily available reduced data. *Energy Build.* **2019**, *203*, 109431. [CrossRef]
105. IEA. *Annex 67—Energy Flexible Buildings*; IEA: Paris, France, 2019.
106. IEA. *Annex 79—Occupant-Centric Building Design and Operation*; IEA: Paris, France, 2018.
107. Jack, R.; Loveday, D.; Allinson, D.; Lomas, K. First evidence for the reliability of building co-heating tests. *Build. Res. Inf.* **2018**, *46*, 383–401. [CrossRef]
108. Lomas, K.; Oliveira, S.; Warren, P.; Haines, V.; Chatterton, T.; Beizaee, A.; Prestwood, E.; Gething, B. Do domestic heating controls save energy? A review of the evidence. *Renew. Sustain. Energy Rev.* **2018**, *93*, 52–75. [CrossRef]
109. Mathieu, J.L.; Price, P.N.; Kiliccote, S.; Piette, M.A. Quantifying Changes in Building Electricity Use, With Application to Demand Response. *IEEE Trans. Smart Grid* **2011**, *2*, 507–518. [CrossRef]
110. Jayaweera, T.; Haeri, H.; Gowans, D. The Uniform Methods Project: Methods for Determining Energy Efficiency Savings for Specific Measures. *Contract* **2013**, *303*, 275–3000.
111. EVO IPMVP New Construction Subcommittee. *International Performance Measurement and Verification Protocol: Concepts and Option for Determining Energy Savings in New Construction*; Efficiency Valuation Organization (EVO): Washington, DC, USA, 2003; Volume 3.
112. FEMP. *Federal Energy Management Program, M&V Guidelines: Measurement and Verification for Federal Energy Projects Version 3.0*; U.S. Department of Energy Federal Energy Management Program: Washington, DC, USA, 2008.
113. CalTRACK Methods. Available online: http://docs.caltrack.org/en/latest/methods.html (accessed on 20 May 2020).

114. Investor Confidence Project. Available online: https://europe.eeperformance.org/ (accessed on 31 August 2020).
115. ISO. *ISO 16346:2013 Energy Performance of Buildings—Assessment of Overall Energy Performance*; ISO: Geneva, Switzerland, 2013.
116. American Society of Heating, Refrigerating and Air-Conditioning Engineers (ASHRAE). *ASHRAE Guideline 14–2014: Measurement of Energy, Demand, and Water Savings*; American Society of Heating, Refrigerating and Air-Conditioning Engineers: Atlanta, GA, USA, 2014.
117. ISO. *ISO 50006:2014 Energy Management Systems—Measuring Energy Performance Using Energy Baselines (EnB) and Energy Performance Indicators (EnPI)—General Principles and Guidance*; ISO: Geneva, Switzerland, 2014.
118. Fabrizio, E.; Monetti, V. Methodologies and Advancements in the Calibration of Building Energy Models. *Energies* **2015**, *8*, 2548–2574. [CrossRef]
119. Paulus, M.T. Algorithm for explicit solution to the three parameter linear change-point regression model. *Sci. Technol. Built Environ.* **2017**, *23*, 1026–1035. [CrossRef]
120. RMV2.0-LBNL M&V2.0 Tool. Available online: https://lbnl-eta.github.io/RMV2.0/ (accessed on 20 May 2020).
121. Masuda, H.; Claridge, D.E. Statistical modeling of the building energy balance variable for screening of metered energy use in large commercial buildings. *Energy Build.* **2014**, *77*, 292–303. [CrossRef]
122. Lin, G.; Claridge, D.E. A temperature-based approach to detect abnormal building energy consumption. *Energy Build.* **2015**, *93*, 110–118. [CrossRef]
123. Kohler, M.; Blond, N.; Clappier, A. A city scale degree-day method to assess building space heating energy demands in Strasbourg Eurometropolis (France). *Appl. Energy* **2016**, *184*, 40–54. [CrossRef]
124. Fazeli, R.; Ruth, M.; Davidsdottir, B. Temperature response functions for residential energy demand—A review of models. *Urban Clim.* **2016**, *15*, 45–59. [CrossRef]
125. Østergård, T.; Jensen, R.L.; Maagaard, S.E. A comparison of six metamodeling techniques applied to building performance simulations. *Appl. Energy* **2018**, *211*, 89–103. [CrossRef]
126. Westermann, P.; Evins, R. Surrogate modelling for sustainable building design—A review. *Energy Build.* **2019**, *198*, 170–186. [CrossRef]
127. Paulus, M.T.; Claridge, D.E.; Culp, C. Algorithm for automating the selection of a temperature dependent change point model. *Energy Build.* **2015**, *87*, 95–104. [CrossRef]
128. Tronchin, L.; Manfren, M.; James, P.A. Linking design and operation performance analysis through model calibration: Parametric assessment on a Passive House building. *Energy* **2018**, *165*, 26–40. [CrossRef]
129. Allard, I.; Olofsson, T.; Nair, G. Energy evaluation of residential buildings: Performance gap analysis incorporating uncertainties in the evaluation methods. *Build. Simul.* **2018**, *11*, 725–737. [CrossRef]
130. Manfren, M.; Nastasi, B.; Tronchin, L. Linking Design and Operation Phase Energy Performance Analysis Through Regression-Based Approaches. *Front. Energy Res.* **2020**, *8*, 288. [CrossRef]
131. Manfren, M.; Nastasi, B. Parametric Performance Analysis and Energy Model Calibration Workflow Integration—A Scalable Approach for Buildings. *Energies* **2020**, *13*, 621. [CrossRef]
132. ISO. *ISO 50001:2018, Energy Management Systems—Requirements with Guidance for Use*; ISO: Geneva, Switzerland, 2018.
133. Miller, C.; Nagy, Z.; Schlueter, A. Automated daily pattern filtering of measured building performance data. *Autom. Constr.* **2015**, *49*, 1–17. [CrossRef]
134. Jalori, S.; Reddy, T.A. A new clustering method to identify outliers and diurnal schedules from building energy interval data. *Ashrae Trans.* **2015**, *121*, 33.
135. Richard, M.A.; Fortin, H.; Poulin, A.; Leduc, M.-A.; Fournier, M. Daily load profiles clustering: A powerful tool for demand side management in medium-sized industries. In Proceedings of the ACEEE Summer Study on Energy Efficiency in Industry, Denver, CO, USA, 15–18 August 2017.
136. Lund, P.D.; Lindgren, J.; Mikkola, J.; Salpakari, J. Review of energy system flexibility measures to enable high levels of variable renewable electricity. *Renew. Sustain. Energy Rev.* **2015**, *45*, 785–807. [CrossRef]
137. Poncelet, K.; Delarue, E.; Six, D.; Duerinck, J.; D'Haeseleer, W. Impact of the level of temporal and operational detail in energy-system planning models. *Appl. Energy* **2016**, *162*, 631–643. [CrossRef]
138. Adhikari, R.S.; Aste, N.; Manfren, M. Multi-commodity network flow models for dynamic energy management–Smart Grid applications. *Energy Procedia* **2012**, *14*, 1374–1379. [CrossRef]
139. Manfren, M. Multi-commodity network flow models for dynamic energy management–Mathematical formulation. *Energy Procedia* **2012**, *14*, 1380–1385. [CrossRef]
140. Kraning, M.; Chu, E.; Lavaei, J.; Boyd, S. Dynamic Network Energy Management via Proximal Message Passing. *Found. Trends®Optim.* **2014**, *1*, 73–126. [CrossRef]
141. Dorfner, J. *Open Source Modelling and Optimisation of Energy Infrastructure at Urban Scale*; Technische Universität München: München, Germany, 2016.
142. Mazzoni, S.; Ooi, S.; Nastasi, B.; Romagnoli, A. Energy storage technologies as techno-economic parameters for master-planning and optimal dispatch in smart multi energy systems. *Appl. Energy* **2019**, *254*, 113682. [CrossRef]
143. Jentsch, M.F.; Bahaj, A.S.; James, P.A. Climate change future proofing of buildings—Generation and assessment of building simulation weather files. *Energy Build.* **2008**, *40*, 2148–2168. [CrossRef]

144. Jentsch, M.F.; James, P.A.B.; Bourikas, L.; Bahaj, A.S. Transforming existing weather data for worldwide locations to enable energy and building performance simulation under future climates. *Renew. Energy* **2013**, *55*, 514–524. [CrossRef]
145. Dias, J.B.; Da Graça, G.C.; Soares, P.M. Comparison of methodologies for generation of future weather data for building thermal energy simulation. *Energy Build.* **2020**, *206*, 109556. [CrossRef]
146. Stadler, P.; Girardin, L.; Ashouri, A.; Maréchal, F. Contribution of Model Predictive Control in the Integration of Renewable Energy Sources within the Built Environment. *Front. Energy Res.* **2018**, *6*, 22. [CrossRef]
147. Orehounig, K.; Mavromatidis, G.; Evins, R.; Dorer, V.; Carmeliet, J. Towards an energy sustainable community: An energy system analysis for a village in Switzerland. *Energy Build.* **2014**, *84*, 277–286. [CrossRef]
148. Orehounig, K.; Evins, R.; Dorer, V. Integration of decentralized energy systems in neighbourhoods using the energy hub approach. *Appl. Energy* **2015**, *154*, 277–289. [CrossRef]
149. Lammers, N.; Kissock, K.; Abels, B.; Sever, F. Measuring Progress with Normalized Energy Intensity. *SAE Int. J. Mater. Manuf.* **2011**, *4*, 460–467. [CrossRef]
150. Hallinan, K.P.; Brodrick, P.; Northridge, J.; Kissock, J.K.; Brecha, R.J. *Establishing Building Recommissioning Priorities and Potential Energy Savings from Utility Energy Data*; ASHRAE: Peachtree Corners, GA, USA, 2011.
151. Hallinan, K.P.; Kissock, J.K.; Brecha, R.J.; Mitchell, A. *Targeting Residential Energy Reduction for City Utilities Using Historical Electrical Utility Data and Readily Available Building Data*; ASHRAE: Peachtree Corners, GA, USA, 2011.
152. Danov, S.; Carbonell, J.; Cipriano, J.; Martí-Herrero, J. Approaches to evaluate building energy performance from daily consumption data considering dynamic and solar gain effects. *Energy Build.* **2013**, *57*, 110–118. [CrossRef]
153. Masuda, H.; Claridge, D.E. Inclusion of building envelope thermal lag effects in linear regression models of daily basis building energy use data. In Proceedings of the 12th International Conference for Enhanced Building Operations, Manchester, UK, USA, 22–26 October 2012; Available online: https://oaktrust.library.tamu.edu/handle/1969.1/148946 (accessed on 25 November 2020).
154. Bynum, J.D.; Claridge, D.E.; Curtin, J.M. Development and testing of an Automated Building Commissioning Analysis Tool (ABCAT). *Energy Build.* **2012**, *55*, 607–617. [CrossRef]
155. Hitchin, R.; Knight, I. Daily energy consumption signatures and control charts for air-conditioned buildings. *Energy Build.* **2016**, *112*, 101–109. [CrossRef]
156. Jalori, S.; Reddy, T.A. A unified inverse modeling framework for whole-building energy interval data: Daily and hourly baseline modeling and short-term load forecasting. *Ashrae Trans.* **2015**, *121*, 156.
157. Abushakra, B.; Paulus, M.T. An hourly hybrid multi-variate change-point inverse model using short-term monitored data for annual prediction of building energy performance, part III: Results and analysis (1404-RP). *Sci. Technol. Built Environ.* **2016**, *22*, 996–1009. [CrossRef]
158. Bauwens, G.; Roels, S. Co-heating test: A state-of-the-art. *Energy Build.* **2014**, *82*, 163–172. [CrossRef]
159. Erkoreka, A.; Garcia, E.; Martin, K.; Teres-Zubiaga, J.; Del Portillo, L. In-use office building energy characterization through basic monitoring and modelling. *Energy Build.* **2016**, *119*, 256–266. [CrossRef]
160. Giraldo-Soto, C.; Erkoreka, A.; Mora, L.; Uriarte, I.; Del Portillo, L.A. Monitoring System Analysis for Evaluating a Building's Envelope Energy Performance through Estimation of Its Heat Loss Coefficient. *Sensors* **2018**, *18*, 2360. [CrossRef]
161. Uriarte, I.; Erkoreka, A.; Giraldo-Soto, C.; Martin-Escudero, K.; Uriarte, A.; Eguia, P. Mathematical development of an average method for estimating the reduction of the Heat Loss Coefficient of an energetically retrofitted occupied office building. *Energy Build.* **2019**, *192*, 101–122. [CrossRef]
162. Busato, F.; Lazzarin, R.M.; Noro, M. Energy and economic analysis of different heat pump systems for space heating. *Int. J. Low Carbon Technol.* **2012**, *7*, 104–112. [CrossRef]
163. Busato, F.; Lazzarin, R.; Noro, M. Two years of recorded data for a multisource heat pump system: A performance analysis. *Appl. Therm. Eng.* **2013**, *57*, 39–47. [CrossRef]
164. Krese, G.; Lampret, Ž.; Butala, V.; Prek, M. Determination of a Building's balance point temperature as an energy characteristic. *Energy* **2018**, *165*, 1034–1049. [CrossRef]
165. Sjögren, J.U.; Andersson, S.; Olofsson, T. Sensitivity of the total heat loss coefficient determined by the energy signature approach to different time periods and gained energy. *Energy Build.* **2009**, *41*, 801–808. [CrossRef]
166. Vesterberg, J.; Andersson, S.; Olofsson, T. Robustness of a regression approach, aimed for calibration of whole building energy simulation tools. *Energy Build.* **2014**, *81*, 430–434. [CrossRef]
167. Meng, Q.; Xiong, C.; Mourshed, M.; Wu, M.; Ren, X.; Wang, W.; Li, Y.; Song, H. Change-point multivariable quantile regression to explore effect of weather variables on building energy consumption and estimate base temperature range. *Sustain. Cities Soc.* **2020**, *53*, 101900. [CrossRef]
168. Oh, S.; Haberl, J.S.; Baltazar, J.-C. Analysis methods for characterizing energy saving opportunities from home automation devices using smart meter data. *Energy Build.* **2020**, *216*, 109955. [CrossRef]
169. Westermann, P.; Deb, C.; Schlueter, A.; Evins, R. Unsupervised learning of energy signatures to identify the heating system and building type using smart meter data. *Appl. Energy* **2020**, *264*, 114715. [CrossRef]
170. Pasichnyi, O.; Wallin, J.; Kordas, O. Data-driven building archetypes for urban building energy modelling. *Energy* **2019**, *181*, 360–377. [CrossRef]
171. Qomi, M.J.A.; Noshadravan, A.; Sobstyl, J.M.; Toole, J.; Ferreira, J.; Pellenq, R.J.-M.; Ulm, F.-J.; Gonzalez, M.C. Data analytics for simplifying thermal efficiency planning in cities. *J. R. Soc. Interface* **2016**, *13*, 20150971. [CrossRef]

172. Afshari, A.; Friedrich, L.A. Inverse modeling of the urban energy system using hourly electricity demand and weather measurements, Part 1: Black-box model. *Energy Build.* **2017**, *157*, 126–138. [CrossRef]
173. Afshari, A.; Liu, N. Inverse modeling of the urban energy system using hourly electricity demand and weather measurements, Part 2: Gray-box model. *Energy Build.* **2017**, *157*, 139–156. [CrossRef]
174. Catalina, T.; Virgone, J.; Blanco, E. Development and validation of regression models to predict monthly heating demand for residential buildings. *Energy Build.* **2008**, *40*, 1825–1832. [CrossRef]
175. Hygh, J.S.; Decarolis, J.F.; Hill, D.B.; Ranjithan, S.R. Multivariate regression as an energy assessment tool in early building design. *Build. Environ.* **2012**, *57*, 165–175. [CrossRef]
176. Asadi, S.; Amiri, S.S.; Mottahedi, M. On the development of multi-linear regression analysis to assess energy consumption in the early stages of building design. *Energy Build.* **2014**, *85*, 246–255. [CrossRef]
177. Al Gharably, M.; Decarolis, J.F.; Ranjithan, S.R. An enhanced linear regression-based building energy model (LRBEM+) for early design. *J. Build. Perform. Simul.* **2015**, *9*, 115–133. [CrossRef]
178. Ipbüker, C.; Valge, M.; Kalbe, K.; Mauring, T.; Tkaczyk, A. Case Study of Multiple Regression as Evaluation Tool for the Study of Relationships between Energy Demand, Air Tightness, and Associated Factors. *J. Energy Eng.* **2017**, *143*, 04016027. [CrossRef]
179. Ligier, S.; Robillart, M.; Schalbart, P.; Peuportier, B. Energy Performance Contracting Methodology Based upon Simulation and Measurement. In Proceedings of the Building Simulation 2017, San Francisco, CA, USA, 7–9 August 2017.
180. Abdelalim, A.; O'Brien, W.; Shi, Z. Data visualization and analysis of energy flow on a multi-zone building scale. *Autom. Constr.* **2017**, *84*, 258–273. [CrossRef]
181. Liu, X.; Mancarella, P. Modelling, assessment and Sankey diagrams of integrated electricity-heat-gas networks in multi-vector district energy systems. *Appl. Energy* **2016**, *167*, 336–352. [CrossRef]
182. Masuda, H.; Claridge, D. *Estimation of Building Parameters Using Simplified Energy Balance Model and Metered Whole Building Energy Use*; Texas A&M University Libraries: Killeen, TX, USA, 2012.
183. Tronchin, L.; Manfren, M.; Tagliabue, L.C. Optimization of building energy performance by means of multi-scale analysis—Lessons learned from case studies. *Sustain. Cities Soc.* **2016**, *27*, 296–306. [CrossRef]
184. Tronchin, L.; Manfren, M.; Nastasi, B. Energy analytics for supporting built environment decarbonisation. *Energy Procedia* **2019**, *157*, 1486–1493. [CrossRef]
185. Manfren, M.; Nastasi, B. From in-situ measurement to regression and time series models: An overview of trends and prospects for building performance modelling. *AIP Conf. Proc.* **2019**, *2123*, 20100. [CrossRef]
186. ISO. *ISO 52000-1:2017, Energy Performance of Buildings—Overarching EPB Assessment–Part 1: General Framework and Procedures*; ISO: Geneva, Switzerland, 2017.
187. Li, F.G.N.; Smith, A.; Biddulph, P.; Hamilton, I.; Lowe, R.J.; Mavrogianni, A.; Oikonomou, E.; Raslan, R.; Stamp, S.; Stone, A.; et al. Solid-wallU-values: Heat flux measurements compared with standard assumptions. *Build. Res. Inf.* **2015**, *43*, 238–252. [CrossRef]
188. Rasooli, A.; Itard, L. In-situ characterization of walls' thermal resistance: An extension to the ISO 9869 standard method. *Energy Build.* **2018**, *179*, 374–383. [CrossRef]
189. Alzetto, F.; Meulemans, J.; Pandraud, G.; Roux, D. A perturbation method to estimate building thermal performance. *Comptes Rendus Chim.* **2018**, *21*, 938–942. [CrossRef]
190. Meulemans, J. An Assessment of the QUB/e Method for Fast In Situ Measurements of the Thermal Performance of Building Fabrics in Cold Climates. In *Cold Climate HVAC Conference*; Springer: Berlin/Heidelberg, Germany, 2018; pp. 317–326.
191. Ahmad, N.; Ghiaus, C.; Thiery, T. Influence of Initial and Boundary Conditions on the Accuracy of the QUB Method to Determine the Overall Heat Loss Coefficient of a Building. *Energies* **2020**, *13*, 284. [CrossRef]
192. Bouchié, R.; Alzetto, F.; Brun, A.; Boisson, P.; Thebault, S. Short methodologies for in-situ assessment ofthe intrinsic thermal performance of the building envelope. In Proceedings of the Sustainable Places 2014 (SP2014), Nice, France, 1–3 October 2014.
193. Thébault, S.; Bouchié, R. Refinement of the ISABELE method regarding uncertainty quantification and thermal dynamics modelling. *Energy Build.* **2018**, *178*, 182–205. [CrossRef]
194. Michalak, P. A thermal network model for the dynamic simulation of the energy performance of buildings with the time varying ventilation flow. *Energy Build.* **2019**, *202*, 109337. [CrossRef]
195. Lundström, L.; Akander, J.; Zambrano, J. Development of a Space Heating Model Suitable for the Automated Model Generation of Existing Multifamily Buildings—A Case Study in Nordic Climate. *Energies* **2019**, *12*, 485. [CrossRef]
196. Fuentenueva, C.; DeNaveros, I.; Ghiaus, C.; Ordoñez, J.; Ruiz, D.P. Thermal networks considering graph theory and thermodynamics. In Proceedings of the 12th International Conference on Heat Transfer, Fluid Mechanics and Thermodynamics, Malaga, Spain, 11–13 July 2016; pp. 1568–1573.
197. Naveros, I.; Ghiaus, C.; Ordoñez, J.; Ruíz, D.P. Thermal networks from the heat equation by using the finite element method. *WIT Trans. Eng. Sci.* **2016**, *106*, 33–43. [CrossRef]
198. Ghiaus, C.; Ahmad, N. Thermal circuits assembling and state-space extraction for modelling heat transfer in buildings. *Energy* **2020**, *195*, 117019. [CrossRef]
199. Ramallo-González, A.P.; Eames, M.E.; Natarajan, S.; Fosas-De-Pando, D.; Coley, D.A. An analytical heat wave definition based on the impact on buildings and occupants. *Energy Build.* **2020**, *216*, 109923. [CrossRef]
200. Michalak, P. The development and validation of the linear time varying Simulink-based model for the dynamic simulation of the thermal performance of buildings. *Energy Build.* **2017**, *141*, 333–340. [CrossRef]

201. Hong, T.; Lee, S.H. Integrating physics-based models with sensor data: An inverse modeling approach. *Build. Environ.* **2019**, *154*, 23–31. [CrossRef]
202. Juricic, S.; Goffart, J.; Rouchier, S.; Foucquier, A.; Cellier, N.; Fraisse, G. Influence of weather natural variability on the thermal characterisation of a building envelope. *arXiv* **2020**, arXiv:2011.13593.
203. Baasch, G.; Wicikowski, A.; Faure, G.; Evins, R. Comparing Gray Box Methods to Derive Building Properties from Smart Thermostat Data. In Proceedings of the 6th ACM International Conference on Systems for Energy-Efficient Buildings, Cities, and Transportation, New York, NY, USA, 13–14 November 2019; ACM: New York, NY, USA, 2019; pp. 223–232.
204. Raillon, L.; Ghiaus, C. Study of Error Propagation in the Transformations of Dynamic Thermal Models of Buildings. *J. Control. Sci. Eng.* **2017**, *2017*, 1–15. [CrossRef]
205. Gustin, M.; McLeod, R.S.; Lomas, K.J. Forecasting indoor temperatures during heatwaves using time series models. *Build. Environ.* **2018**, *143*, 727–739. [CrossRef]
206. Gustin, M.; McLeod, R.S.; Lomas, K.J. Can semi-parametric additive models outperform linear models, when forecasting indoor temperatures in free-running buildings? *Energy Build.* **2019**, *193*, 250–266. [CrossRef]
207. Panão, M.J.O.; Santos, C.A.; Mateus, N.M.; Da Graça, G.C. Validation of a lumped RC model for thermal simulation of a double skin natural and mechanical ventilated test cell. *Energy Build.* **2016**, *121*, 92–103. [CrossRef]
208. Naveros-Mesa, I.; Ghiaus, C.; Ruiz, D.; Castaño, S. Physical parameters identification of walls using ARX models obtained by deduction. *Energy Build.* **2015**, *108*, 317–329. [CrossRef]
209. Andriamamonjy, A.; Klein, R.; Saelens, D. Automated grey box model implementation using BIM and Modelica. *Energy Build.* **2019**, 209–225. [CrossRef]
210. Lehmann, B.; Gyalistras, D.; Gwerder, M.; Wirth, K.; Carl, S. Intermediate complexity model for Model Predictive Control of Integrated Room Automation. *Energy Build.* **2013**, *58*, 250–262. [CrossRef]
211. Raillon, L.; Ghiaus, C. An efficient Bayesian experimental calibration of dynamic thermal models. *Energy* **2018**, *152*, 818–833. [CrossRef]
212. Raillon, L.L.; Ghiaus, C. Sequential Monte Carlo for states and parameters estimation in dynamic thermal models. In Proceedings of the Building Simulation 2017 Conference, San Francisco, CA, USA, 7–9 August 2017; pp. 988–997.
213. Serale, G.; Fiorentini, M.; Capozzoli, A.; Bernardini, D.; Bemporad, A. Model Predictive Control (MPC) for Enhancing Building and HVAC System Energy Efficiency: Problem Formulation, Applications and Opportunities. *Energies* **2018**, *11*, 631. [CrossRef]
214. Ahmad, M.W.; Mourshed, M.; Mundow, D.; Sisinni, M.; Rezgui, Y. Building energy metering and environmental monitoring—A state-of-the-art review and directions for future research. *Energy Build.* **2016**, *120*, 85–102. [CrossRef]
215. Carstens, H.; Xia, X.; Yadavalli, S. Measurement uncertainty in energy monitoring: Present state of the art. *Renew. Sustain. Energy Rev.* **2018**, *82*, 2791–2805. [CrossRef]
216. Senave, M.; Roels, S.; Verbeke, S.; Saelens, D. Analysis of the influence of the definition of the interior dwelling temperature on the characterization of the heat loss coefficient via on-board monitoring. *Energy Build.* **2020**, *215*, 109860. [CrossRef]
217. Lydon, G.; Caranovic, S.; Hischier, I.; Schlueter, A. Coupled simulation of thermally active building systems to support a digital twin. *Energy Build.* **2019**, *202*, 109298. [CrossRef]
218. Hamilton, I.; Summerfield, A.; Lowe, R.J.; Ruyssevelt, P.; Elwell, C.A.; Oreszczyn, T. Energy epidemiology: A new approach to end-use energy demand research. *Build. Res. Inf.* **2013**, *41*, 482–497. [CrossRef]
219. Hamilton, I.; Summerfield, A.; Oreszczyn, T.; Ruyssevelt, P. Using epidemiological methods in energy and buildings research to achieve carbon emission targets. *Energy Build.* **2017**, *154*, 188–197. [CrossRef]
220. Junker, R.G.; Azar, A.G.; Lopes, R.; Lindberg, K.B.; Reynders, G.; Relan, R.; Madsen, H. Characterizing the energy flexibility of buildings and districts. *Appl. Energy* **2018**, *225*, 175–182. [CrossRef]
221. Kohlhepp, P.; Harb, H.; Wolisz, H.; Waczowicz, S.; Müller, D.; Hagenmeyer, V. Large-scale grid integration of residential thermal energy storages as demand-side flexibility resource: A review of international field studies. *Renew. Sustain. Energy Rev.* **2019**, *101*, 527–547. [CrossRef]
222. Vijay, A.; Hawkes, A. Demand side flexibility from residential heating to absorb surplus renewables in low carbon futures. *Renew. Energy* **2019**, *138*, 598–609. [CrossRef]
223. Bee, E.; Prada, A.; Baggio, P.; Psimopoulos, E. Air-source heat pump and photovoltaic systems for residential heating and cooling: Potential of self-consumption in different European climates. *Build. Simul.* **2019**, *12*, 453–463. [CrossRef]
224. Psimopoulos, E.; Bee, E.; Widén, J.; Bales, C. Techno-economic analysis of control algorithms for an exhaust air heat pump system for detached houses coupled to a photovoltaic system. *Appl. Energy* **2019**, *249*, 355–367. [CrossRef]
225. Facci, A.L.; Krastev, V.K.; Falcucci, G.; Ubertini, S. Smart integration of photovoltaic production, heat pump and thermal energy storage in residential applications. *Sol. Energy* **2019**, *192*, 133–143. [CrossRef]
226. Oliveira-Lima, J.A.; Delgado-Gomes, V.; Martins, J.; Lima, C. Standard-based service-oriented infrastructure to integrate intelligent buildings in distributed generation and smart grids. *Energy Build.* **2014**, *76*, 450–458. [CrossRef]
227. Sun, M.; Djapic, P.; Aunedi, M.; Pudjianto, D.; Strbac, G. Benefits of smart control of hybrid heat pumps: An analysis of field trial data. *Appl. Energy* **2019**, *247*, 525–536. [CrossRef]
228. Gui, E.M.; MacGill, I. *Consumer-Centric Service Innovations in an Era of Self-Selecting Customers. Consumer, Prosumer, Prosumager: How Service Innovations will Disrupt the Utility Business Model*; Academic Press: Cambridge, MA, USA, 2019; pp. 127–151. [CrossRef]

229. Clauß, J.; Finck, C.; Vogler-Finck, P.; Beagon, P. Control strategies for building energy systems to unlock demand side flexibility—A review. In Proceedings of the IBPSA Building Simulation 2017, San Francisco, CA, USA, 7–9 August 2017.
230. Péan, T.; Salom, J.; Costa-Castelló, R. Review of control strategies for improving the energy flexibility provided by heat pump systems in buildings. *J. Process. Control.* **2019**, *74*, 35–49. [CrossRef]
231. Clauß, J.; Georges, L. Model complexity of heat pump systems to investigate the building energy flexibility and guidelines for model implementation. *Appl. Energy* **2019**, *255*, 113847. [CrossRef]
232. Drgoňa, J.; Arroyo, J.; Figueroa, I.C.; Blum, D.H.; Arendt, K.; Kim, D.; Ollé, E.P.; Oravec, J.; Wetter, M.; Vrabie, D.L.; et al. All you need to know about model predictive control for buildings. *Annu. Rev. Control.* **2020**, *50*, 190–232. [CrossRef]
233. De Coninck, R.; Baetens, R.; Saelens, D.; Woyte, A.; Helsen, L. Rule-based demand-side management of domestic hot water production with heat pumps in zero energy neighbourhoods. *J. Build. Perform. Simul.* **2013**, *7*, 271–288. [CrossRef]
234. Klein, K.; Kalz, D.; Herkel, S. Grid impact of a net zero energy building with BiPV using different energy management strategies. In Proceedings of the International Conference CISBAT 2015 Future Buildings and Districts Sustainability from Nano to Urban Scale, Lausanne, Switzerland, 9–11 September 2015; pp. 579–584.
235. Le Dréau, J.; Heiselberg, P. Energy flexibility of residential buildings using short term heat storage in the thermal mass. *Energy* **2016**, *111*, 991–1002. [CrossRef]
236. Dar, U.I.; Sartori, I.; Georges, L.; Novakovic, V. Advanced control of heat pumps for improved flexibility of Net-ZEB towards the grid. *Energy Build.* **2014**, *69*, 74–84. [CrossRef]
237. Reynders, G.; Diriken, J.; Saelens, D. A generic quantification method for the active demand response potential of structural storage in buildings. In Proceedings of the 14th International Conference of IBPSA-Building Simulation, Hyderabad, India, 7–9 December 2015; pp. 1986–1993.
238. Turner, W.J.N.; Walker, I.S.; Roux, J. Peak load reductions: Electric load shifting with mechanical pre-cooling of residential buildings with low thermal mass. *Energy* **2015**, *82*, 1057–1067. [CrossRef]
239. Esfehani, H.H.; Kriegel, M.; Madani, H. Load balancing potential of ground source heat pump system coupled with thermal energy storage: A Case Study for Berlin. In Proceedings of the CLIMA 2016, 12th REHVA World Congress, Aalborg, Denmark, 22–25 May 2016.
240. Alimohammadisagvand, B.; Jokisalo, J.; Kilpeläinen, S.; Ali, M.; Sirén, K. Cost-optimal thermal energy storage system for a residential building with heat pump heating and demand response control. *Appl. Energy* **2016**, *174*, 275–287. [CrossRef]
241. Salpakari, J.; Lund, P. Optimal and rule-based control strategies for energy flexibility in buildings with PV. *Appl. Energy* **2016**, *161*, 425–436. [CrossRef]
242. Masy, G.; Georges, E.; Verhelst, C.; Lemort, V.; André, P. Smart grid energy flexible buildings through the use of heat pumps and building thermal mass as energy storage in the Belgian context. *Sci. Technol. Built Environ.* **2015**, *21*, 800–811. [CrossRef]
243. Panão, M.J.O.; Mateus, N.M.; Da Graça, G.C. Measured and modeled performance of internal mass as a thermal energy battery for energy flexible residential buildings. *Appl. Energy* **2019**, *239*, 252–267. [CrossRef]
244. Vivian, J.; Chiodarelli, U.; Emmi, G.; Zarrella, A. A sensitivity analysis on the heating and cooling energy flexibility of residential buildings. *Sustain. Cities Soc.* **2020**, *52*, 101815. [CrossRef]
245. De Coninck, R.; Helsen, L. Quantification of flexibility in buildings by cost curves—Methodology and application. *Appl. Energy* **2016**, *162*, 653–665. [CrossRef]
246. Halvgaard, R.; Poulsen, N.K.; Madsen, H.; Jorgensen, J.B. Economic Model Predictive Control for building climate control in a Smart Grid. In Proceedings of the 2012 IEEE PES Innovative Smart Grid Technologies (ISGT), Washington, DC, USA, 16–20 January 2012; pp. 1–6.
247. Maasoumy Haghighi, M. *Controlling Energy-Efficient Buildings in the Context of Smart Grid: A Cyber Physical System Approach*; Technical Report No. UCB/EECS-2013-244; University of California: Berkley, CA, USA, 2014.
248. Corbin, C.D.; Henze, G. Predictive control of residential HVAC and its impact on the grid. Part I: Simulation framework and models. *J. Build. Perform. Simul.* **2016**, *10*, 294–312. [CrossRef]
249. Corbin, C.; Henze, G. Predictive control of residential HVAC and its impact on the grid. Part II: Simulation studies of residential HVAC as a supply following resource. *J. Build. Perform. Simul.* **2017**, *10*, 365–377. [CrossRef]
250. Lindelöf, D.; Afshari, H.; Alisafaee, M.; Biswas, J.; Caban, M.; Mocellin, X.; Viaene, J. Field tests of an adaptive, model-predictive heating controller for residential buildings. *Energy Build.* **2015**, *99*, 292–302. [CrossRef]
251. Garnier, A.; Eynard, J.; Caussanel, M.; Grieu, S. Predictive control of multizone heating, ventilation and air-conditioning systems in non-residential buildings. *Appl. Soft Comput.* **2015**, *37*, 847–862. [CrossRef]
252. Kandler, C.; Wimmer, P.; Honold, J. Predictive Control and Regulation Strategies of Air-to-Water Heat Pumps. *Energy Procedia* **2015**, *78*, 2088–2093. [CrossRef]
253. Blum, D.H.; Arendt, K.; Rivalin, L.; Piette, M.; Wetter, M.; Veje, C. Practical factors of envelope model setup and their effects on the performance of model predictive control for building heating, ventilating, and air conditioning systems. *Appl. Energy* **2019**, *236*, 410–425. [CrossRef]
254. Dong, B.; Yan, D.; Li, Z.; Jin, Y.; Feng, X.; Fontenot, H. Modeling occupancy and behavior for better building design and operation—A critical review. *Build. Simul.* **2018**, *11*, 899–921. [CrossRef]
255. Naylor, S.; Gillott, M.; Lau, T. A review of occupant-centric building control strategies to reduce building energy use. *Renew. Sustain. Energy Rev.* **2018**, *96*, 1–10. [CrossRef]

256. Carpino, C.; Mora, D.; Arcuri, N.; De Simone, M. Behavioral variables and occupancy patterns in the design and modeling of Nearly Zero Energy Buildings. *Build. Simul.* **2017**, *22*, 860–888. [CrossRef]
257. Caucheteux, A.; Sabar, A.E.; Boucher, V. Occupancy measurement in building: A litterature review, application on an energy efficiency research demonstrated building. *Int. J. Metrol. Qual. Eng.* **2013**, *4*, 135–144. [CrossRef]
258. Naspi, F.; Arnesano, M.; Stazi, F.; D'Orazio, M.; Revel, G.M. Measuring Occupants' Behaviour for Buildings' Dynamic Cosimulation. *J. Sens.* **2018**, *2018*, 2756542. [CrossRef]
259. Cecconi, F.R.; Manfren, M.; Tagliabue, L.C.; Ciribini, A.L.C.; De Angelis, E. Probabilistic behavioral modeling in building performance simulation: A Monte Carlo approach. *Energy Build.* **2017**, *148*, 128–141. [CrossRef]
260. Tagliabue, L.C.; Manfren, M.; Ciribini, A.L.C.; De Angelis, E. Probabilistic behavioural modeling in building performance simulation—The Brescia eLUX lab. *Energy Build.* **2016**, *128*, 119–131. [CrossRef]
261. De Menezes, A.C.K.; Cripps, A.; Bouchlaghem, D.; Buswell, R.A. Predicted vs. actual energy performance of non-domestic buildings: Using post-occupancy evaluation data to reduce the performance gap. *Appl. Energy* **2012**, *97*, 355–364. [CrossRef]
262. Aragon, V.; Gauthier, S.; Warren, P.; James, P.A.B.; Anderson, B. Developing English domestic occupancy profiles. *Build. Res. Inf.* **2019**, *47*, 375–393. [CrossRef]
263. Zangheri, P.; Pagliano, L.; Armani, R. How the comfort requirements can be used to assess and design low energy buildings: Testing the EN 15251 comfort evaluation procedure in 4 buildings. In Proceedings of the ECEEE 2011 Summer Study "Energy Efficiency First: The Foundation of a Low-Carbon Society", Hyeres, France, 6–11 June 2011; pp. 1569–1579.
264. Fabbri, K.; Tronchin, L. Indoor Environmental Quality in Low Energy Buildings. *Energy Procedia* **2015**, *78*, 2778–2783. [CrossRef]
265. Manfren, M.; Nastasi, B.; Piana, E.A.; Tronchin, L. On the link between energy performance of building and thermal comfort: An example. *AIP Conf. Proc.* **2019**, *2123*, 20066. [CrossRef]
266. Sarfraz, O.; Bach, C.K. Equipment power consumption and load factor profiles for buildings' energy simulation (ASHRAE 1742-RP). *Sci. Technol. Built Environ.* **2018**, *24*, 1054–1063. [CrossRef]
267. Gunay, B.; Shen, W.; Yang, C. Characterization of a building's operation using automation data: A review and case study. *Build. Environ.* **2017**, *118*, 196–210. [CrossRef]
268. Saini, J.; Dutta, M.; Marques, G. Indoor Air Quality Monitoring Systems Based on Internet of Things: A Systematic Review. *Int. J. Environ. Res. Public Health* **2020**, *17*, 4942. [CrossRef]
269. Martín-Garín, A.; Millán-García, J.; Baïri, A.; Millán-Medel, J.; Sala-Lizarraga, J. Environmental monitoring system based on an Open Source Platform and the Internet of Things for a building energy retrofit. *Autom. Constr.* **2018**, *87*, 201–214. [CrossRef]
270. Lucchi, E. Environmental Risk Management for Museums in Historic Buildings through an Innovative Approach: A Case Study of the Pinacoteca di Brera in Milan (Italy). *Sustainability* **2020**, *12*, 5155. [CrossRef]
271. Pauwels, P.; Zhang, S.; Lee, Y.-C. Semantic web technologies in AEC industry: A literature overview. *Autom. Constr.* **2017**, *73*, 145–165. [CrossRef]
272. Corry, E.; Pauwels, P.; Hu, S.; Keane, M.; O'Donnell, J. A performance assessment ontology for the environmental and energy management of buildings. *Autom. Constr.* **2015**, *57*, 249–259. [CrossRef]

energies

MDPI

Article

Utilising Open Geospatial Data to Refine Weather Variables for Building Energy Performance Evaluation—Incident Solar Radiation and Wind-Driven Infiltration Modelling

Kristian Skeie * and **Arild Gustavsen**

Department of Architecture and Technology, Norwegian University of Science and Technology, Alfred Getz vei 3, 7491 Trondheim, Norway; arild.gustavsen@ntnu.no
* Correspondence: kristian.skeie@ntnu.no

Abstract: In building thermal energy characterisation, the relevance of proper modelling of the effects caused by solar radiation, temperature and wind is seen as a critical factor. Open geospatial datasets are growing in diversity, easing access to meteorological data and other relevant information that can be used for building energy modelling. However, the application of geospatial techniques combining multiple open datasets is not yet common in the often scripted workflows of data-driven building thermal performance characterisation. We present a method for processing time-series from climate reanalysis and satellite-derived solar irradiance services, by implementing land-use, and elevation raster maps served in an elevation profile web-service. The article describes a methodology to: (1) adapt gridded weather data to four case-building sites in Europe; (2) calculate the incident solar radiation on the building facades; (3) estimate wind and temperature-dependent infiltration using a single-zone infiltration model and (4) including separating and evaluating the sheltering effect of buildings and trees in the vicinity, based on building footprints. Calculations of solar radiation, surface wind and air infiltration potential are done using validated models published in the scientific literature. We found that using scripting tools to automate geoprocessing tasks is widespread, and implementing such techniques in conjunction with an elevation profile web service made it possible to utilise information from open geospatial data surrounding a building site effectively. We expect that the modelling approach could be further improved, including diffuse-shading methods and evaluating other wind shelter methods for urban settings.

Keywords: thermal building performance; satellite-based solar radiation data; meteorological reanalysis data; ISO 52016-1; single-zone infiltration

Citation: Skeie, K.; Gustavsen, A. Utilising Open Geospatial Data to Refine Weather Variables for Building Energy Performance Evaluation—Incident Solar Radiation and Wind-Driven Infiltration Modelling. *Energies* **2021**, *14*, 802. https://doi.org/10.3390/en14040802

Academic Editor: Benedetto Nastasi

Received: 16 December 2020
Accepted: 22 January 2021
Published: 3 February 2021

1. Introduction

Meteorological data like temperature, wind speed, and solar radiation are essential input for characterising buildings' thermal performance. Ideally, these elements are measured locally using a well-maintained weather station near the building site or on the building itself. The increasing availability of high-resolution geospatial data, gridded weather data and adequate modelling techniques (including web-services) can provide an alternative approach to estimating local climatic building boundary conditions in the built environment [1,2]. Using assimilated data-sources has several advantages, e.g., making it possible to supplement low-cost air temperature observations, that are relatively common to measure on-site, with other weather variables that are more difficult to capture or predict in a simple way. Such as solar irradiance data from services built on remote sensing of sky conditions, or wind speed estimations from numerical weather prediction (NWP)-models in forecast or reanalysis-mode. Reanalysis is a method to reconstruct the past weather by combining modelling of the atmospheric dynamics and physics of the earth climate systems with historical observations. Daily updated information about the past weather and historical climate is available via Copernicus Climate Change Service (C3S) Climate Data

Store (CDS) such as the ECMWF (European Centre for Medium-Range Weather Forecasts) fifth-generation global reanalysis (ERA5) and the soon to be released Copernicus Regional Reanalysis for Europe (CERRA). Advancements in temporal and spatial resolutions and dedicated land surface analysis, like ERA5-Land [3], may extend the popularity and application to many fields [4]. Some meteorology institutions are also developing hourly surface analysis products on a regional level, combining their operational mesoscale models for weather forecasting with observations and making the assimilated products available as gridded datasets free-of-charge [5]. However, with the use of any gridded weather data product for building energy performance evaluation, comes a need to adjust the data to fit local building boundary conditions [1,2].

Plenty of methods exist to downscale or bias-correct gridded weather data and to include local effects from the terrain, vegetation and buildings in solar and wind assessments. Downscaling techniques reaching the meso and micro-scale span from simple analytical or statistical methods to running high-resolution NWP-models informed by global reanalysis [6], or even computational fluid dynamics (CFD) codes uncoupled or coupled with atmospheric models [7]. When it comes to including local sheltering effects, tools vary significantly in the overall approach and temporal and spatial resolution. With increasing interest in local renewable energy generation and urban scale modelling, numerous research efforts have been initiated to develop and refine tools and methods to support wind energy analysis, façade and rooftop solar potential assessments, urban building energy modelling and urban micro-climate studies [8–10]. The diversity of methods and tools logically reflect the wide variety of use cases, but also evident are the multiple ways to represent surfaces and other features in geospatial datasets (e.g., terrain, tree canopies, roofs and facades). Workflows integrated with graphical information system (GIS)-tools operating on two-dimensional raster maps that supply the surface elevation are prevalent [10]. Modern toolkits offer to automate geoprocessing tasks through Python scripts and web-mapping services [11,12]. To encompass height information, e.g., point-clouds, into full 3D-processing, often requires time-consuming manual work and expertise [13,14]. Although more and more mapping agencies and local authorities are releasing point cloud data or 3D-building models according to Open Geospatial Consortium (OGC) standards, these are mostly limited to city-scale, province, or municipality levels. 3D-building models and point clouds also risk being outdated if not updated at frequent intervals.

In the following, we use high-resolution height data from airborne laser scanning which is becoming widely available in the form of pre-processed digital surface and terrain models covering large land surface areas on a regional or national scale [15]. Despite that these datasets may suffer from the same problems as above, they may easily be served in a web-service and supplemented by building footprints with user-specified heights in the nearby area of interest. Footprints are broadly available as open data from authorities or volunteered geoinformation and usually produced at more frequent intervals [13].

The application of geospatial techniques combining multiple open datasets is not common in the often-scripted workflows of data-driven building thermal performance characterisation.

This work aims to investigate ways to adapt site-specific climate data for building thermal energy analysis, by identifying suitable open geospatial datasets that can be served in a web-service and demonstrate a scripted workflow that can be implemented to calculate solar and wind effects on buildings facades. We present a method for processing time-series from climate reanalysis and satellite-derived solar irradiance services, by implementing land-use, and elevation raster maps served in an elevation profile web-service. Building footprints from OpenStreetMap Overpass API complement the analysis by separating buildings and trees in the vicinity.

1.1. Outline

This paper first addresses how much we can adapt gridded weather data to local building boundary conditions using only building location (latitude, longitude) and a selection of free/open geospatial datasets covering Europe. A comparison between observations and the ERA5 reanalysis and CAMS-Rad satellite service is shown for a low-rise building localised in a relatively open landscape in the south of Germany. Including a surface wind downscaling method using weighted surface roughness derived from land-use maps. Next, we investigate what detailed geospatial data was found for our three other residential buildings in Norway, UK and Belgium. These are located in more urban settings, where shading and wind sheltering of nearby obstructions have a more significant impact. The four case studies are used to evaluate the proposed methods to assess solar radiation distribution on the facades and wind and temperature-dependent infiltration; Two factors influenced by nearby topography and obstructions that are not always considered physically in data-driven building energy performance evaluation. We focus on the influence of sheltering of nearby obstacles, buildings, and trees in either case, using the methods:

(1) The Alberta single-zone air infiltration model [16], including a concept of wind shadow on building facades, projected downstream by upwind obstacles [17].
(2) Solar irradiance modelling according to EN-ISO 52010:2017, including direct and circumsolar beam façade surface shading from obstacles as outlined in EN-ISO 52016:2017.

1.2. Identifying Suitable Open Geospatial Datasets and Previous Works

Open geospatial datasets are growing in diversity, from crowdsourcing efforts to data produced by authorities and scientific collaborations [18]. Developments around open spatial infrastructures ease access to meteorological data and other relevant spatial information. Open data and modelling are valuable, as building monitoring data is often limited. Buildings are complex systems because their energy use and indoor conditions vary dynamically under the influence of weather, occupancy and component performance [19]. The drive towards a more sustainable built environment and low carbon transition of the energy system give rise to challenges that can only be met by multi-disciplinary knowledge [20,21]. Interaction of open data and models may become fundamental for monitoring, verifying and tracking performance at multiple levels [21].

Many geospatial datasets are available via the EU Copernicus Earth observation programme which has operated a policy of open data since its inception. INSPIRE (Infrastructure for Spatial Information in Europe) is a related EU initiative that aims to ease access to public data through standardisation of spatial data among member states. Two of Copernicus regional products, the Digital Elevation Model (DEM) over Europe (EU-DEM), and the CORINE land cover (CLC) maps are relevant examples of pan-European cooperation [22,23]. CLC maps have been used to derive surface roughness classes in numerous wind resource studies [24,25]. The EU-DEM is a hybrid product based on the larger SRTM and ASTER GDEM datasets produced by NASA Earthdata and Japan Space Systems [26], two of many global and freely available DEM's. These are satellite sensor-based models representing the first-return earth surface (including trees, buildings) at a relatively coarse resolution (from ca. 30 m) and accuracy (ca. 5 m to several hundred) [27]. Still, they have been used in solar resource map creation and operating a service that returns a site's horizon profile, available from JRC's PVGIS website [28].

Lately, more and more light detection and ranging (LiDAR) data obtained from state-funded airborne laser scanning (ALS) are published around the world under free licenses. Pre-built digital elevation models are distributed in high resolution as digital surface models (DSMs) and digital terrain models (DTMs) covering entire regions and countries. In Europe, the INSPIRE Geoportal keeps track of downloadable elevation data [15]. For example, in Norway, both post-processed LiDAR point cloud data and digital surface model data are made openly available by the Norwegian Mapping Authority under CC-BY licensing [29]. Pre-built DSM and DTM-tiles in 1-m resolution covering the whole country can be downloaded amounting to a combined download of ca. 2.5 TB in GeoTIFF format. This national model is updated sector-wise when new surveys are produced. For the other countries in the study, DSM's and DTM's in 1 m resolution published under open government licenses were available for the UK, and the region of Flanders, Belgium. In parts of Germany, datasets are still proprietary and come at a cost [30].

Many researchers have shown that high-resolution LiDAR data in its point-cloud form is enabling determination of building geometry and shading from the surrounding environment. Nonetheless, composed DSMs (of 1 m resolution) have proven up to the task to capture the slope and aspect of basic roof shapes (without variation in the architecture of the roof) required to estimate solar potential on rooftops [31,32]. At least two different 2.5D solar models are previously published detecting vertical façades from 1 m DSM pixels and either estimating wall irradiances under clear-sky [33,34] or based on observations of global horizontal radiation [35]. Another feature of the second model is the inclusion of vegetation which is found to be crucial when modelling irradiance on walls in an urban setting, especially where building heights are relatively low [35].

A model resolution of 1 m is interesting, as it allows handling large areas, but still maintain a file-size that is easy to store and work with by splitting data in raster tiles. Another reason to consider digital surface and terrain models over higher-resolution LiDAR point cloud data is that considerable work and expertise [36], has gone into creating the DSM's and DTM's to meet the requirements of the commissioner. Airborne imagery, stereography or orthophotos are also often overlaid in the creation. Likewise, imagery integration is recommended for creating 3D building geometry from LiDAR point clouds to capture objects more accurately [18–20], underlining that expertise in data fusion, processing, and acquisition is needed. Other recent developments show that micro-drones [37] or mobile ground-based laser surveying can be used for detailed building shape and façade mapping [38], indicating new workflows and applications to the building industry.

Other works make use of building ground plans or roof perimeters obtained through user-contributed data such as OpenStreetMap or derived from administrative databases to extrude lower detail building models from the ground and up, or DSM's with or without terrain and surface model data available [10]. Some cities and local authorities are openly distributing 3D city-scale models at higher detail level or building cadastral data at greater accuracy then what can be expected by crowdsourced content [39,40].

However, all of the datasets above suffers from the same challenges: acquisition and that the most recent dataset available may be outdated. Evaluating design or as-built building energy performance as part of commissioning requires recent height-data. There are also other applications than energy evaluation where better assessments of local climatic conditions can positively influence the building design process or operation-phase. Building information and geospatial assessment techniques can be used to reduce climate-induced damages on buildings, improve user quality, and improve the balance between climatic adaptation demands and other demands [41]. When it comes to design-studies, the buildings under consideration may not even be built yet. Therefore, the proposed approach will need to be easily updated when new DSM/DTM's are available for an area. It will also need to have multiple ways to input building footprint and height information and separate between ground, buildings and other tall obstacles like trees to overcome the identified gaps.

2. Methods

This section gives a brief overview of the methods, the code design and implementation. Calculation details and input data are provided in the Appendices A and B. A public Github repository will be published with the full code when the paper is published. Existing packages and scripts used in the workflow include:

- Obtaining reanalysis data from the Copernicus CDS using the ecmwf R package [42]
- Solar radiation from the CAMS-Rad service using the rOpenSci camsRad client [43]
- Self-hosting elevation and land cover data in an Open Topo Data server [44]
- Modules to create horizon profiles from a viewpoint by calling elevation services [45]
- Solar irradiance transposition model according to the ISO 52010 standard [46]
- Wind speed interpolation using key portions of the R-code printed in [47] (p. 45)

- The code implemented in the workflow relies on many additional popular Python and R packages such as rcpp, ncdf4, gdal, pyProj, shapely and netCDF4.

Table 1 provides an overview of the different datasets, spatialisation techniques and analytical models used in this study to process weather data sourced from the climate reanalysis and satellite irradiance service in order to evaluate local solar and wind effects on buildings facades.

Table 1. An overview of open geospatial data sources that cover different spatial scales.

Description	Weather Data Acquisition	Downscaling	Area of Local Study
Scale	Large scale (>10 km)	Medium scale	Small scale (<1 km)
User input	Latitude, longitude	Latitude, longitude	Latitude, longitude and building information
Datasets	Climate reanalysis, Satellite irradiance	Land cover maps, Satellite DEM	Building footprints, LiDAR DSM/DTM
Data resolution	5 to 30 km	30 to 100 m	<1 m
Data sources	Copernicus Climate & Atmosphere Data Store	Copernicus programme, and the JRC (PVGIS)	National authorities and crowdsourced (OSM)
Modelling techniques	Bilinear interpolation, nearest neighbour selection	2-layer wind model [48], Perez transposition model	Wind shadow method [17], ISO shading method

The workflow needed to obtain the local data has largely been automated (Figure 1). First, the area of the local study is defined by the geographic position. LiDAR DSM and DTM raster maps in local projections were downloaded from national mapping services and stored on the server. New rasters containing only DSM data within building footprints were created from Open Street Map (OSM) building layers by calling the Overpass Application Programming Interfaces (API) and an open map layer from the UK Ordnance survey (OSM building outlines were not available for the area of interest in the UK at the time of study (mid-2020)). The local area LiDAR raster datasets were stored as .tif and the building footprints as .shp files. The rasters were served together with the EU-DEM and CLC land cover maps via the elevation API configured to return JSON strings of height (or land type classes) along paths resolved to sets of latitude and longitude points. The service "Open Topo Data REST API" relies on Python's gdal and pyProj packages for conversion between latitude, longitude, and local map projections in meter [44]. Scripting was adjusted to handle specifying latitude and longitude in decimal degrees with six decimal places precision (translating to 0.11 m at the equator) and to include datum shifts, ensuring more accurate conversions between latitude, longitude and local map projections (in meter).

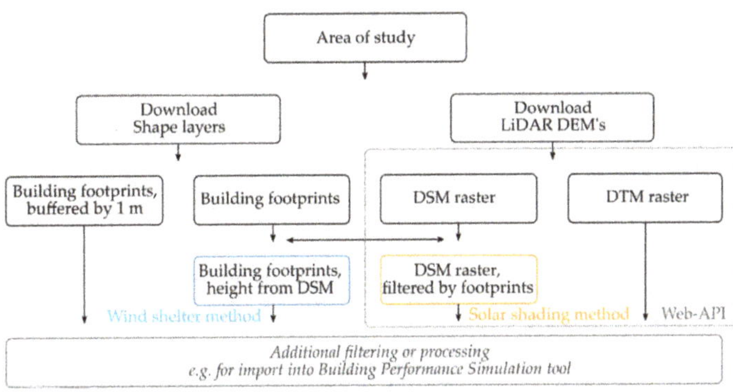

Figure 1. Proposed workflow to assess local sheltering using the wind shadow method and the solar shading model in Section 2.4.

Before running the path profiling scrips, we define the footprint of the case building from the centre point in decimal degrees (lat, lon) by the length of the building envelope in the x-direction, length in the y-direction (in meter) and building rotation (in degrees). We specify the intermediate distance of mapping points for each façade and calculate their respective projection lines in all directions. Figure 2 shows the building and façade centre viewpoints used for the path profiling calculations (in green). In this way, it is irrelevant if the building is represented on the surface raster and building footprint shape layers or not.

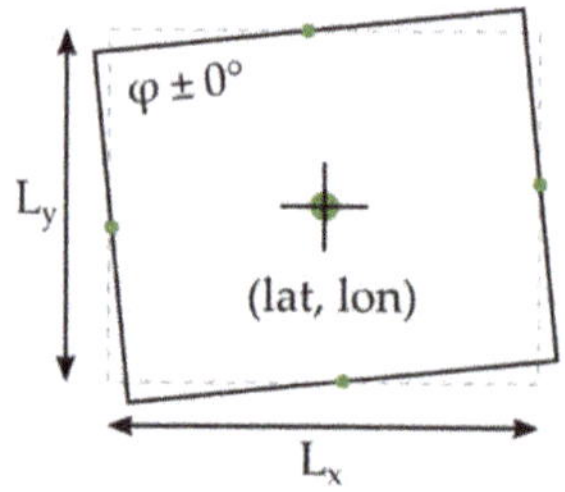

Figure 2. The building façade geometry for solar shading and wind shelter calculations was created from three inputs: length in the x-direction, length in the y-direction (in meter) and building rotation. The latitude and longitude were selected to be the centre-point that the envelope is rotated around.

2.1. *Environmental Variables*

2.1.1. Environmental Variables Derived from Reanalysis Data

Meteorological data (Table 1) were derived from the ERA5 atmospheric reanalysis of the global climate released by the European Centre for Medium-Range Weather Forecasts (ECMWF) as part of the Copernicus Climate Change Service (CDS). A comprehensive description of ERA5 is provided by Hersbach et al. [49]. ERA5 provides validated estimates for each hour of the day, worldwide, with a two to three months delay and leading back to 1979, or as recent as up to a couple of days ago through the preliminary dataset ERA5T.

Table 2 also shows selected variables from "ERA5-Land" another current dataset produced by ECMWF with around four times finer spatial resolution (~9 km grid spacing compared to the ~31 km grid of ERA5). It is a simulation of the land surface components of ERA5 forced by ERA5's lower atmospheric fields, currently without coupling or additional data assimilation, meaning that observations only influence the simulation indirectly through the forcing [50].

Table 2. Climate variables acquired from ERA5 and ERA5-land reanalysis and the model transformation. The name of each variable used in the comparison study is shown on the same line as the sourced variables according to their short names in the Copernicus CDS.

Description	Name	ERA5	$ERA5_{land}$	Transformation
External air temperature at 2 m	θ_e	2t	2t	Kelvin to centigrade
Wind speed at 10 m	U_{10m}	10u, 10v	10u, 10v	$U_{10m} = \sqrt{10u^2 + 10v^2}$
Wind from direction at 10 m	D_{10m}	10u, 10v	10u, 10v	$D_{10m} = \text{atan2}(10u, 10v) + \pi$
Forecasted surface roughness	$z_{0;M}$	fsr		
Ground albedo without snow cover	α_{gr}	ssr, ssrd	fal *	$\alpha_{gr} = \max(1 - ssr/ssrd, fal)$
Snow cover	f_{sn}		snowc	
Ground albedo with snow cover	$\alpha_{gr;sn}$		asn	$\alpha_{gr;sn} = f_{sn} \cdot (asn - \alpha_{gr}) + \alpha_{gr}$
Surface thermal radiation downwards	φ_{strd}	strd		Joule to Watt-hours
Sky temperature	θ_{sky}			$\theta_{sky} = (\varphi_{strd}/\sigma)^{0.25} - 273.15$ K

* fal is a diagnostic broadband albedo, whereas the true ground value is calculated by: $\alpha_{gr} = 1$—ssr/ssrd [51]. To account for increased reflectance when the ground is covered by snow, the ERA5-Land surface model parameters: snow cover and snow albedo were included: $\alpha_{gr;sn} = snowc \cdot asn + (1 - snowc) \cdot \alpha_{gr}$.

We used nearest-neighbour and bilinear interpolation to create hourly time-series for each site-location. Figure 3 shows the four case studies' location relative to the ERA5 grid points and each grid tile's relative weighting. The ERA5-land data were also interpolated to the site (not shown).

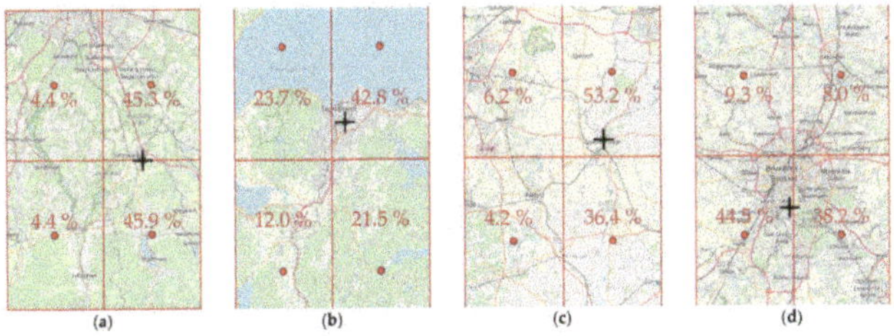

Figure 3. Locations of cases (**a**) Holzkirchen DE, (**b**) Trondheim NO, (**c**) Gainsborough UK, and (**d**) Brussels BE, and nearest ERA5 grid cells (red lines) overlaid over local area maps. The percentages illustrate the resulting bilinear weighting of the adjacent cells. Map data from OpenStreetMap.

2.1.2. Environmental Variables Derived from Remote Sensing

For solar radiation, it is possible to use surface downwelling radiation from the reanalysis, but several studies show that better products exist to account for clouds' variability. Services that combine solar models with remote sensing techniques provide greater temporal and spatial resolution. An overestimation of solar radiation is often observed in reanalysis, and an underestimation is observed in satellite methods [52].

Solar irradiance data were acquired from Copernicus Atmosphere Monitoring Service: CAMS Radiation Service (CAMS-Rad) version 3.2 [53]. CAMS-Rad's satellite-based solar irradiance data are available at a spatial resolution of ~5 km over central Europe in 15-min time steps from 2004 until the present time (up to two days ago) and covers the field of view of the Meteosat satellite (Europe, Africa and the Middle East). An account of the radiative transfer scheme in CAMS-Rad is provided by Qu et al. [54]. We used the camsRad R-package [43] to obtain 1-min time-series of direct normal, global horizontal and diffuse horizontal irradiance for clear-sky and cloudy conditions. The dataset was then resampled to 10-min intervals for the shading calculations.

2.2. Downscaling

2.2.1. Local Wind Speed Estimation

The ERA5 reanalysis is already being applied in wind energy assessment, showing improvements over previously released global and regional reanalysis datasets [55–57]. In several recent studies, dynamical downscaling of ERA5 data using the high-resolution Weather Research and Forecasting (WRF) model has demonstrated an added value of introducing a higher spatial resolution [6,57,58]. The near-surface wind speeds in the reanalysis are advised not to be used directly to indicate surface wind conditions at a site [56], as the relatively low spatial resolution lacks the local representativity of the site surroundings. This recommendation is reflected by a word of caution in the API-documentation:

> *"Care should be taken when comparing this variable with observations because wind observations vary on small space and time scales and are affected by the local terrain, vegetation and buildings that are represented only on average in the ECMWF Integrated Forecasting System."*

In fact, the 10-m near-surface winds in ERA5 are parametrised as the potential wind in open terrain [59], which can differ substantially from the model representation over the whole grid cell [60]. Overland, exposure correction in ERA5 is done taking the lowest model level winds (at a height above the surface that is less influenced by underlining terrain) and applying vertical interpolation to 10 m through a logarithmic wind profile including stability indices (Monin-Obhukov theory). An open-terrain surface roughness (0.03 m) is used in the transformation [60].

In an attempt to better represent surface conditions, a simple analytical downscaling method is used. First presented by Wieringa [59], the "2L-method" consists of a two-layer model of the atmospheric boundary level to account for the difference in surface roughness from one location to another. It is to be used in combination with roughness lengths obtained through anemometric analysis or a surface roughness map. We used parts of the code as printed in [47] and a surface roughness conversion table (see Appendix C) for our implementation. Other more sophisticated models reported in the literature are included in the widely used commercial software WAsP, WindSim, and windPRO. There are similarities between the procedures implemented in WAsP, the 2L-method, and another simple model by De Rooy and Kok [61–63]. The latter was recently used to create a 1 km gridded dataset for Germany of hourly surface variables using station records and a regional climate reanalysis model [4]. Other authors have successfully combined WAsP and statistical approaches [6,64]. These simple downscaling methods are not claimed to represent the full complexity of the boundary layer. The 2L-method has mainly been used to create wind resource maps and determining extreme open-water winds [47,48,62,65]. It was first developed as an interpolation method for surface wind measurements [59] and has later been developed by Verkaik [66,67] and by Wever and Growen [68]. Evaluations over land have revealed mixed results, outlining that the roughness lengths significantly impact model performance and that the use of uniform (non-directional) roughness values leads to large errors [47,65–67].

We follow the approach of Verkaik [66,67] using a high-resolution land-use map with derived surface roughness values together with a simplified footprint model to downscale model wind. In the lower level, the surface wind is transformed into the so-called blending height, where local disturbances have been blended out (Figure 4). Next, the wind speed in the upper layer is determined by using the NWP-model grid cell roughness value and geostrophic resistance laws [59]. At the height of the boundary layer, the wind speed is interpolated between model grid points to site. The wind speed at the blending height is calculated using regional surface roughness length and transformed back to surface height (10-m) using local roughness length in the given wind direction.

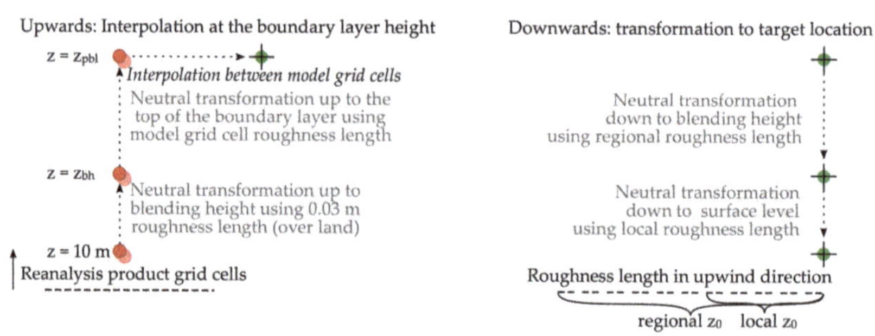

Figure 4. The 2-layer downscaling model concept applied to the reanalysis model surface wind.

If the surface roughness used in the upwards and downwards transformation in the upper layer are identical (the model grid cell roughness and the regional roughness is set to the same value), the 2-layer model reduces to a neutral logarithmic wind profile conversion via blending height [59]. In the result section, we refer to this common conversion as the 1L-method:

$$U_{loc} = U_{10\mathrm{m}} \frac{\ln\left(\frac{z_{\mathrm{bh}}}{z_{0;\mathrm{WMO}}}\right) \ln\left(\frac{z_{\mathrm{loc}}}{z_{0;\mathrm{loc}}}\right)}{\ln\left(\frac{10}{z_{0;\mathrm{WMO}}}\right) \ln\left(\frac{z_{\mathrm{bh}}}{z_{0;\mathrm{loc}}}\right)}, \ z_{0;\mathrm{WMO}} = 0.03 \ \mathrm{m} \tag{1}$$

where z_{loc} is the local height to use in the conversion (10 m or roof height), z_{bh} is the blending height, U_{10m} is the wind speed from reanalysis, $z_{0;loc}$ is the local roughness length at the target location and $z_{0;WMO}$ is the open-terrain surface roughness (Equation (1)).

For the land cover classification, we use the CORINE land use (CLC) map covering the entire Europe. The map was stored in the elevation path profile web-service. The returned land cover classes along paths (with an intermediate spacing of 100 m) were used to assign their relative roughness using tables from literature (see Appendix C). The details of the regional and local roughness length calculation and the footprint model is given in the Appendix C.

2.2.2. Transforming Solar Irradiance Data Using a Satellite DEM

In CAMS Radiation Service, the irradiance calculations are done under the assumption of a flat terrain within the satellite pixel, without considering the diffuse parts masked or reflected by surrounding slopes [53]. In this study, no attempts are made to calculate the diffuse part reflected by the surrounding slopes, but it could be possible to use the satellite-derived DEM to calculate reflections. For the direct irradiance, to account for shading from local hills or mountains, we combined the matrices with the terrain shading angle from the JRC's PVGIS API, selecting the two's maximum shading angle in each sector. The PVGIS horizon angles were interpolated to match sectors of 2.5°, from the native 7.5° sectors corresponding to half-hour intervals. A distance of 10 km was assumed. Except for the PVGIS horizon profile API, no existing services were found to consider local hill-shading effects. Still, many web-based height map services provide tools to create elevation profiles manually along user-defined paths. We also want to test to what extent services that calculate terrain shading angle from satellite-derived DEM's can supplement or substitute higher-resolution DSM's/DTM's covering the nearby building vicinity.

2.3. *Transformations to Local Building Boundary Conditions Using Detailed Surface Models*

2.3.1. Wind Shadow Sheltering on Facades by Nearby Upwind Obstacles

The wind shadow model to calculate wind sheltering effects on building air-infiltration is implemented into a Python script in the following work. The main concept of this simple empirical model first presented in Walker, Wilson and Forests 1996 paper [17], is a wind shadow projected downstream by upwind obstacles to determine the effect of wake velocity on the building surfaces. It applies a Gaussian-shaped weighting reduction, projected and weighted on the facades, that extends beyond the width of the obstacle in the far wake region. In the following, we apply the calculated directional sheltering factor to scale the wind speed in the infiltration model as intended in the original paper, not trying to solve the full wind profile with urban canyon effects, localised flow accelerations, vertical spread and other effects, which would be possible with CFD-simulation or with the three-dimensional diagnostic urban wind models described in the Appendix B that share the empirical parameterisation for far wakes [69,70]. In complexity level, the analytical model implemented here is more similar to the extensive work focusing on deriving wind conditions in urban environments using morphometric approaches [71].

The method calculates an effective mean wind speed U_λ based on the unobstructed wind speed U, multiplied by the shelter factor λ_w which takes a value between 1 (no shelter) to 0 (complete shelter), or 1 to 0.3 in a physical setting where large buildings are immediately adjacent [16]:

$$U_\lambda = U \cdot \lambda_w(\theta) \tag{2}$$

where the shelter factor $\lambda_w(\theta)$ is expressed as a function of wind direction angle θ.

The authors of the wind shadowing method make it clear that the coefficients to find λ_w and U_λ (Equation (2)), were not based on measured wake velocities, but on measured sheltered and unsheltered façade surface pressures. See Appendix B for further discussion.

The other unique feature of the model is a flapped notch wake used to simulate the effect of wind direction fluctuations on near wake spread and growth. The notch wake indicated by notch centreline velocity in Figure 5, is flapped over a range of wind angles assuming a Gaussian distribution of wind direction about the mean angle θ:

$$\lambda_w(\theta) = \sum_{j=1}^{61} f(\phi_j, \theta, \sigma_\theta) \cdot \lambda_w(\phi_j) \tag{3}$$

where the standard deviation of the wind distribution σ_θ is estimated based on a function that changes with averaging time, e.g., 10° for a time-step of 1 h [17], which translates into a range ϕ_j of +/− 30° for each side of the mean wind angle θ (Equation (3)).

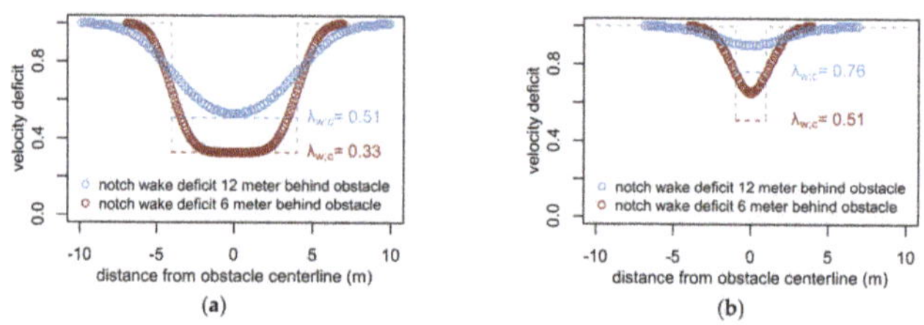

Figure 5. The concept of a Gaussian-shaped wake extending beyond a (**a**) wide obstacle of 8 × 8 × 10 m (**b**) narrow obstacle of 2 × 2 × 10 m.

Figure 5 illustrates how narrow and wide objects differ in notch wake velocity deficit and how fast the wake assumes a Gaussian profile behind the obstacle. A standard deviation of $\sigma_\theta = 10°$ was used for the calculations (Equation (3)).

As described in Appendix B, to determine the scaling length, the obstacles' aspect ratio is considered (calculating a characteristic dimension in the wind direction), but not including the roof pitch or edges of the obstacle relative to the wind direction, or the geometrical relationship between the two objects in consideration. These and other simplifications of the three-dimensional flows are discussed in the paper [17] and wind-tunnel experiments that evaluate the model [72]. Other studies show that although the assumption of wake symmetry may be a reasonable approximation for simple cubes oriented normal to the wind, for more complex geometries the functional form of the velocity deficit in the far wake is neither symmetric nor Gaussian [73,74]. For situations where the obstacle has protruding edges in the wind direction, standing vortices are formed that may lead to velocity deficits that differ significantly from what is predicted by a simple wake model [74].

In our implementation, the effective distance between the obstacle and the facade is calculated differently for the wind shadow coming into the facade, fully immersed, and out of the facade. When the projection line intersects with the façade, the distance is calculated from the intersection point to the obstacle's edge. For the particular case, when the wind direction is perpendicular to the façade, the mid façade point is used in the distance calculation, which is the situation described in the paper.

For each façade, j, the effective shelter $\lambda_{w;j}$ is found by a weighting of sheltered and unsheltered portions of the wall:

$$\lambda_w = 1 - (1 - \lambda_{w;c})\left(\frac{L_S}{L_W}\right) \tag{4}$$

where $\lambda_{w;c;j}$ is the shelter factor on the wake centerline, $L_{W;j}$ is the length of the façade, and $L_{S;j}$ is the sheltered façade length (Equation (4)). The sheltered façade length and distance between the obstacle and the façade (the wake distance) will differ for each wall and wind angle. Every façade and obstacle are considered independently, and upwind walls do not shelter downwind building walls on the same building, as these effects are accounted for in the pressure coefficients [17].

The footprint of the building under consideration was defined as described in Figure 2. The nearby building footprints (including calculated mean building height information) were imported from building shape layers and analysed with Python's shapely library. Trees and vegetation can be included as points (narrow obstacles), but this was not tested as there were no tall trees located within 2–3 building heights in the prevailing wind direction of either case building. Finally, we evaluate the maximum sheltering factor on each façade for every wind direction resulting in a combined directional sheltering factor for the building.

2.3.2. Approach to Calculate Sheltering from Surface Digital Elevation Models (DEM)

For the shading study, we created path profiling scripts using the gdal library for Python and later found Python modules of [45] and shifted to use these modules together with the elevation API. Calling the API repeatedly, one can use a façade sub-division, yet we restricted the evaluation to a single point per façade. For each case building, the digital elevation maps hosted in the elevation API covered a radius of at least 100 m around the building. Focusing on obstacles in the close surroundings is in agreement with Lingfors [31] who found that for roof surfaces a radius of 50 m is satisfactory with little impact on annual direct irradiance beyond 75 m. Further procedure:

(1) Surface height and distance were evaluated extending at least e = 100 m outwards in each façade direction, using a spacing n of 1 m and a sector angle s of 2.5 degrees, creating matrices of dimension $(e/n) \cdot (360/s)$, see also Figure 2 for illustration.
(2) In the next step, the terrain reference height (above mean sea level) for the building and the façade height was used to calculate each sector's maximum obstacle angle. We return the height and distance to this obstacle along with the obstacle height angle based on projection lines from mid-façade height, creating three arrays of (360/s) values for each façade. Knowing the façade orientations, the length of these arrays can optionally be reduced by half (to 180/s sectors).
(3) To account for shading from local hills or mountains, we combined the matrices with the terrain shading angle from the EU-DEM mapping (up to 10 km) and the PVGIS API, selecting the maximum shading angle in each sector. The PVGIS horizon angles were interpolated to match sectors of 2.5°, from the native 7.5° sectors corresponding to half-hour intervals. A distance of 10 km was assumed to calculate PVGIS terrain height.
(4) A fixed sky view factor was calculated for each façade orientation based on the mid-façade height horizon angle. Finally, the variable percentage of façade surface shaded by obstacles was calculated based on the full façade height, solar height and solar azimuth position according to the ISO52010 methodology. The selected 2.5° sectors correspond to 10 min intervals, making it straight forward to apply to 10-min time-series.

We start evaluating the horizon angle a few meters away from the façade to avoid heightmap slope artefacts or exclude trees or other nearby obstructions that are not shielding the entire façade. For the lower resolution DEMs, starting the evaluation, e.g., 5 m away also help to reduce how precisely the façade lines need to be defined relative to the underlying surface DEMs. Following the methodology laid out in the EN ISO 52016-1:2017 standard, it is suggested that building self-shading or "side-fins" are assessed separately.

2.3.3. Solar Shading by Nearby Objects

We follow the procedure of EN ISO 52016-1:2017 for calculation of solar shading reduction factors of nearby (and distant) objects. The standard differentiates between shading reduction factor by objects and on or close by the building itself like overhangs, sides and rebates. In the present work, we did not consider building self-shading. The standard gives two methods to assess shading of diffuse radiation; either disregard shading of the diffuse or use a sector-based evaluation applied to the Perez transposition model. In this anisotropic model, the diffuse part is separated into sky diffuse, circumsolar, horizon band and ground reflected irradiance (EN ISO 52010:2017 calculation method). The technical report accompanying the standard also outlines how some sky patch dome models are compatible with the Perez model. An example of implementing a sky patch model with Perez surface transposition and LiDAR data for shading analysis is found in [35]. As objects may not only block solar irradiance on a surface, but may also reflect solar radiation (e.g., hills, trees, other buildings, or other parts of the same building), the three-dimensional problem quickly calls for more advanced methods, e.g., ray-tracing supported by GPU acceleration. For simplicity, we follow the first approach in the standard, which is equivalent to a situation where the radiation reflected or transmitted by objects in the environment is equal to the diffuse radiation blocked by these object.

The direct radiation (including circumsolar) is fully or partially blocked by a factor F_{sh} if the object is between the sun and surface (Figure 6).

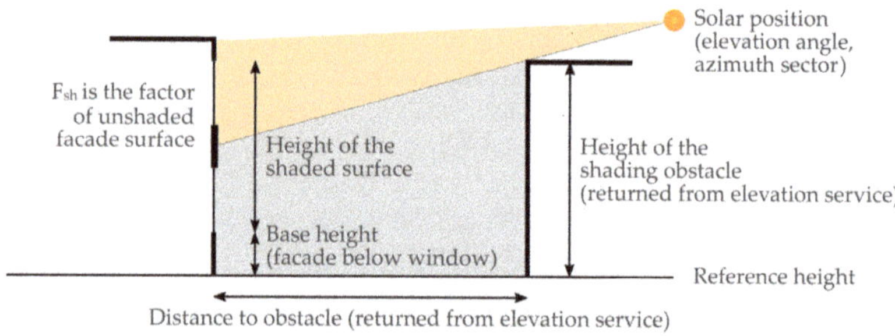

Figure 6. Shading of direct and circumsolar beam irradiance and the vertical shading factor.

2.4. Including Environmental Variables and Local Sheltering Effects

In the following section, we discuss models that can be used pre-process weather variables in order to capture the thermal tie between indoor temperature, and boundary conditions, in this case, incident solar insolation and temperature- and wind-dependent air-leakages across the building envelope which result in infiltration losses.

2.4.1. Infiltration Losses

Infiltration is the uncontrolled air leakage through cracks and other unintentional openings in the building envelope introducing outdoor air into a building. Many infiltration models for residential buildings have been developed based on statistical fits of infiltration data. By considering that weather is the dominant driving force, infiltration flow can be assumed to be linearly dependent on the outside-inside temperature difference and wind-speed [75]. However, the simplicity of regression is not without limitations. The fitted coefficients carry little physical meaning, and the collinearity between heat transmission across the building envelope and infiltration losses driven by the indoor-outdoor temperature difference may lead to identifiability issues.

An empirical single-zone infiltration model accounts for infiltration, relying on physical parameters and building information, the AIM-2 model. We use the model form presented by Lundström in [58], where the calculated potential specific infiltration flow rate $Q*_{inf}$ [Pa^n] multiples with the infiltration coefficient C_{inf} [$1/(s\ Pa^n m^2)$] which can be estimated by inverse approaches or obtained from fan pressurisations tests. A literature review of studies using the model and the differences in our implementation to [58] are presented in Appendix B. The single-zone infiltration models' performance is mainly sensitive to the highly uncertain distribution of air leakages across the envelope and the parameters used for converting wind data measured at a weather station to the building site and local wind shelter effects from typography and nearby buildings. This uncertainty can be reduced by wind measurements on-site. However, shelter coefficient may still apply as a simplified approach to account for direct wind shielding caused by trees and neighbouring buildings located within 2–3 building heights of the building facades [16,17]. To estimate local wind velocity U_{loc}, AIM-2 uses unobstructed wind speed transformed to building eave height at the building site. A power law wind profile conversion was used in the original AIM-2 model, whereas a logarithmic wind profile conversion can be found in the implementation in the popular building energy simulation software ESP-r [76]. Furthermore, the wind shelter coefficient λ_W of 0–1 is multiplied by the wind speed at roof height. Both the authors of AIM-2 and the LBL infiltration method recommends making this shelter effect directional based on wind direction [16,75].

An engineering approach to make the wind sheltering directional is proposed by Walker and Wilson [16] and can be found reprinted in the ASHRAE Handbook of design guidelines [77]:

$$\lambda_w(\theta) = 0.5 \cdot \left((\lambda_{w;1} + \lambda_{w;3}) \cdot \cos^2\theta + (\lambda_{w;1} - \lambda_{w;3}) \cdot \cos\theta + (\lambda_{w;2} + \lambda_{w;4}) \cdot \sin^2\theta + (\lambda_{w;2} - \lambda_{w;4}) \cdot \sin\theta \right), \quad (5)$$

where $\lambda_w(\theta)$ is the shelter factor for the particular wind direction θ, and $\lambda_{w;j}$ is the shelter factor when the wind direction is normal to a wall *j* (estimated perpendicular to each side building side) which can be estimated from sheltering class tables in literature [78].

We compare this interpolation approach (Equation (5)) to the wind sheltering model in the result section.

2.4.2. Solar Heat Irradiance on Facades

The solar irradiance is calculated as a weighted mean vertical input according to the proportions of total solar gains expected for each façade orientation. This input can be used in simplified thermal models that are suitable to determine building heat transmission losses (HTC) [79]. When measuring and accounting for solar gains in steady-state whole building heat loss experiments, one approach is weighting each façade by their respective glazing proportions. Stamp et al. found that for north-south oriented dwellings, vertical south-facing or weighted means provide the most accurate results to determine heat transmission losses (HTC), whilst estimating solar gains from global horizontal measurements overestimated HTC [80]. For east-west facing dwellings, mean or weighted means may provide more accurate results than a single vertical measurement in the dominant direction, particularly where there are local shading effects [80]. The presence and operation of solar shading devices represent a considerable uncertainty.

3. Results and Discussion

The proposed method is applied to four buildings:

- The TWIN detached house oriented directly towards south at the Fraunhofer IBP test site in Holzkirchen, Germany.
- The ZEBLL Living Lab detached house oriented south with 4° westward tilt on the main campus of NTNU, Trondheim, Norway.
- The GBORO south-facing apartment end-unit oriented 12° eastwards in Gainsborough, UK.
- The UKULE townhouse oriented 71° westwards from the south in a historic part of Brussels, Belgium.

Figure 7 illustrates that nearby buildings do not shade the east, south, and west façades of the TWIN house, at the 15. of February ca. 9:30 in the morning, 12:30 mid-day and 15.30-afternoon local time (or any other time in winter). The ZEBLL house at the NTNU campus, on the other hand, is shaded in the afternoon on this day from a nearby building located west of the house. We evaluate the shading model by using data from a pyranometer on the south façade in Section 3.2.

(a) (b)

Figure 7. Shading of east, south and west facades on the 15. February with solar position 135° (SE), 180° (S) and 225° (SW). (**a**) TWIN, Holzkirchen, DE; (**b**) ZEBLL, Trondheim, NO.

3.1. Comparison of Sourced Weather Data to Observations at the Holzkirchen Site

The key weather variables sourced from the reanalysis and the satellite irradiance service are presented, by comparing a two and a half month-long winter period to observations at the Fraunhofer IBP test site in Holzkirchen. The weather data was collected at the IBP's weather station at 1-min intervals, provided as 10-min averages for the period 7 December 2018 to 28 February 2019 as part of IEA EBC Annex 71. The 10-min data were processed to hourly observations for the following analysis.

3.1.1. Air Temperature and Sky Longwave Irradiance

The air temperature and sky longwave irradiance from reanalysis are seen to represent the diurnal cycle (Figure 8). The ERA5-Land temperature at 2-m scores somewhat better on central performance metrics (0.33 °C, 2.04 °C and 1.36 °C) compared to ERA5's (1.32 °C, 2.27 °C and 1.61 °C) for mean bias difference, root mean square error and mean absolute difference respectively. The errors are largest under cold spells. Both products underpredict the temperature under cold conditions, which is a feature of NWP models, they struggle to represent cold temperatures in stable conditions well [81].

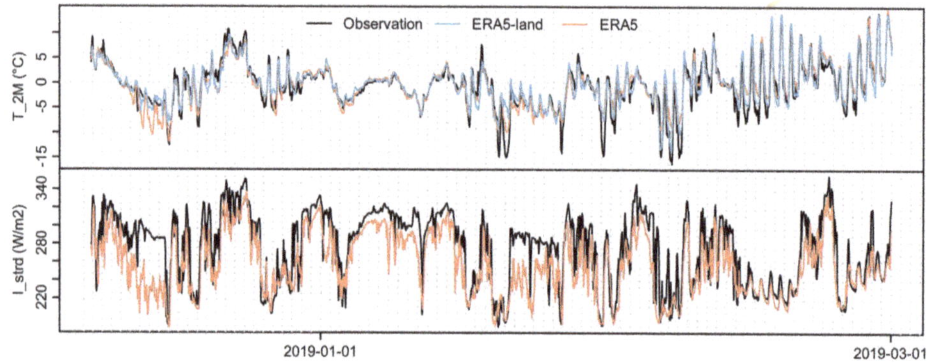

Figure 8. Observed outdoor temperature and longwave sky radiation at the Holzkirchen site compared to the ERA5 reanalysis for the period 7 December 2018 to 28 February 2019.

The longwave sky irradiance and the sky temperature, calculated from the ERA5 longwave sky irradiance and air temperature (using the conversion in Table 2) correlate to local measurements on most of the days. In Figure 9, the distribution and sorted values in ascending order (black) reveal a bias across the distribution. These quantile-quantile plots are helpful to determine if the distributions are similar. The actual hourly values aligned in time are shown as coloured point samples. The reanalysis does not match observed hourly temperature in the lower end, but the distributions are similar, so more confidence can be placed for longer periods (e.g., monthly averages).

3.1.2. Wind Speed and Direction

As described in the method section, the near-surface 10-m wind in the reanalysis is not itself a direct output of the model: instead, the lowest predicted model level wind is post-processed using an exposure correction to better represent observed 10-m wind in open terrain [60]. In this case, the wind mast's local surroundings match the condition of open unobstructed terrain in the prevailing wind direction, making a direct comparison possible. When comparing the reanalysis data to site-observations, there is a clear bias in the reanalysis (Figure 10). Both products generally capture the hourly fluctuations reasonably but a consistent underprediction effect is observed.

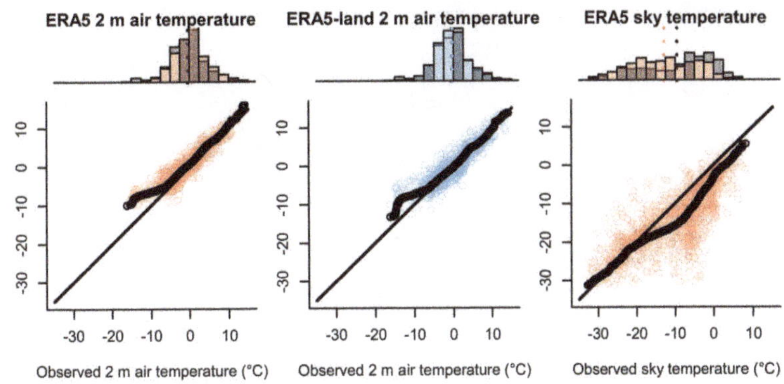

Figure 9. Observed outdoor temperature and sky temperature (computed from longwave radiation) at the Holzkirchen site compared to the ERA5 reanalysis for the period 7 December 2018 to 28 February 2019.

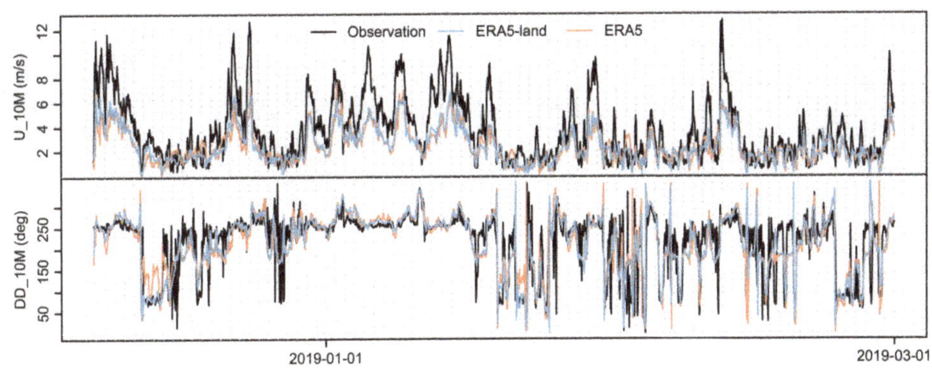

Figure 10. Observed wind speed and direction 10 m above ground at the Holzkirchen site compared to the ERA5 reanalysis for the period 7 December 2018 to 28 February 2019.

By applying the 2L-method, the 10 m near-surface wind from ERA5 is bias-adjusted, leading to a better match with the observed wind (Figure 11). More details on the correction factors are presented in the downscaling chapter. As with the temperature comparison, the actual hourly values aligned in time are shown as coloured point samples and sorted in ascending order (black) to see model bias across the theoretical distribution (Q-Q plot). A Weibull distribution is overlaid on top of each scatter plot, displaying the observations series in grey and the reanalysis in colours.

After the bias correction, the Weibull scale parameter is increased from 3.1 m/s to 3.8 m/s, compared to the observed 4.0 m/s. The mean bias difference is reduced from −0.85 m/s to −0.2 m/s, and the mean absolute difference is improved by 0.4 m/s from 1.2 to 0.8 m/s.

3.1.3. Global Horizontal and Diffuse Irradiance

The global horizontal and diffuse irradiance observed at the site compares quite well to the satellite-derived irradiance from the CAMS-Rad service (Figure 12) on most days. There is a series of days in February, towards the end of the period, where diffuse irradiance is overpredicted, and the global horizontal is underpredicted. These are clear-sky days not interpreted as such by the satellite-model product. Earlier in the period, the cloud cover appears to be predominantly overcast (global horizontal and diffuse irradiance are equal), but some day-to-day variability can be spotted in both observations and satellite irradiances.

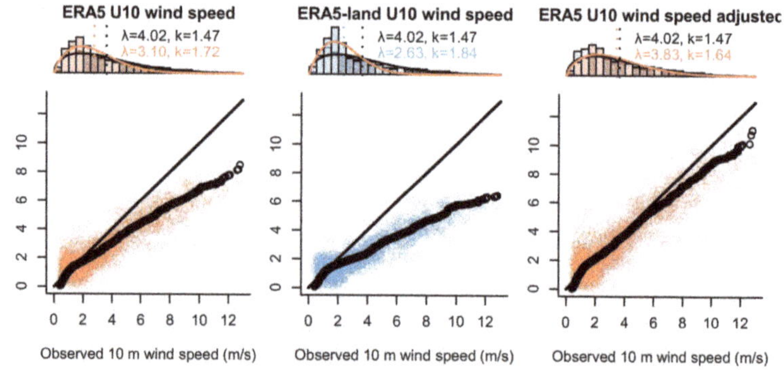

Figure 11. Observed wind speed and direction 10 m above ground at the Holzkirchen site compared to the ERA5 reanalysis for the period 7 December 2018 to 28 February 2019.

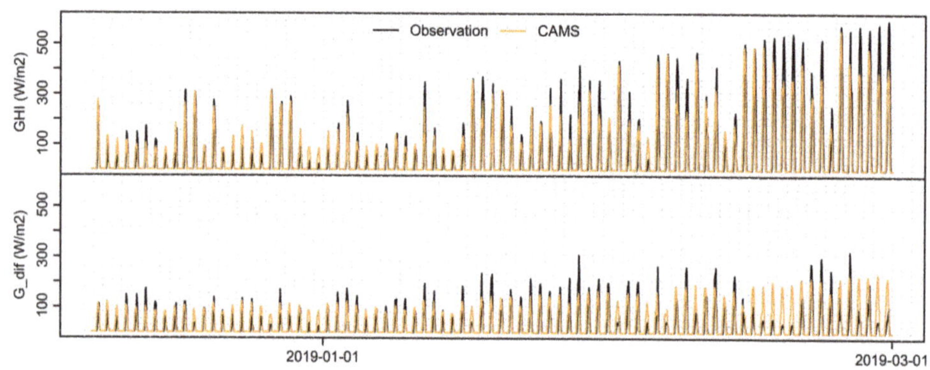

Figure 12. Observed global horizontal (GHI) and diffuse irradiance (G_{dif}) at the Holzkirchen site compared to the variables from CAMS radiation service for the period 7 December 2018 to 28 February 2019.

3.1.4. Snow Depth and Ground Surface Albedo

The snow model depth in the ERA5-Land grid cell matches the observed depth quite well (Figure 13). The calculated snow broadband albedo adjusted for snow cover (equation in Table 2) is compared to the measurements from two pyranometers on-site, filtered by solar azimuth height (only displaying values when the sun is 5° above the horizon) to account for uncertainty at sunrise and sunset. The uncertainty in the measurements is likely to be high at low radiation intensity, so applying more precise criteria could improve the correlation.

3.1.5. Vertical Solar Irradiance on Facades

The TWIN buildings and weather station at the Holzkirchen site lie unobstructed in a flat open terrain making it impossible to evaluate the shading model with measurements from this site. It is still included to show the calculated horizontal and vertical solar irradiance compared to observations, as TWINS is the only case with measurements in all four façade orientations. On the particular day selected for analysis, clear sky conditions can be observed from ca. 10:00 in the morning (Figure 14). Even if the global horizontal irradiance from CAMS-Rad matches observations well, the calculated irradiance on each façade orientation is underestimated. We can observe that the diffuse fraction on the vertical appears to be underestimated in every facade direction even when using clear-sky irradiance as input to the solar transposition model. Possible explanations are due to how diffuse radiation is parametrised in the Perez anisotropic sky model (including model attenuation coefficients), the assumption that diffuse sky irradiance is isotropic in the shading calculations (each surface sees 50% of the sky) and physical effects caused by terrestrial reflectance or measurement errors. Due to fresh snow, the modelled ground reflectance was set to a high value of 0.70 at this day (calculated from the ERA5-Land snow cover).

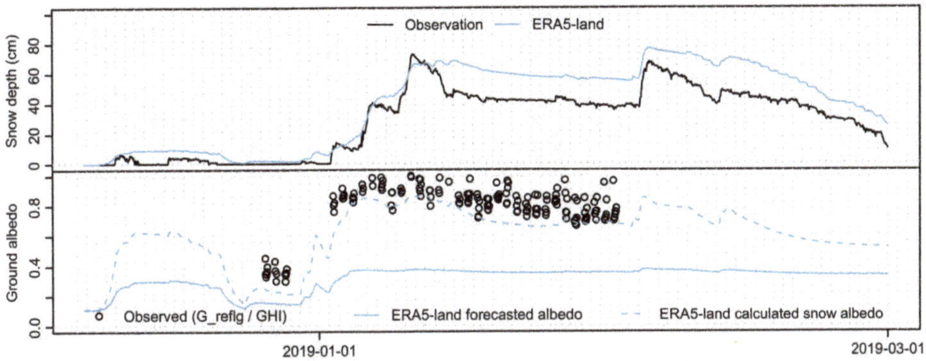

Figure 13. Observed snow depth and ground reflectance (calculated from ground reflected short wave radation) at the Holzkirchen site compared to the ERA5 reanalysis forecasted albedo for the period 7 December 2018 to 28 February 2019. The dashed blue line shows the calculated snow albedo of Table 2.

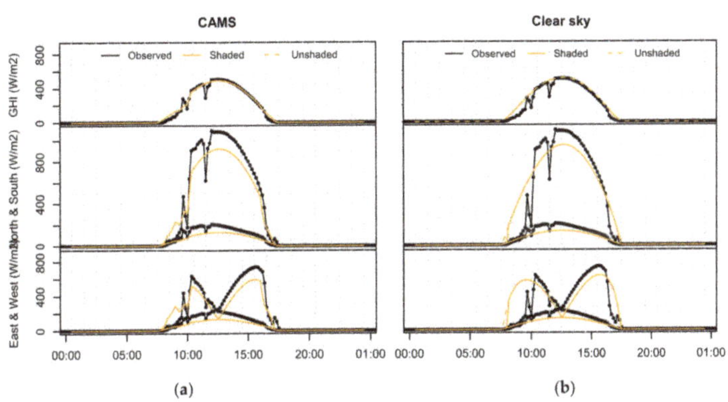

Figure 14. Observed global horizontal (GHI) and total vertical irradiance measured in each direction for a sunny day in winter at the Holzkirchen site compared to the calculated surface irradiances using the (**a**) CAMS radiation service as input. The (**b**) clear sky data are based on the CAMS-Rad McClear service.

3.2. *Comparison of Sourced Weather Data to Observations at the NTNU Campus*

Vertical Solar Irradiance on Facades

The ZEBLL campus building also features pyranometers measuring global horizontal irradiance on the roof and total vertical irradiance on the south façade. This building has its largest windows towards south, and the view from the wall-mounted sensor is shown in Figure 15. In winter, the afternoon sun is obstructed by large buildings towards west, well-captured by the shading model (Figure 16).

Figure 15. Panoramic view from a vertical sensor positioned south, ZEB Living Lab, NTNU Campus.

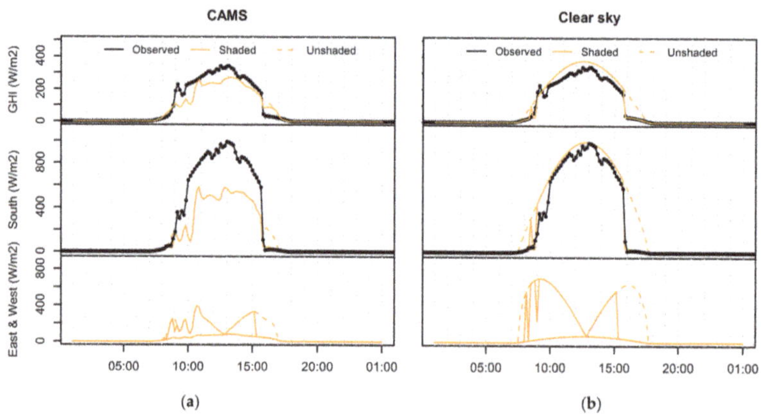

Figure 16. Observed global horizontal (GHI) and total vertical irradiance measured from south sensor position, ZEB Living Lab, NTNU campus compared to the calculated surface irradiances using the CAMS radiation service as input. The CAMS-Rad irradiance over-evaluates cloud cover (**a**) and clear sky (**b**).

The CAMS-Rad irradiance over-evaluates cloud cover on this particular day (Figure 16a, but a better match can be seen in the clear sky irradiance from the CAMS-Rad McClear model (Figure 16b). The discrepancy on how vertical solar irradiance is calculated is likely due to surfaces partially shaded by trees before noon (Figure 15), because only buildings and terrain are considered in this particular diagram. The difference between observed hourly global horizontal irradiance and what is interpreted by the satellite product as mostly diffuse sky irradiance on this particular morning, also exemplifies how surface shading is left unaffected when cloud cover is predominant over the hour in the satellite product. Improving the shading calculations to include diffuse shading and not only effects on direct surface radiation.

3.3. Horizon Angle and Solar Radiation

3.3.1. Horizon Profile Using a Satellite-Derived Pan-European Surface Height Model

The proposed workflow for solar and wind assessment has in common that the same datasets are used on a local level, but in different ways. Moreover, that available dataset for downscaling to medium scale covers all of Europe (or global datasets), whereas the local effects are only assessable using local area height maps or building footprint vector map-layers with user contributed height-specification. Although there are initiatives of standardisation and contribution within the INSPIRE framework, including OSM contribution, and a community-led project to derive a pan-European terrain model online [82], high-resolution models involve more complex issues that are less pronounced in lower-resolution models due to the already high uncertainty. The EU-DEM v.1 product is evaluated to a vertical accuracy of 2.9 m RMSE, with higher values for the Nordic countries (e.g., 5.75 m RMSE for Norway) [83]. EU-DEM v1.1 used in this study is an improvement over the first version, but it has not been validated yet [23]. In order to use different datasets together, we implemented datum conversion and height adjustment to the different national height systems [82]. For Norway, the difference between the national height and the one used in EU-DEM was less than 1 cm, but for Belgium, the offset is as much as 2.31 m.

We tested combinations of height information from the EU-DEM model and the local high-resolution DSM's. However, when mapping from the EU-DEM model surface height, which may very well be above the building height in steep terrain or in places with low buildings and heigh vegetation, we did not find a significant benefit to use the 30-m resolution of EU-DEM compared to the existing horizon profile API of PVGIS, which is a service that relies on a pre-processed global SRTM of 3 arc-seconds (around 90 m). However, we found that if we have actual information about the building elevation height, and create a buffer around the viewpoint (e.g., 50 to 100 m), we can create a horizon profile that in some situations is roughly similar to the detailed DSM's.

3.3.2. Horizon Profile Using Local High-Resolution Surface Height Models

Figure 17, shows the calculated horizon angle at mid-façade height using the maximum local terrain profile of the PVGIS API (thin black line) and the high-resolution digital surface models filtered to buildings only (thick black line) or without filter (green), for the various case buildings. The TWINS building (a) is located in an open flat area. ZEBLL (b) is located in Norway where solar height at the summer solstice (red line) is lower than the others, and in this case, vegetation towards north-east and larger campus buildings westward block morning and afternoon sun. GBORO (c) is shaded by a neighbouring townhouse distanced 7 m from the south façade, and UKULE (d) is also located in an urban environment facing a street and a backyard with trees in the backyard.

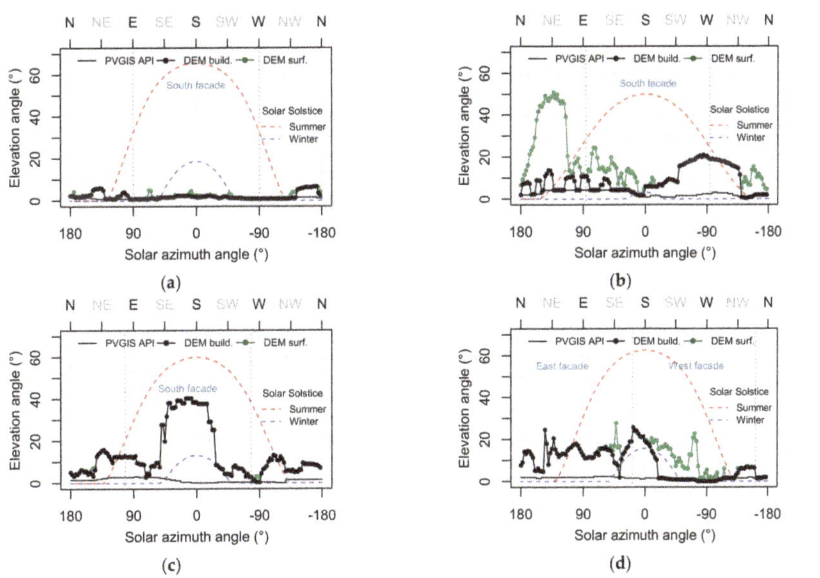

Figure 17. Calculated horizon angle at mid-façade height using the maximum local terrain profile of the PVGIS API (thin black line) and the digital elevation model DEM filtered to buildings only (thick black line) or without filter (green) for the (**a**) TWINS, (**b**) ZEBLL (**c**) GBORO and (**d**) UKULE buildings.

Table 3 summarises the average calculated horizon height per façade, separated by considering only buildings obstacles (abbreviated "Build.") or all obstacles including vegetation (abbreviated "Surf.").

Table 3. Calculated average shading angle at mid-façade height filtered by terrain and building outlines, or without surface filtering (including all trees or other obstacles from the surface DEM).

Mean Horizon	TWINS		ZEBLL		GBORO		UKULE	
Angle from DEM	Build.	Surf.	Build.	Surf.	Build.	Surf.	Build.	Surf.
North façade (°)	3.1	3.1	8.2	28.6	-	-	-	-
East facade (°)	1.8	2.0	6.8	29.2	11.2	11.2	11.8	12.6
South façade (°)	1.4	1.8	8.9	12.7	19.4	19.4	-	-
West façade (°)	2.6	2.7	13.8	15.9	8.1	8.3	5.3	10.3

Figure 18 shows the clear-sky shade index per winter month by [31] defined as the ratio of shaded and unshaded clear-sky irradiance (only direct and circumsolar beam is included in the following). The impact of vegetation is presented as a range from fully opaque (by an x mark) to fully transparent (by a cross mark) and a mix of the two (dot).

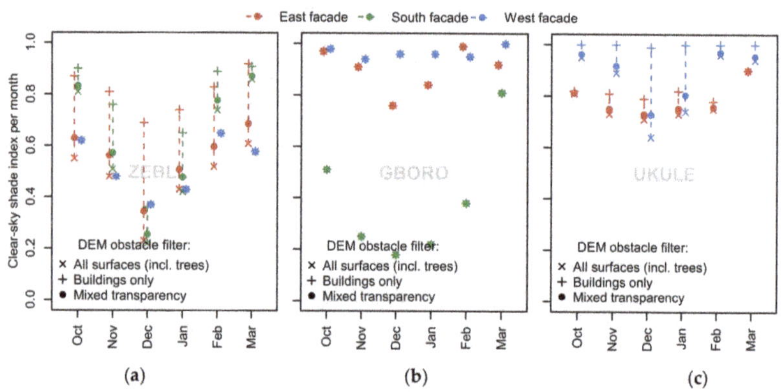

Figure 18. Monthly clear-sky shading index per façade for the winter months presented per façade from left to right: (**a**) ZEBLL (**b**) GBORO (**c**) UKULE.

- For ZEBLL, the incoming direct radiation on the south and west façades is reduced substantially (up to 60%) by buildings in winter, and if trees are included, they may block beam insolation from south and east.
- For the GBORO case building, trees have no shading impact, but the south façade is almost entirely in shadow from December to February, rapidly diminishing in March as the mid-day solar angle climbs.
- For UKULE, which on average has moderate shading of vertical beam irradiance (ca. 20% reduction) on the two facades, the west façade is only shaded from vegetation.

By weighting the calculated surface radiation by the glazing ratios, we can estimate the solar heat gains on the window facades in the winter months and estimate a total solar aperture for the building.

3.4. Surface Roughness and Wind Sheltering

3.4.1. Unobstructed Height Adjusted Wind Speed

ERA5 has a model resolution of approximately 31 km and lack the local representativity of the site surroundings. To better represent surface conditions upwind influence, a downscaling method is used in an attempt to increase the local accuracy. The method consists of a high-resolution land-use map with derived surface roughness values from tables (Appendix A) and a simple two-layer model of the atmospheric boundary layer.

The analysis in Section 1 illustrated that 10-m surface wind speed at the Holzkirchen site (TWINS) is underestimated in the reanalysis. An explanation can be found in the high forecasted model surface roughness, indicating that the model grid box surface-average does not represent the local site conditions. The forecasted $z_{0;M}$ = 1.5 m (Table 4) is a value typically representative for forested areas. The site is located in an open agricultural area bordering the nearby town of Holzkirchen to the north-west and a golf course towards the south (Figure 9). The prevailing wind direction is west-south-west, and the first km up-wind are open landscape. Further out forest surrounds the farmland, influencing the regional scale weighted mean surface roughness plotted in Figure 17.

Table 4. Surface roughness from ERA5 grid cells and local value derived from the land cover maps.

Surface Roughness z_0 (in Meter)	TWINS	ZEBLL	GBORO	UKULE
$z_{0;M}$ ERA5 forecasted in the nearest grid cell (31 km)	1.52	1.17	0.23	0.34
$z_{0;WMO}$ ERA5 open terrain roughness for U10 wind	0.03	0.03	0.03	0.03
$z_{0;CLC}$ Land cover map grid cell closest to site (100 m)	0.03	1.00	1.00	1.00

First, the forecasted surface roughness for each grid cell was extracted from the ERA5 reanalysis. Table 4 show the values for each site, together with the open terrain roughness used to compute 10-m wind in ERA and the derived surface roughness from the closest grid cell of the CORINE Land Cover (CLC) 2018 map paired with a table of roughness values (Appendix C). The roughness length for the land use type "Sport and leisure facilities" was adjusted to a lower value from 0.5 m to 0.03 m to represent the snow-covered golf course.

Next, the directional surface roughness from the footprint model applied to local scale (up to 1.8 km), and regional scale (up to 9 km) scale are plotted (Figures 19–22). For TWINS we can see that on the local level, the surface roughness is well approximated to 0.03 m in the prevailing wind direction (SW), but the model level of 1.5 is many classes higher than the regional level z_0, except in the north direction (due to the proximity to forest). The difference between the model average forecasted in the nearest grid cell ($z_{0;M}$), and the regional footprint model is less pronounced for cases (b) to (d) (Figures 20–22). However, all display a considerably higher surface roughness on local-level than the open terrain roughness ($z_{0;WMO}$).

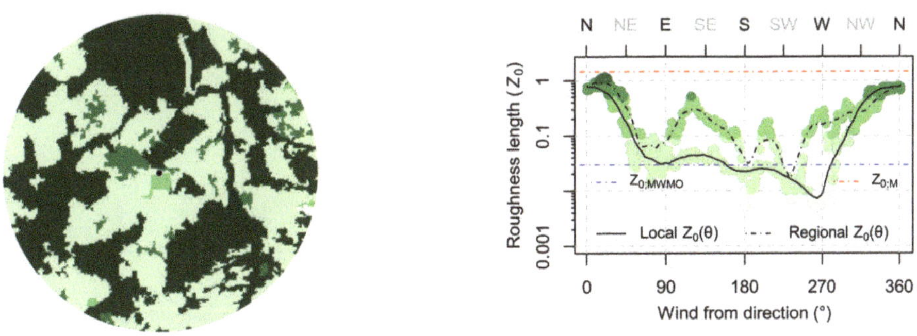

(a) (b)

Figure 19. Derived surface roughness for TWINS site (**a**) in 10 km radius adapted from CLC map (© European Union, Copernicus Land Monitoring Service 2018, European Environment Agency (EEA)); (**b**) Directional local and regional surface roughness weighted by distance to building-site. Darker green colour indicates higher surface roughness values.

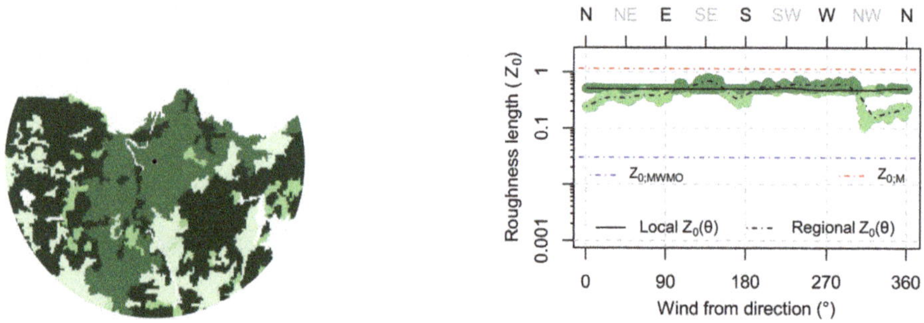

(a) (b)

Figure 20. Derived surface roughness for ZEBLL site (**a**) in 10 km radius adapted from CLC map[2]; (**b**) Directional local and regional surface roughness weighted by distance to building-site. Darker green colour indicates higher surface roughness values.

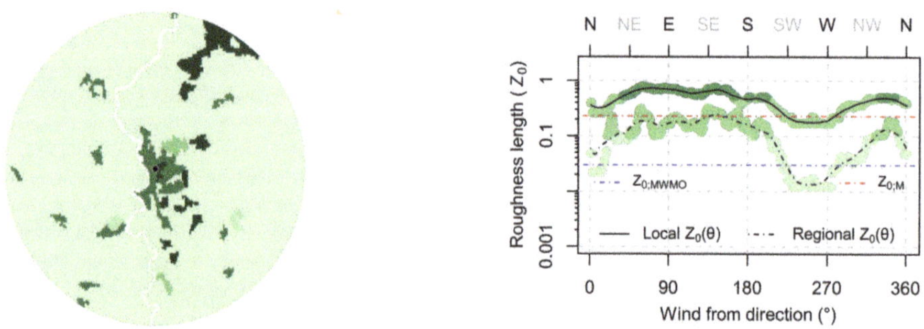

(a) (b)

Figure 21. Derived surface roughness for GBORO site (**a**) in 10 km radius adapted from CLC map[2]; (**b**) Directional local and regional surface roughness weighted by distance to building-site. Darker green colour indicates higher surface roughness values.

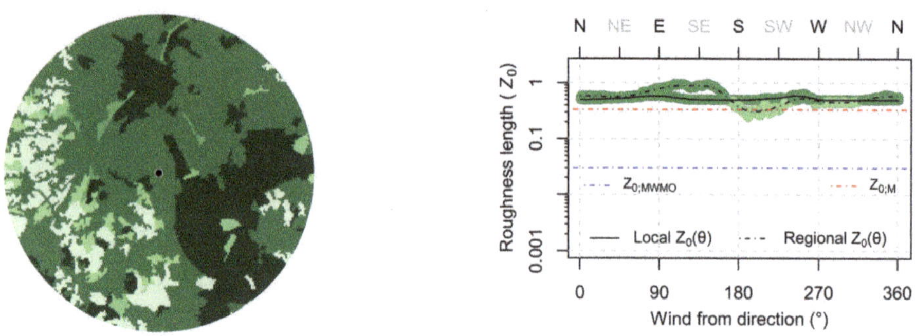

(a) (b)

Figure 22. Derived surface roughness for UKULE site (**a**) in 10 km radius adapted from CLC map[2]; (**b**) Directional local and regional surface roughness weighted by distance to building-site. Darker green colour indicates higher surface roughness values.

The resulting unobstructed surface winds transformed to roof height are used together with the wind sheltering method in the estimation of infiltration loss (see Section 3.4.3).

3.4.2. Wind Sheltering of Nearby Obstructions

The wind sheltering model results are shown below for the three cases (b) to (d) where nearby buildings wake significantly impact surface conditions in the dominating wind direction. For the campus building (Figure 23), obstacles downwind coincide with the dominating south-west wind direction. Likely, these buildings will also influence the wind direction and lead to more complex airflow patterns than captured by the model.

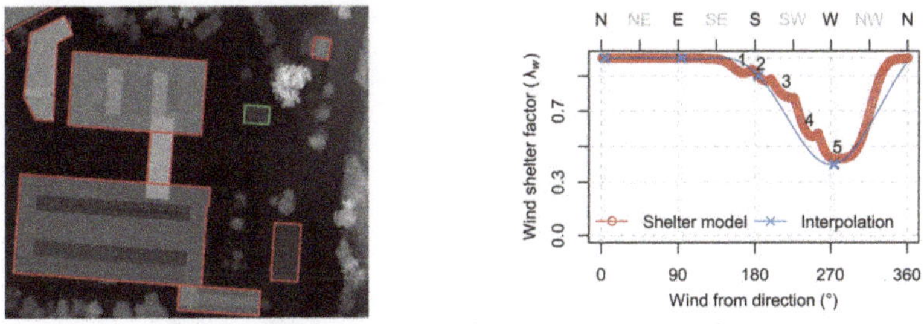

(a) (b)

Figure 23. Air infiltration sheltering for the NTNU Living Lab (ZEBLL) case (**a**) Upwind obstacle geometries in red (OSM building footprints) and building exterior facades in green overlaid on the local DSM height map by Kartverket/CC-BY 4.0; (**b**) The combined directional shelter factor for the four facades using the shelter model (red) and the table-values obtained for each building side (blue).

For the two other cases, GBORO and UKULE, located in urban/semi-urban setting, the calculated maximum sheltering coefficient from building obstacles are more irregular (Figure 24). Therefore, it fits less to the interpolated values (using Equation (5)) estimated qualitatively for each building side from sheltering classes in literature [78]. Instead of a selecting the sheltering factor manually from a table, one could use the model to approximate the sheltering factor from obstacles located perpendicular to each façade orientations (when the wind direction is normal to a wall).

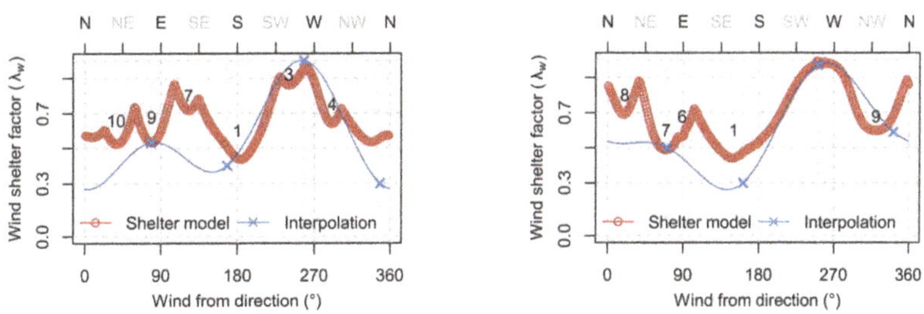

(a) (b)

Figure 24. Air infiltration sheltering (**a**) for the GBORO case-building; (**b**) and for the UKULE case building. The combined directional shelter factor for the three exterior facades of either plotted over the wind direction. Only the obstacles that yielded an impact to the shelter factor are enumerated.

In the presence of more tightly-spaced building obstacles, the limitations of the shelter-model implementation become evident. First, overlapping sheltering is not considered, which may be detrimental to performance. We have no way to assess skimming flow that is usually accounted for in urban wind morphometric approaches by increasing the zero-plane displacement height [71]. Secondly, with more complex and overlapping geometries, the functional form of the velocity deficit in the far wake is not Gaussian [72–74]. In the urban/semi-urban situation below, limiting the assessment to approximate the sheltering factor in the four façade directions and interpolating using Equation (5) may be a more physically sound and robust simplification.

3.4.3. Infiltration Loss

In this section, the infiltration model results are plotted against the potential wind speed of ERA5 at 10-m = height to illustrate the effect of the wind exposure and shelter correction. The AIM-2 models physical model output, "infiltration potential" (a scalable time-dependent variable) is multiplied by the results of air-pressurisation tests n50 value. It could be replaced by either a design-value or a parameter estimation in an inverse modelling approach. All the four cases (which are well-insulated and air-tight residential buildings) have measured air-pressurisation n50 values of 1.0 air change rate per hour at 50 Pa pressure difference, except the last case d) with a lower value of 0.5 h^{-1}.

The slope of the infiltration air change rate at 1 atm. pressure (Figure 25) is mainly influenced by the n50 value, the shelter coefficient used in AIM-2 and what kind of exposure correction is made from ERA5 10 m potential wind speed (in open terrain) to building roof height, U_{rf}.

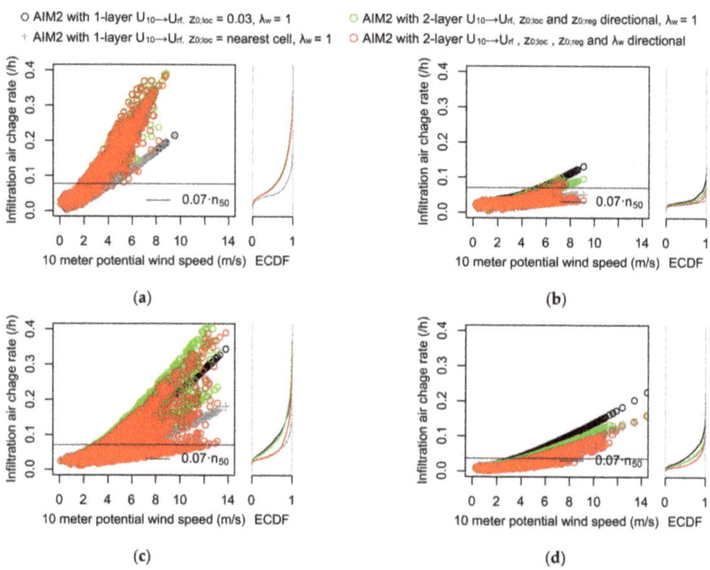

Figure 25. Calculated infiltration rate (air change per hour) over six months in winter versus 10-m wind speed in open terrain from reanalysis and estimated cumulative distribution (ECFD) **(a)** TWINS; **(b)** ZEBLL; **(c)** GBORO; **(d)** UKULE.

- The black horizontal line is a common conversion of the n50 infiltration rate to 1 atm. pressure, simply multiplying the n50 values by a constant of 0.07, a rule-of-thumb conversion factor representative of a moderately sheltered building with more than one exposed façade.

The first two AIM-2 model variants uses a 1-layer (1L) logarithmic transformation of the wind speed (up to a blending height of 60 m) which is a common conversion method in many Building Energy Simulation (BES) tools.

- The black dots show the infiltration rate with a neutral logarithmic correction to roof height using the same surface roughness value of 0.03 m present in the reanalysis surface wind.
- The grey crosses show the infiltration rate using a transformation from 10-m potential wind to roof height using a local roughness value. The local surface roughness obtained from the CLC-classification is 0.03 m for case (a) TWINS and 1.0 m for the other cases (Table 4).

In the last two AIM-2 model variants, the full 2-layer (2L) downscaling method was used with and without the directional surface sheltering method. In the 2L-method, the directional surface roughness upwind to the site is obtained for the required regional and local footprint using derived surface roughness from the CLC-classification.

- The green points show the infiltration rate using the full 2L transformation of reanalysis wind.
- The red points show the infiltration rate using the same 2L transformation with the directional surface sheltering method to account for nearby obstructions.

For the first two model runs using fixed surface roughness, the variability of air change comes from the outdoor indoor temperature difference, which is more pronounced at calm wind conditions. This effect is explained by the AIM2-models sub-addition of stack and wind-driven infiltration, where the stack effect loses significance at higher wind speeds. At higher wind speeds, the wind direction plays a role when the two-layer downscaling method is applied and through the wind shelter method in cases (b) to (d) where there are buildings located in the prevailing wind direction (Figure 25).

The resulting infiltration loss over the six-month winter period is presented in the table below in W/m^2 K indoor-outdoor temperature difference and W/m^2 assuming a constant indoor temperature of 21 °C in the period (Table 5). Consistent with the hourly infiltration air change rate (Figure 25), including shading from nearby obstacles significantly impacts the average infiltration heat loss over the winter period for cases (b) to (d). Comparing the average infiltration loss with nearby sheltering effect (final line) to the "rule-of-thumb" approach of scaling n50 by a constant of 0.07 (first line), the mean differences between the two are less than 30 % except for case b) ZEBLL.

Table 5. Infiltration loss calculated mean over the six month winter period per floor area in W/Km^2 and assuming an indoor temperature of 21 °C in W/m^2 heated floor area.

Infiltration Heat Loss	**(a) TWINS**		**(b) ZEBLL**		**(c) GBORO**		**(d) UKULE**	
Average Per Floor Area	**W/m^2 K**	**W/m^2**	**W/m^2 K**	**W/m^2**	**W/m^2 K**	**W/m^2**	**W/m^2 K**	**W/m^2**
Constant $n_{50} \cdot 0.07$	0.08	1.35	0.08	1.57	0.07	0.89	0.03	0.47
AIM2-1L, $z_{0;WMO}$, $\lambda = 1$	0.06	1.00	0.04	0.87	0.09	1.11	0.04	0.60
AIM2-1L, $z_{0;CLC}$, $\lambda = 1$	0.06	1.00	0.03	0.60	0.05	0.66	0.03	0.43
AIM2-2L, $z_0(\theta)$, $\lambda = 1$	0.10	1.68	0.04	0.76	0.09	1.10	0.03	0.46
AIM2-2L, $z_0(\theta)$, $\lambda(\theta)$	0.10	1.63	0.03	0.67	0.06	0.78	0.02	0.33

4. Conclusions

In this paper we examine how open geospatial data can be used to refine weather variables for building energy performance evaluation with focus on incident solar radiation and wind-driven infiltration modelling. By using only building location (latitude, longitude) and a selection of free/open geospatial datasets covering Europe, we were able to acquire and adapt gridded weather data variables to local building boundary conditions. The nearly three-month-long winter comparison of a building test-site in South-Germany indicates that hourly surface variables from climate reanalysis and satellite-based solar radiation can become a feasible supplement to local observations for heating season building performance modelling and evaluation. However, the air temperature, vital to heating analysis, did not fully capture the extreme lows. In Europe, local observations of air-temperature and to some extent other weather variables are commonly available closer to site than the resolution of the current global reanalysis datasets (as was the case for the four case buildings). New regional surface products are expected to take advantage of observation stations' density and reduce the need for site-correction techniques by featuring higher spatial resolution.

To include local effects from the terrain, vegetation and buildings in solar and wind assessments, 1-m resolution DSM's and DTM's from airborne laser scanning were acquired for each case building. The separation of nearby vegetation and buildings is essential to model incident solar insolation on low-rise building facades. The impact of distinguishing between the two was clear on the building-sites with trees. The separation was obtained by pre-processing the DSM into a new raster based on building outlines, but this operation could be made more efficient by using the Overpass API directly within the elevation profile web service to obtain building footprints and filter between building obstacles and vegetation. The building footprints (with building heights obtained from the DSM) were also used to include wind sheltering effects in the infiltration calculations. Including wind sheltering from nearby building obstacles in the AIM-2 model significantly impacts the average infiltration heat loss over the winter period for three of the four cases. Based on the four case studies, it seems like the approach work better for more open situations, other approaches may be better suited for buildings situated in a more dense urban setting.

Overall, we found that using scripting tools to automate geoprocessing tasks in conjunction with an elevation profile web service made it possible to utilise information from open geospatial data surrounding a building site effectively. However, there are needs for improvements to the methodology and risk of oversimplification. A next step could be to include diffuse-shading models and evaluate other wind conversions to site and shelter methods for urban settings. It appears that the often-scripted data-driven building thermal evaluation workflows can benefit from using climatological and spatial tools and datasets, especially to include local effects, but more practical evaluation studies are needed.

Author Contributions: Data curation, K.S.; Investigation, K.S.; Methodology, K.S.; Writing—original draft, K.S.; Writing—review and editing, K.S. and A.G.; Visualization, K.S.; Supervision, A.G. Both authors have read and agreed to the published version of the manuscript.

Funding: This paper has been written within the Research Centre on Zero Emission Neighbourhoods in Smart Cities (FME ZEN) funded by the Research Council of Norway (RCN), the research partners NTNU and SINTEF, and the user partners from the private and public sector.

Data Availability Statement: Produced using Copernicus data and information funded by the European Union-EU-DEM layers and Copernicus Land Monitoring Service, Copernicus Atmosphere Monitoring Service (CAMS) radiation service information and Copernicus Climate Change Service information–ERA5 and ERA5-Land global reanalysis. Contains DSM and DTM data produced by The Norwegian Mapping Authority/CC-BY 4.0, DSM and DTM data produced by the Environment Agency UK/Open Government Licence, DSM and DTM data produced by Information Flanders/Open Data Commons Attribution license 1.0, OpenStreetMap data © OpenStreetMap contributors/Open Data Commons Open Database License and Ordnance Survey data © Crown copyright and database right 2020.

Acknowledgments: The case studies were made available to participants in the IEA–EBC Annex 71 project Building Energy Performance Assessment Based on In-Situ Measurements.

Conflicts of Interest: The authors declare no conflict of interest.

Appendix A

Table A1. Building information and assumptions used to estimate solar irradiance on the facades. Façade glazing distribution and azimuth orientation are given in a list form where the four cardinal directions (N E S W) are (180 90 0 -90) degrees.

Parameter (N, E, S, W)	TWINS	ZEBLL	GBORO	UKULE
Facade height h_k	4 m	2.8 m	5.0 m	6.5 m
Aperture above gr. $h_{0;k}$	1.2 m	0.6 m	0.5 m	1.7 m
Façade azimuth az (°)	(180 90 0 -90)	(176 86 -4 -94)	(-168 112 12 -78)	(-161 109 19 -71)
Glazing distribution f_{gl}	(.07 .21 .49 .23)	(.19 .25 .40 .16)	(0 .22 .39 .39)	(0 .41 0 .59)
Window area A_{wi}	23.9 m^2	39.3 m^2	14.7 m^2	40.1 m^2
Frame factor $F_{fr;wi}$	0.23	0.40	0.33	0.31
Transmittance $g_{gl;n;wi}$	0.67	0.5	0.67	0.67

Table A2. Building information and assumptions used in the building infiltration model.

Parameter	Case 1	Case 2	Case 3	Case 4
Floors N_{fl}	2	1.5	2	4
Int. building height h_b	5.2 m	3.4 m	5.5 m	11.3 m
Air tightness n_{50}	1.0 h^{-1}	1.0 h^{-1}	1.0 h^{-1}	0.5 h^{-1}
Flow coefficient n	0.67	0.67	0.67	0.67
Indoor temperature	21	21	21	21

Appendix B

B.1. The AIM-2 Infiltration Model

A single zone infiltration model is used to account for building air-infiltration. Empirical single-zone infiltration models were developed in the 1980s and have seen some renewed interest in later years [84–88]. In most simplified infiltration models, the stack and wind-induced infiltration rates are assessed and derived separately and then superpositioned for a total infiltration rate. Two of the most established models are the LBL and AIM-2 models, adapted into simplified and advanced form in the ASHRAE Fundamentals Handbook. AIM-2 was developed by Walker and Wilson (1990) for houses [16], and a model implementation can be found in the BES software ESP-r [76].

The accuracy of the AIM-2 model to predict infiltration rates in dwellings can be excellent (±10%) when the model parameters are well known according to validations by the authors [78]. A separate validation study found a mean error of 16–27% assessing ten single-family homes [89]. Another study found an average error of about 19% predicting air infiltration rates for 16 detached houses under a wide range of weather conditions [84]. More recently, the AIM-2 model was utilised to predict infiltration rates in three stone-churches in Sweden [85]. The median absolute prediction error was 25%. Considering the model was not developed for large structures, a correction factor of 0.8 to account for overprediction was shown to reduce the error from 25 to 11%. In another recent study, the infiltration model was validated on a single building, obtaining a mean absolute value error of 17–35% by using different parameters for envelope leakage distribution [86]. A methodology is presented in successive work to determine the air change rate in near-real-time by combining the AIM-2 model with a tracer gas decay test method, reducing the error to 10% [90].

Lundström implemented the AIM-2 model in a building energy model [87], recently transformed to stochastic state-space form followed by a Bayesian calibration procedure [91]. The stochastic approach includes a logistic function to model occupant induced manual venting during heating and cooling season. Lundström presents the following version to calculate infiltration loss ϕ_{inf} in [87], where the calculated potential specific infiltration flow rate Q^*_{inf} [Pa^n] multiples with the infiltration coefficient C_{inf} [$l/(s\ Pa^n m^2)$] which can be estimated or obtained from fan pressurisations tests.

$$\phi_{inf} = C_{inf} \cdot Q^*_{inf} \cdot \kappa \cdot \rho_a \cdot (\theta_e - \theta_i), \quad Q^*_{inf} = Q/C_{inf} \tag{A1}$$

$$Q^*_{inf} = \left((Q^*_s)^{\frac{1}{n}} + (Q^*_w)^{\frac{1}{n}} - 0.33 \cdot (Q^*_s \cdot Q^*_w)^{\frac{1}{2n}} \right)^n \tag{A2}$$

where, ρ_a is the density of outdoor air, κ is the heat capacity, and exponent n is the building leakage flow coefficient of the orifice power law. Like many simple infiltration models, AIM-2 uses a superposition technique where the infiltration flow rates due to wind and stack effects, Q^*_s and Q^*_w, are added non-linearly in addition to an interaction term (Equations (A1) and (A2)).

In [87], building height H is adjusted for buildings taller than two floors to H^* (m), resembling the correction factor used by [85] to account for over-prediction. Potential infiltration rates due to stack and wind effects can be pre-calculated by either using the calculated indoor temperature from the previous time-step in [91] or by assuming a constant pre-defined set-point temperature in [87].

$$Q_s^* = f_s(\Delta P_s)^n = f_s\left(\frac{9.806 \cdot H^* \cdot \rho_a \cdot |\theta_e - \theta_i|}{\theta_i + 273.15}\right)^n \quad \text{(A3)}$$

$$Q_w^* = f_w(\Delta P_w)^n = f_w\left(0.5 \cdot U_{loc}^2 \cdot \lambda_w^2 \cdot \rho_a\right)^n \quad \text{(A4)}$$

Assuming evenly distributed envelope leakage, no flue (chimney), and basement or slab-on-grade foundation, the wind and stack factors reduces to f_s = 0.25 [(Pa/K)^n] and f_w = 0.22 [(Pa s^2/m^2)]. However, in this simplified form AIM-2 loses some of its flexibility to utilise building and site-specific data. The leakage distribution input values is a major source of uncertainty influencing AIM-2 and determining the values through measurements is difficult [78]. Noting a lack of reliable data, one of the validation studies shows that minor improvement is achievable (over the default uniform values) if certain building characteristics are taken into consideration [84]. Based on the optimisation of the leakage distribution (between ceilings, floors, and walls) on different groups of houses, their study recommends using values provided in a guideline for estimating leakage distribution parameters according to house types, number of storeys, and foundation type by Lew [92]. The leakage distribution tables by Lew were implemented as cited in the ESP-r source-code [76]. The full equations for f_s and f_w can be found in [78] including model forms for building flues or crawl spaces.

B.2. The Wind Shadow Method

The concept of a Gaussian-shaped wake in Walker, Wilson and Forest's wind shadow method [17] is similar to a more commonly used shelter model "WEMOD" for far wake effects by Taylor and Salmon [65] that can be found implemented in the QUICK-URB and SkyHelios urban wind models [69,70]. Both QUICK-URB and SkyHelios are light-weight diagnostic urban wind models that do not solve the full Navier-Strokes equations but are based on empirical parameterisations, and mass conservation principles first compiled by Röckle [93] and later improved with updated parameterisations like the WEMOD wake model [69,70]. A difference between the wake model by Walker et al. and the WEMOD model is that the latter was developed to adjust wind speed measurements in a single point in space as opposed to being used on whole building facades. However, being of the same family of models many of the assumptions and limitations apply to both.

The sheltering factors were derived from measured sheltered and unsheltered surface pressures [17]. Therefore, the authors suggest that the surface pressure coefficients C_p of a building shielded by adjacent obstacles could be predicted by correcting the C_p obtained for an isolated building. This was later investigated in wind tunnel experiments using scale building models with different shapes and surrounding conditions by Sawachi et al. [72]. Their results indicate that the influence of the upwind building on the C_p distribution of the downwind building is clearly different inside and outside of the wind shadow. The best correlations are shown when the distance of an adjacent building (obstacle) is more than twice the obstacle's height and width. It is suggested that when the adjacent building obstacle is closer, the width of the wind shadow should be given a wider area, depending on the depth of the shielding obstacle. Still, it is unclear from the paper whether they applied the flapping technique or simply projected a shadow of constant width. Another proposition is that the distance beyond where the shielding effect is negligible could be defined more clearly by the full three- dimensional size and geometrical relationship between the two objects in consideration [72].

For the scaling length in the model, a characteristic dimension of the obstacle is considered which is an empirical relationship between the smallest and the largest dimensions in projected width or height (cast in the direction of the wind). The definition is supported by former experimental studies. We refer to the original paper for the theoretical explanation of model assumptions and the functional form of the wake decay [17].

Appendix C

The roughness length table for each CLC type used in this work is published on the Finish wind atlas website with distinctive values for summer and winter. The values for winter are used.

To calculate the regional and local roughness length, a surface drag coefficient $C_{d;i}$ is averaged at the blending height which is a method to give weighting to the larger roughness values [59,66]:

$$C_{d;i} = \left(\frac{\kappa}{ln(z_{bh}/z_0)}\right)^2, \kappa = 0.4 \quad \text{(A5)}$$

where z_{bh} is the blending height and z_0 is the roughness computed for each map point i (100 m). A simple footprint model is used to scale the significance of roughness further away from the site:

$$C_d = \frac{W_n C_{d;i}}{\sum W_n}, W_n = \exp\left(-\frac{x_n}{D}\right) \quad \text{(A6)}$$

where the scaling factor D is 600 m for the local footprint and 3 km for regional footprint [66]. A source area of up to 3 times D is considered, resulting in an effective evaluation length of 1.8 km for the local scale and 9 km for the regional, which accounts for a total of 80% of the integral of W_n (Equation (A6)). Other studies using this footprint model have reported other distance weightings [48,63].

References

1. Roth, M. The Use of Reanalysis in ASHRAE Applications. *ASHRAE Trans.* **2020**, *126*, 580–592.
2. Qiu, X.; Roth, M.; Corbett-Hains, H.; Fuquan, Y. Mesoscale Climate Modeling Procedure Development and Performance Evaluation. *ASHRAE Trans.* **2016**, *122*, 186–201.
3. Pelosi, A.; Terribile, F.; D'Urso, G.; Chirico, G.B. Comparison of ERA5-Land and UERRA MESCAN-SURFEX Reanalysis Data with Spatially Interpolated Weather Observations for the Regional Assessment of Reference Evapotranspiration. *Water* **2020**, *12*, 1669. [CrossRef]
4. Krähenmann, S.; Walter, A.; Brienen, S.; Imbery, F.; Matzarakis, A. High-resolution grids of hourly meteorological variables for Germany. *Theor. Appl. Climatol.* **2018**, *131*, 899–926. [CrossRef]
5. MET Norway NWP Wiki; GitHub: San Francisco, CA, USA. 2020. Available online: https://github.com/metno/NWPdocs/wiki/Post-processed-products (accessed on 12 October 2020).
6. Dörenkämper, M.; Olsen, B.; Witha, B.; Hahmann, A.; Davis, N.; Barcons, J.; Ezber, Y.; García-Bustamante, E.; González Rouco, J.F.; Navarro, J.; et al. The Making of the New European Wind Atlas—Part. 2: Production and Evaluation. *Geosci. Model Dev.* **2020**, *13*, 5079–5102. [CrossRef]
7. Udina, M.; Montornès, À.; Casso, P.; Kosović, B.; Bech, J. WRF-LES Simulation of the Boundary Layer Turbulent Processes during the BLLAST Campaign. *Atmosphere* **2020**, *11*, 1149. [CrossRef]
8. Resch, B.; Sagl, G.; Törnros, T.; Bachmaier, A.; Eggers, J.-B.; Herkel, S.; Narmsara, S.; Gündra, H. GIS-Based Planning and Modeling for Renewable Energy: Challenges and Future Research Avenues. *ISPRS Int. J. Geo Inf.* **2014**, *3*, 662–692. [CrossRef]
9. Biljecki, F.; Stoter, J.; Ledoux, H.; Zlatanova, S.; Çöltekin, A. Applications of 3D City Models: State of the Art Review. *ISPRS Int. J. Geo Inf.* **2015**, *4*, 2842–2889. [CrossRef]
10. Lindberg, F.; Grimmond, C.S.B.; Gabey, A.; Huang, B.; Kent, C.W.; Sun, T.; Theeuwes, N.E.; Jrvi, L.; Ward, H.C.; Capel-Timms, I.; et al. Urban Multi-scale Environmental Predictor (UMEP). *Environ. Model. Softw.* **2018**, *99*, 70–87. [CrossRef]
11. Altaweel, M. Python and Geospatial Analysis. Available online: https://www.gislounge.com/python-and-geospatial-analysis/ (accessed on 12 January 2021).
12. ESRI ArcGIS REST API. Available online: https://developers.arcgis.com/rest/ (accessed on 10 December 2020).
13. Biljecki, F.; Ledoux, H.; Stoter, J. Generating 3D city models without elevation data. *Comput. Environ. Urban Syst.* **2017**, *64*, 1–18. [CrossRef]
14. Wendel, J.; Murshed, S.M.; Sriramulu, A.; Nichersu, A. Development of a Web-Browser Based Interface for 3D Data—A Case Study of a Plug-in Free Approach for Visualizing Energy Modelling Results. In *Progress in Cartography*; Springer: New York, NY, USA, 2016; pp. 185–205.
15. European Commission. The INSPIRE Geoportal. Available online: https://inspire-geoportal.ec.europa.eu/ (accessed on 12 November 2020).
16. Walker, I.S.; Wilson, D. *The Alberta Air Infiltration Model: AIM-2*; Department of Mechanical Engineering, University of Alberta: Edmonton, AL, Canada, 1990.

17. Walker, I.S.; Wilson, D.J.; Forest, T.W. Wind Shadow Model for Air Infiltration Sheltering by Upwind Obstacles. *HVAC&R Res.* **1996**, *2*, 265–282.
18. Coetzee, S.; Ivánová, I.; Mitasova, H.; Brovelli, M.A. Open Geospatial Software and Data: A Review of the Current State and A Perspective into the Future. *ISPRS Int. J. Geo Inf.* **2020**, *9*, 90. [CrossRef]
19. Clarke, J.A. Why Tools for Buildings and Cities Performance Simulation Need to Evolve. Available online: https://www.buildingsandcities.org/insights/commentaries/tools-for-buildngs-and-cities-performance.html (accessed on 28 December 2020).
20. Clarke, J. IOP Simulation-Based Procedure for the Holistic Resilience Testing of Building Performance. In *Conference Series: Earth and Environmental Science*; IOP Publishing: Bristol, UK, 2019; p. 12027.
21. Manfren, M.; Nastasi, B.; Groppi, D.; Astiaso Garcia, D. Open data and energy analytics—An analysis of essential information for energy system planning, design and operation. *Energy* **2020**, *213*, 118803. [CrossRef]
22. Copernicus Land Monitoring Service. CORINE Land Cover. 2018. Available online: https://land.copernicus.eu/pan-european/corine-land-cover/clc2018 (accessed on 15 May 2020).
23. Copernicus Land Monitoring Service. EU-DEM v1.1. Available online: https://land.copernicus.eu/imagery-in-situ/eu-dem (accessed on 15 May 2020).
24. Tabas, D.; Fang, J.; Porté-Agel, F. Wind Energy Prediction in Highly Complex Terrain by Computational Fluid Dynamics. *Energies* **2019**, *12*, 1311. [CrossRef]
25. Silva, J.; Ribeiro, C.; Guedes, R.; Rua, M.-C.; Ulrich, F. Roughness length classification of Corine Land Cover classes. In Proceedings of the EWEC, Milano, Italy, 7–10 May 2007; Citseer: Princeton, NJ, USA, 2007.
26. NASA Land Processes Distributed Active Archive Center (LP DAAC). Available online: https://lpdaac.usgs.gov/ (accessed on 12 November 2020).
27. Szypuła, B. Quality assessment of DEM derived from topographic maps for geomorphometric purposes. *Open Geosci.* **2019**, *11*, 843–865. [CrossRef]
28. Huld, T. PVMAPS: Software tools and data for the estimation of solar radiation and photovoltaic module performance over large geographical areas. *Sol. Energy* **2017**, *142*, 171–181. [CrossRef]
29. Statens Kartverk Høydedata. Available online: https://hoydedata.no/ (accessed on 10 September 2020).
30. Bayerische Vermessungsverwaltung. Gebühren- und Preisliste für Geobasisdaten der Bayerischen Vermessungsverwaltung. 2020. Available online: ldbv.bayern.de/file/pdf/1269/Preisliste_aktuell.pdf (accessed on 15 April 2020).
31. Lingfors, D.; Bright, J.M.; Engerer, N.A.; Ahlberg, J.; Killinger, S.; Widén, J. Comparing the capability of low- and high-resolution LiDAR data with application to solar resource assessment, roof type classification and shading analysis. *Appl. Energy* **2017**, *205*, 1216–1230. [CrossRef]
32. Buffat, R.; Grassi, S.; Raubal, M. A scalable method for estimating rooftop solar irradiation potential over large regions. *Appl. Energy* **2018**, *216*, 389–401. [CrossRef]
33. Redweik, P.; Catita, C.; Brito, M. Solar Energy potential on roofs and facades in an urban landscape. *Sol. Energy* **2013**, *97*, 332–341. [CrossRef]
34. Brito, M.C.; Redweik, P.; Catita, C.; Freitas, S.; Santos, M. 3D Solar Potential in the Urban Environment: A Case Study in Lisbon. *Energies* **2019**, *12*, 3457. [CrossRef]
35. Lindberg, F.; Jonsson, P.; Honjo, T.; Wästberg, D. Sol. Energy on building envelopes—3D modelling in a 2D environment. *Sol. Energy* **2015**, *115*, 369–378. [CrossRef]
36. Chen, Z.; Gao, B.; Devereux, B. State-of-the-Art: DTM Generation Using Airborne LIDAR Data. *Sensors* **2017**, *17*, 150. [CrossRef] [PubMed]
37. Nex, F.; Remondino, F. UAV for 3D mapping applications: A review. *Appl. Geomat.* **2014**, *6*. [CrossRef]
38. Puente, I.; Gonzalez, H.; Arias, P.; Armesto, J. Land-Based Mobile Laser Scanning Systems: A review. *Int. Arch. Photogramm. Remote Sens. Spat. Inf. Sci.* **2011**, *38*. [CrossRef]
39. Brovelli, M.A.; Zamboni, G. A new method for the assessment of spatial accuracy and completeness of OpenStreetMap building footprints. *ISPRS Int. J. Geo Inf.* **2018**, *7*, 289. [CrossRef]
40. Fan, H.; Zipf, A.; Fu, Q.; Neis, P. Quality assessment for building footprints data on OpenStreetMap. *Int. J. Geo Inf. Sci.* **2014**, *28*, 700–719. [CrossRef]
41. Hjelseth, E.; Thiis, T. Use of BIM and GIS to enable climatic adaptations of buildings. In *Ework and Ebusiness in Architecture, Engineering and Construction*; Christodoulou, S., Scherer, R., Eds.; CRC Press Taylor & Francis Group: Boca Raton, FL, USA, 2008; pp. 409–417.
42. Hufkens, K.; Reto, S.; Campitelli, E. *ECMWFR: Programmatic Interface to the Two European Centre for Medium-Range Weather Forecasts API Services*; Version 1.2.0; Zenodo: Meyrin, Switzerland, 2019.
43. Lundström, L. *CamSrad: Client for CAMS Radiation Service*; Version 0.3. 0; R Package; R Core Team: Vienna, Austria, 2016.
44. Nisbet, A. *Open Topo Data*; GitHub: San Francisco, CA, USA, 2020. Available online: https://github.com/ajnisbet/opentopodata/ (accessed on 11 November 2020).
45. Peronato, G. *getHorizon*; Laboratory of Integrated Performance in Design (LIPID), Ecole Polytechnique Fédérale de Lausanne (EPFL): Lausanne, Switzerland, 2017.
46. Lundström, L. *Total Solar Irradiance According to ISO 52010-1:2017*; GitHub: San Francisco, CA, USA, 2018. Available online: https://github.com/lukas-rokka/solarCalcISO52010 (accessed on 2 May 2020).

47. Stepek, A.; Wijnant, I.L. *Interpolating Wind Speed Normals from the Sparse Dutch Network to a High Resolution Grid Using Local Roughness from Land Use Maps*; Royal Netherlands Meteorological Institute: De Bilt, The Netherlands, 2011.
48. Caires, S.; de Waal, H.; Groeneweg, J.; Groen, G.; Wever, N.; Geerse, C.; Bottema, M. Assessing the uncertainties of using land-based wind observations for determining extreme open-water winds. *J. Wind Eng. Ind. Aerodyn.* **2012**, *110*, 70–85. [CrossRef]
49. Hersbach, H.; Bell, B.; Berrisford, P.; Hirahara, S.; Horányi, A.; Muñoz-Sabater, J.; Nicolas, J.; Peubey, C.; Radu, R.; Schepers, D.; et al. The ERA5 global reanalysis. *Q. J. R. Meteorol. Soc.* **2020**, *146*. [CrossRef]
50. Muñoz Sabater, J. First ERA5-Land dataset to be released this spring. *ECMWF Newslett.* **2019**, *159*, 8–9.
51. Hogan, R. Radiation Quantities in the ECMWF Model and MARS. Technical Report ECMWF. 2015. Available online: https://www.ecmwf.int/node/18490 (accessed on 10 April 2020).
52. Babar, B.; Graversen, R.; Boström, T. Solar radiation estimation at high latitudes: Assessment of the CMSAF databases, ASR and ERA5. *Sol. Energy* **2019**, *182*, 397–411. [CrossRef]
53. Schroedter-Homscheidt, M.; Hoyer-Klick, C.; Killius, N.; Betcke, J.; Lefèvre, M.; Wald, L.; Wey, E.; Saboret, L. User's Guide to the CAMS Radiation Service (CRS): Status December 2018; Copernicus Atmosphere Monitoring Service. 2019. Available online: http://www.soda-pro.com/documents/10157/326332/CAMS72_2015SC3_D72.1.3.1_2018_UserGuide_v1_201812.pdf/95ca8325-71f6-49ea-b5a6-8ae4557242bd (accessed on 10 April 2020).
54. Qu, Z.; Oumbe, A.; Blanc, P.; Espinar, B.; Gesell, G.; Gschwind, B.; Klüser, L.; Lefèvre, M.; Saboret, L.; Schroedter-Homscheidt, M.; et al. Fast radiative transfer parameterisation for assessing the surface solar irradiance: The Heliosat-4 method. *Meteorol. Z.* **2017**, *26*, 33–57. [CrossRef]
55. Ramon, J.; Lledó, L.; Torralba, V.; Soret, A.; Doblas-Reyes, F.J. What global reanalysis best represents near-surface winds? *Q. J. R. Meteorol. Soc.* **2019**, *145*, 3236–3251. [CrossRef]
56. Olauson, J. ERA5: The new champion of wind power modelling? *Renew. Energy* **2018**, *126*, 322–331. [CrossRef]
57. Vortex. *Vortex ERA5 Downscaling: Validation Results*; VORTEX FdC S.L.: Barcelona, Spain, 2017; p. 11.
58. Wang, X.; Tolksdorf, V.; Otto, M.; Scherer, D. WRF-based Dynamical Downscaling of ERA5 Reanalysis Data for High Mountain Asia: Towards a New Version of the High Asia Refined Analysis. *Int. J. Climatol.* **2020**. [CrossRef]
59. Wieringa, J. Roughness-dependent geographical interpolation of surface wind speed averages. *Q. J. R. Meteorol. Soc.* **1986**, *112*, 867–889. [CrossRef]
60. European Centre for Medium-Range Weather Forecasts. Part IV: Physical Processes. In *IFS Documentation CY41R2*; ECMWF: Reading, UK, 2016.
61. De Rooy, W.C.; Kok, K. A combined physical–statistical approach for the downscaling of model wind speed. *Weather Forecast.* **2004**, *19*, 485–495. [CrossRef]
62. Baas, P.; Bosveld, F.; Burgers, G. The impact of atmospheric stability on the near-surface wind over sea in storm conditions. *Wind Energy* **2015**, *19*. [CrossRef]
63. Troen, I.; Petersen, E.L. *European Wind Atlas*; Office of Scientific and Technical Information, U.S. Department of Energy: Oak Ridge, TN, USA, 1989.
64. Barrachina, A.B.; Beek, J.V.; Alzamora, F.M.; Jiménez, P. Using downscaled NCEP/NCAR reanalysis data for wind resource mapping. *Int. J. Energy Environ.* **2014**, *5*, 305–316.
65. Van Ackere, S.; Van Eetvelde, G.; Schillebeeckx, D.; Papa, E.; Van Wyngene, K.; Vandevelde, L. Wind Resource Mapping Using Landscape Roughness and Spatial Interpolation Methods. *Energies* **2015**, *8*, 8682–8703. [CrossRef]
66. Verkaik, J.W. On Wind and Roughness over Land. Ph.D. Thesis, Wageningen Universiteit, Wageningen, The Nederlands, 2006.
67. Verkaik, J.W.; Jacobs, A.; Tijm, A.; Onvlee, J. Local Wind Speed Estimation by Physical Downscaling of Weather model forecasts. *J. Wind Eng. Ind. Aerodyn.* **2005**, submitted.
68. Wever, N.; Groen, G. *Improving Potential Wind for Extreme Wind Statistics*; Koninklijk Nederlands Meteorologisch Institut: De Bilt, The Netherlands, 2009.
69. Fröhlich, D.; Matzarakis, A. Spatial Estimation of Thermal Indices in Urban Areas—Basics of the SkyHelios Model. *Atmosphere* **2018**, *9*, 209. [CrossRef]
70. Singh, B.; Pardyjak, E.; Brown, M. Testing of a Far-wake Parameterization for a Fast Response Urban Wind Model. In Proceedings of the Sixth Symposium on the Urban Environment/14th Joint Conference on the Applications of Air Pollution Meteorology with the Air and Waste Management Association, Atlanta, GA, USA, 27 January–3 February 2006.
71. Grimmond, C.; Oke, T.R. Aerodynamic properties of urban areas derived from analysis of surface form. *J. Appl. Meteorol.* **1999**, *38*, 1262–1292. [CrossRef]
72. Sawachi, T.; Maruta, E.; Takahashi, Y.; Ken-ichi, S. Wind Pressure Coefficients for Different Building Configurations with and without an Adjacent Building. *Int. J. Vent.* **2006**, *5*, 21–30. [CrossRef]
73. Brunskill, A.; Lubitz, W. A neural network shelter model for small wind turbine siting near single obstacles. *Wind Struct.* **2012**, *15*, 43–64. [CrossRef]
74. Hansen, A.; Peterka, J.A.; Cermak, J.E. *Wind-Tunnel Measurements in the Wake of a Simple Structure in a Simulated Atmospheric Flow*; Libraries, Colorado State University: Fort Collins, CO, USA, 1975.
75. Sherman, M.; Grimsrud, D. Measurement of Infiltration Using Fan Pressurization and Weather Data. In Proceedings of the First Air Infiltration Centre Conference, Windsor, UK, 6–8 October 1980; p. 81.
76. Strachan, P. *ESP-r: Summary of Validation Studies*; Energy Systems Research Unit, University of Strathclyde: Glasgow, UK, 2002.

77. American Society of Heating, Refrigerating and Air-Conditioning Engineers. *ASHRAE Handbook, Edition SI*; American Society of Heating, Refrigerating and Air-Conditioning Engineers: Atlanta, GA, USA, 2017.
78. Walker, I.S.; Wilson, D.J. Field validation of algebraic equations for stack and wind driven air infiltration calculations. *HVAC&R Res.* **1998**, *4*, 119–139.
79. Madsen, H.; Bacher, P.; Bauwens, G.; Deconinck, A.-H.; Reynders, G.; Roels, S.; Himpe, E.; Lethé, G. *Thermal Performance Characterization Using Time Series Data—IEA EBC Annex 58 Guidelines*; DTU Compute-Technical Report-2015, 8; Technical University of Denmark: Copenhagen, Denmark, 2015.
80. Stamp, S.; Altamirano-Medina, H.; Lowe, R. Measuring and accounting for solar gains in steady state whole building heat loss measurements. *Energy Build.* **2017**, *153*, 168–178. [CrossRef]
81. Sandu, I.; Beljaars, A.; Bechtold, P.; Mauritsen, T.; Balsamo, G. Why is it so difficult to represent stably stratified conditions in numerical weather prediction (NWP) models? *J. Adv. Model. Earth Syst.* **2013**, *5*, 117–133. [CrossRef]
82. Over, M. *OpenDEM Europe*; Universität Bonn Geographie, FOSSGIS: Bonn, Germany, 2018.
83. Gras, D. *EU-DEM Statistical Validation Report*; European Environment Agency: Copenhagen, Denmark, 2014.
84. Wang, W.; Beausoleil-Morrison, I.; Reardon, J. Evaluation of the Alberta air infiltration model using measurements and inter-model comparisons. *Build. Environ.* **2009**, *44*, 309–318. [CrossRef]
85. Hayati, A.; Mattsson, M.; Sandberg, M. Evaluation of the LBL and AIM-2 air infiltration models on large single zones: Three historical churches. *Build. Environ.* **2014**, *81*, 365–379. [CrossRef]
86. Tirfe, A.; Zhang, J. A Novel Approach to Near-Real Time Monitoring of Ventilation Rate and Indoor Air Quality in Residential Houses. In Proceedings of the 7th International Building Physics Conference, Syracuse, NY, USA, 23–26 September 2018; Syracuse University: Syracuse, NY, USA, 2018; pp. 841–846.
87. Lundström, L.; Akander, J.; Zambrano, J. Development of a Space Heating Model Suitable for the Automated Model Generation of Existing Multifamily Buildings—A Case Study in Nordic Climate. *Energies* **2019**, *12*, 485. [CrossRef]
88. Wills, A.D. On the Modelling and Analysis of Converting Existing Canadian Residential Communities to Net-Zero Energy. Ph.D. Thesis, Carleton University, Ottawa, ON, Canada, April 2018.
89. Francisco, P.W.; Palmiter, L. Modeled and Measured Infiltration in Ten Single-Family Homes. In *Proceedings of the ACEEE Summer Study on Energy Efficiency in Buildings*; American Council for an Energy-Efficiency Economy: Washington, DC, USA, 1996.
90. Tirfe, A. A Novel Approach to Near-Real Time Monitoring of Ventilation Rate and Indoor Air Quality in Residential Houses. Ph.D. Thesis, Syracuse University, Syracuse, NY, USA, August 2018.
91. Lundström, L.; Akander, J. Bayesian Calibration with Augmented Stochastic State-Space Models of District-Heated Multifamily Buildings. *Energies* **2020**, *13*, 76. [CrossRef]
92. Lew, L. *Evaluation of AIM-2*; Natural Resources Canada: Ottawa, ON, Canada, 1993.
93. Röckle, R. Bestimmung der Strömungsverhältnisse im Bereich Komplexer Bebauungsstrukturen. Ph.D. Thesis, Darmstadt Technische Hochschule, Darmstadt, Germany, 1990.

energies

MDPI

Article

Cyber-Physical Systems Improving Building Energy Management: Digital Twin and Artificial Intelligence

Sofia Agostinelli [1,*], Fabrizio Cumo [1], Giambattista Guidi [2] and Claudio Tomazzoli [3]

1 CITERA Interdepartmental Centre, Sapienza University of Rome, 00197 Rome, Italy; fabrizio.cumo@uniroma1.it
2 National Agency for New Technologies, Energy and Sustainable Economic Development, 00123 Rome, Italy; giambattista.guidi@enea.it
3 Computer Science Department, University of Verona, 37129 Verona, Italy; claudio.tomazzoli@univr.it
* Correspondence: sofia.agostinelli@uniroma1.it

Abstract: The research explores the potential of digital-twin-based methods and approaches aimed at achieving an intelligent optimization and automation system for energy management of a residential district through the use of three-dimensional data model integrated with Internet of Things, artificial intelligence and machine learning. The case study is focused on Rinascimento III in Rome, an area consisting of 16 eight-floor buildings with 216 apartment units powered by 70% of self-renewable energy. The combined use of integrated dynamic analysis algorithms has allowed the evaluation of different scenarios of energy efficiency intervention aimed at achieving a virtuous energy management of the complex, keeping the actual internal comfort and climate conditions. Meanwhile, the objective is also to plan and deploy a cost-effective IT (information technology) infrastructure able to provide reliable data using edge-computing paradigm. Therefore, the developed methodology led to the evaluation of the effectiveness and efficiency of integrative systems for renewable energy production from solar energy necessary to raise the threshold of self-produced energy, meeting the nZEB (near zero energy buildings) requirements.

Keywords: digital construction; artificial intelligence; digital twin; nZEB; energy management; energy efficiency; edge computing

check for updates

Citation: Agostinelli, S.; Cumo, F.; Guidi, G.; Tomazzoli, C. Cyber-Physical Systems Improving Building Energy Management: Digital Twin and Artificial Intelligence. *Energies* **2021**, *14*, 2338. https://doi.org/10.3390/en14082338

Academic Editor: Ioan Sarbu

Received: 9 March 2021
Accepted: 14 April 2021
Published: 20 April 2021

1. Introduction

The energy management of building systems and urban areas such as residential districts is assuming an increasingly relevant role in the control and assessment of urban development and refurbishment processes.

Digital predictive technologies and sensor-based control systems are becoming fundamental tools [1] supporting policies to reach near-zero requirements and targets for buildings and urban districts. Nowadays, the integration of information communication technologies (ICT) has an important role in the configuration of smart cities and in defining digital strategies addressing social, public health, economic, environmental, and safety issues [2].

The success of such digital transformations requires the ability to meet and manage new emerging challenges [3]. Deep interactions between humans, infrastructures, and technologies are increasingly created over time by the global consequences of urbanization and the growth of human activities. Dealing with complexities related to sustainability matters, cities are implementing technological improvements achieving smarter performances through the definition of smart cities that adhere to a smart growth agenda [4].

According to the above mentioned, it can be introduced the urban intelligence [5] concept, providing insights into a number of issues currently faced by modern cities (i.e., air pollution, communication network demand, congested traffic, water floods, etc.) through the introduction of data from Internet of Things (IoT) sensors processed by intelligent and real-time advanced analytics. According to the United Nations prediction, 60% of cities will have at least half a million inhabitants by 2030, leading to issues in cities such as the increasing of network demand and crowd congestions [6].

In the future, progressively current problems in cities will be necessarily managed through intelligent urban reasoning algorithms and suitable deployment data-model based on urban intelligence systems, pervasive computing, communication, big data management technologies, and artificial intelligence (AI), leading to a strong evolution in the management of urban environments as well as in the quality of life in smart cities [7,8].

The configuration of city digital twins represents a giant leap forward for urban sustainability from design to construction and maintenance basing on the implementation of Industry 4.0 principles [9,10]. It is defined as a digital replica of a physical asset, collecting information from sensors, drones, or other sensive IoT devices, applying advanced analytics, machine learning (ML), and AI obtaining real-time processed data about the lifecycle process of physical assets.

In particular, digital twin (DT) ecosystems are related to three main entities: a physical object, its virtual replica, and the connection between them in terms of collecting and connecting real-time information. Such a digital ecosystem can effectively contribute to the lifecycle management of both vertical and horizontal systems, in order to store, manage and process big data about the urban environment in a three-dimensional data model as a structured information system connected to the physical.

In this paper, the applications of such ICT-based digital approaches are related to energy management systems, in order to predict real time situations, enriching and leading to more effective decisions, obtaining the automation of repetitive tasks, and providing added value with the optimization of decision-making processes.

In particular, the objective concerns the configuration of a solid methodology for an increasingly intelligent system where the potential of ICT, IoT, big data and AI are combined interacting with BIM (building information modeling) models (Figure 1), defining three-dimensional information and predictive systems for energy management.

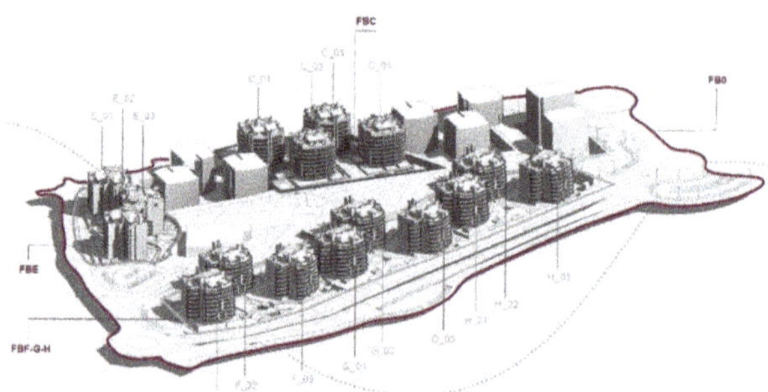

Figure 1. Rione Rinascimento III three-dimensional BIM model overview, consisting of encoded functional blocks (FB) and building models (from C_0n to H_0n).

In fact, the connection between IoT devices, digital information models (BIM), and AI defines an advanced smart-city ecosystem as an intelligent, ubiquitous, and sustainable digital urban context [2] where real-time monitoring systems allow data connections and processing anytime and anyplace [3,4].

More specifically, the project developed by CITERA Interdepartmental Centre of Sapienza University of Rome explores the potential of digital-twin models integrated with AI systems finding a specific application as an opportunity to apply the developed methodology. The case study is related to the configuration of an effective DT model of a residential district in Rome, increasing energy efficiency and identifying a cost-optional solution for which both consumption and costs are expected to be reduced.

Therefore, the 3D information model was developed gradually from the territorial, infrastructural (using Autodesk InfraWorks for geographic information systems) up to the building scale (using Autodesk Revit for building information modeling). The model resulted both as a microscopic and macroscopic digital database, containing static, dynamic, geometric, and semantic data about buildings and their functional interactions.

As mentioned, a BIM approach was carried out focusing on energy management model-uses and leveraging interoperability using IFC (industry foundation classes) models for energy diagnosis purposes. Basing on such analysis, a smart-energy-grid management system was developed combining BIM as-built models with IoT and AI obtaining a substantial as-performed and up-to-date city digital twin.

2. Background

The objective of bringing the virtual and physical worlds together is focused to better support decision-making, reducing risks and configuring a citizen engagement tool, improving urban sustainability [9]. The introduction of DT in construction processes addresses the improvement of decision-making focusing on well-informed and advanced real-time "what-if" scenario assessments, reducing wastes of time and resources that are typical in construction.

In this regard, the Newcastle University created a DT of the city dedicated to incidents and disasters responding and prevention, running simulations of incidents such as burst pipes, heavy rainfall or floods to evaluate the potential impact on communities over a 24 h period [10].

Another effective example of smart-city DT currently ongoing is virtual Singapore, which provides capabilities from virtual experimentations, test-bedding, and decision-making up to research and development [11].

Moreover, a relevant experience is carried out by the Centre for Digital Built Britain (CDBB) delivering a "smart digital economy for infrastructure and construction", as a transformation of the UK AEC (architecture engineering and construction) industry's approach about planning, building, maintenance and utilization of social and economic infrastructures [12].

In addition, the ongoing project for the city digital twin of Atlanta creates a virtual reality (VR)-based platform (built basing on the unity interactive and data-driven cross-platform game engine) which contains a three-dimensional fully modeled city of Atlanta, reproducing the entire city into a virtual space, facilitating spatial-temporal feedbacks and interactions between the human/infrastructure systems and their virtual representations [13].

Focusing on the energy implementations, three significant experiences related to DT developments integrated with AI systems can be mentioned, in order to define a systemic approach for the present study, aiming at integrating the objectives of the single experiences reported below.

The first concerns a microclimatic study on urban scale carried out in the Kalasatama district by the Municipality of Helsinki, in which it is important to highlight the "Energy and Climate Atlas", defined as a city information model for studying and developing strategies for the mitigation of climate changes and improving energy efficiency. The atlas includes a number of specific information about the buildings, such as heating systems, energy certification, electricity consumption, district heating, and water distribution. As configured, the model helps to analyze a series of technological scenarios, allowing users to define the solar energy potential of buildings, evaluating the possibility for reducing carbon dioxide emissions or outlining cost-impact scenarios for different interventions [14].

In addition, it is important to investigate the behavior of energy-smart-grid systems serving differentiated users managed by ML. As known, the main issue to be resolved concerns the need to implement storage systems due to the characteristics of discontinuity of renewable energy production.

The ESS (energy storage system) management realized through a DT integrated with ML systems can bring significant improvements leading to consequent bill savings, if compared with the current systems based on predefined control systems of the electrical power supply from the batteries.

In addition, the development of an energy management system (EMS) is fundamental. As reported by Park, Byeon et al. [15] "an EMS reinforces operational functions such as adjusting the amount and schedule of charging and discharging through the efficient control of the ESS and power conditioning system (PCS) and manages the overall power flow". Moreover, it is connected with sensors and measurement equipment able to analyze and monitor consumption patterns, managing information about power activities and optimizing the overall efficiency.

Another extremely significant energy application of DT is the simulation and testing of scenarios for energy-efficiency interventions aiming to achieve nZEB (near zero energy buildings) requirements on buildings. Since most buildings today are already built, it is necessary to underline the essential application of nZEB parameters on existing built environments through the use of BIM-oriented 5D and 6D digital approaches [16].

The fifth and sixth dimensions of BIM are used and developed to promote stakeholder's collaboration, visualizing and evaluating different options with the configuration of nZEBs, in terms of sustainability and energy efficiency parameters (6D), estimating associated costs (5D) and technical issues [16].

From there, the advances in building data interoperability both at a technical and organizational level enable relevant innovation in end–user energy delivery and optimization [17] beside to open data availability, leveraging on technologies [18] such as the IoT and cyber–physical systems.

3. Material and Methods

The case study of the present research analyzes digital ICT-based energy management techniques applied to a 16 eight-floor buildings residential district called Rione Rinascimento III, located in Rome, which represents the most significant Italian residential implementation of a geothermal source heat pump (GSHP) system, that is currently the largest in Europe.

3.1. The Urban Context

Rinascimento III (Figure 2) is configured as a building intervention characterizing an energetically self-sufficient new portion of the city, integrated as much as possible with the surrounding areas in terms of urban planning and services, and it is considered of relevant significance since it is powered by a still not-commonly-deployed kind of renewable energy system.

In the urban planning agreement between the Municipality of Rome and the private owner, primary and secondary public works were planned, as well as the completion of the Talenti Park area in front of the district. According to the Italian regulations, the new district is included in the category of bioenergetic improvement interventions, which aim at improving the bioclimatic performance of the settlement.

Moreover, the introduced Italian energy policies (such as Decree Law no. 63 of 4 June 2013) aim at a partial refunding up to 65% of the amount for energy requalification expenses, consistently improving the use of renewable sources such as the geothermal one.

The geological characteristics of the Italian territory are particularly favorable for the development of geothermal energy systems and could allow one to exploit low-enthalpy resources at different depths and in numerous areas of the country.

According to the above mentioned, a research activity was developed by the CNR (National Centre for Research) with a pilot project promoted in four Italian regions (Calabria, Campania, Apulia, and Sicily), contributing to the increase of knowledge about the use of geothermal resources, with the aim of providing useful information to start activities of exploration for the improvement of geothermal energy uses in the south of Italy [19].

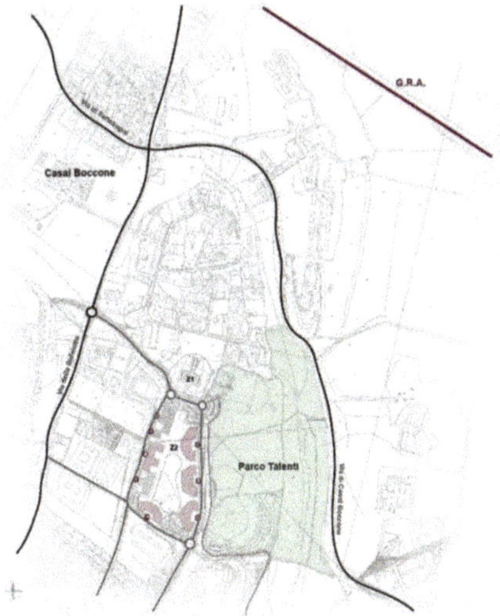

Figure 2. Master plan of the considered area (Rinascimento III) in Rome.

3.2. *Linking Virtual to Physical*

The concept of Construction 4.0 defines a framework where data-driven systems are able to manage physical processes by configuring a virtual replica of the physical world and achieving decentralized decision-making processes based on self-learning mechanisms [20].

Therefore, BIM models containing data and information useful for processing assessments become able to communicate with the real systems using data from sensors, developing learning capabilities, and being able to process the received information.

The collaboration between 3D information models and IoT devices is highly necessary for a successful implementation of real-time DT purposes, as well as for energy management optimizations. However, the implementation of IoT in real-world environments configuring smart, ubiquitous, and live-interconnected systems (Figure 3) is currently still restricted by technical barriers such as device battery life, network capacity, and maintenance costs.

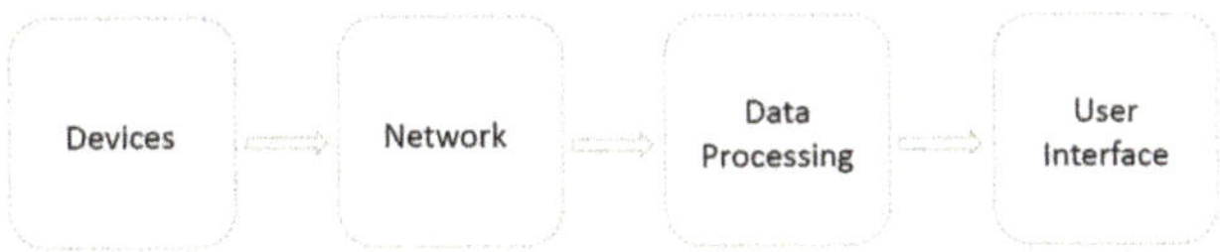

Figure 3. Block diagram of the IoT system.

The core functionality of IoT devices is to reliably collect and share data (such as flow rates, temperatures, pressures, physical movements, distance, mass, etc.) from its designated environment to the virtual world.

The hardware elements consist of a battery-powered sensor, an actuator, and a network communication system in which the collected data are processed and consequently sent to remote servers.

In the present application, the connection between the physical and virtual model is made through sensors [21] able to monitor and communicate electrical power data such as power energy voltmeter ammeter for lighting and heating, ventilation and air conditioning (HVAC) systems and smart plugs for electromotive equipment such as computers, televisions, washing machines, and so forth (Table 1) [22].

Table 1. Review of the implemented IoT devices.

Communication Technology	Functions	Technical Parameters	Real-Time Monitoring
Smart plugs	Communication of the overall profile of energy consumption to AI systems, in order to provide data learning on every single socket, defining a hierarchy of energy priorities to be attributed to the different zones of the apartment in case of deficit in energy production systems.	220–240 V ~10 A, Max 2300 W WiFi connection	Power supply
Energy Power Meter	Functionality as above, the dual relay switch with dual power metering, can be installed into the wall under the power socket or a standard light switch.	110~230 V AC 50 Hz~60 Hz 0~10 A. 0–2300 W WiFi connection	Power supply
Temperature and Humidity Monitors	Temperature and humidity monitoring, WiFi connected with other smart devices enabling smart appliances through app platforms.	0–60 °C 0–99% RH QB/WSDJ2401-2019 Bluethoot 4.2 BLE	Temperature and humidity

In this case, AI systems allow the DT to develop predictive capabilities, learning from the events and improving outputs, ultimately taking and implementing autonomous decisions based on the analysis carried out without human interventions.

Moreover, the AI system achieves a balanced condition between energy consumption and energy production system's performance parameters [23], adapting itself to the environment in order to achieve the predefined objectives.

In other words, the system takes data from sensing devices, and it generates appropriate and specific actions through reasoning systems, modifying the behavior of the equipment in order to optimize energy consumptions. Specifically, it takes information from IFC-BIM and CityGML-GIS (geographic information systems) models, constantly updating them with real-time data as described in Figure 4.

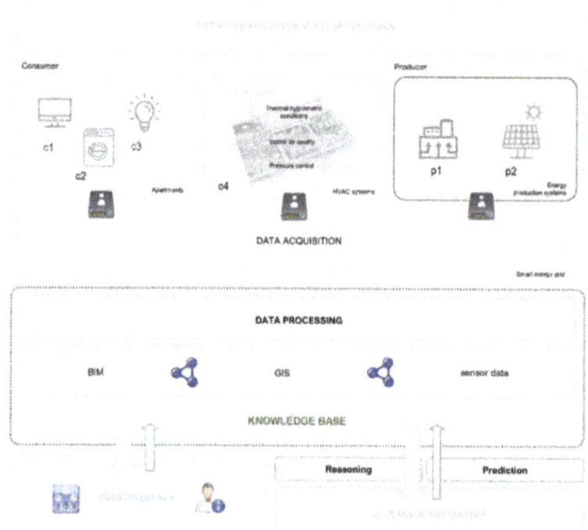

Figure 4. Data flow and processing for digital-twin-based energy optimization.

3.3. *Data Interoperability*

Principles of Industry 4.0 and data interoperability in the AEC sector are extensively applicable on linking GIS and BIM models, providing data for real-time multiscale object-oriented simulations of the built environment. As configured, GIS-BIM 3D city information models and applications require common communication standards introducing problems related to information integration and data interoperability at different domains and scales [24].

In the specific case of information management in construction processes based on BIM methodologies, interoperability consists in exchanging data from models to different software and application platforms, implemented for different purposes and functionalities throughout to the whole lifecycle.

The main objective of interoperability is to facilitate the interaction between different and nonhomogeneous information systems, minimizing errors and aiming at reliability, effectiveness, and optimization of resources.

For the above mentioned, different levels and approaches on interoperability, are defined by the Information Technology Vocabulary (ISO/ISO/IEC 2382) [25] as the "capability to communicate, execute programs, or transfer data among various functional units in a manner that requires the user to have little or no knowledge of the unique characteristics of those units" [26,27].

Industry foundation classes (IFC) were defined as a reference standard format for the building industry to develop different advanced processes based on spatial data relations between building components of a BIM model.

In the present application, specific processes can be scheduled for different activities, objectives and domains (Table 2) since objects are connected to data entities and properties such as name, geometry, identifications, material parameters, etc.

Table 2. Data domains and collection of the interoperability process.

Domain	Data Collection	Software	Interoperability
1. Building information modeling 1.1 Building energy modeling 1.2 Computational fluid dynamics (CFD) simulations	BIM objects, LOD 400	Autodesk Revit MC4 Suite for Revit Autodesk CFD	IFC Standards
2. Geographic information systems 2.1 City information model	BIM/GIS objects	Autodesk InfraWorks	IFC/City GML Standards

In the GIS field, CityGML was developed as a model standard representing geometric and information relationships between geographic entities, being defined as the most appropriate territorial modeling standard in different levels of detail. In addition, IFC and CityGML standard were used, as they are currently the two semantic models dedicated to the configuration of object-oriented information management systems, even though research is still focused on information exchanging, linking IFC and CityGML toward an advanced 3D city information model [28].

3.4. *6D BIM for Sustainability and Energy Efficiency*

The study focuses on the Rinascimento III district (about 85,000 m^2) which is a part of Rione Rinascimento, consisting of 16 eight-floor buildings hosting about 900 apartment units with 2500 inhabitants.

A significant part of the energy supplied to the building complex is self-produced using renewable geothermal sources. For this reason, the following case study is considered to be extremely relevant for approaching digital methodologies integrating DT and AI systems for an efficient energy-smart-grid management.

According to the BIM Use Classification System developed by Penn State University [29] which basically categorizes BIM Uses (Figure 5) as the main purpose to be achieved when implementing BIM in construction processes, specific purposes and objectives for BIM models were identified.

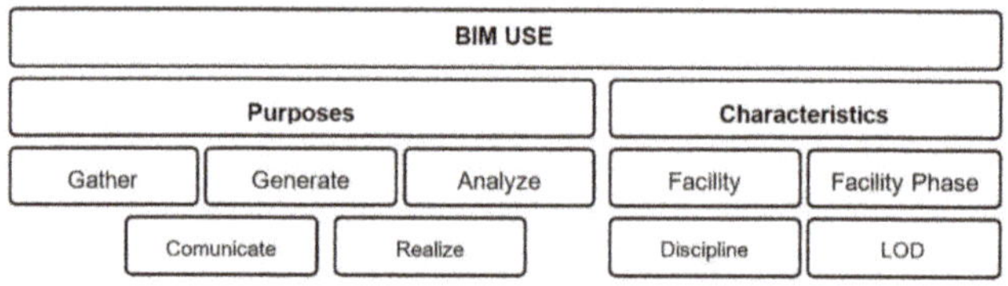

Figure 5. The components of a BIM use, adapted from ref. [29].

The definition of the main BIM purposes led to the identification of specific requirements for data implementation and model configuring.

Since the current application is based on the use of BIM and GIS models for energy management purposes, priority was given to the implementation of specific data such as well-defined technical parameters of the building envelope, thermal zones, rooms, HVAC systems, and equipment, as well as specific data about localization, climate [30], boundary conditions, etc., as information coming directly from the BIM system in the interoperability process.

Moreover, BIM models can have different level of depth both geometrically and informatively, depending on the BIM Uses and related objectives. According to the ISO 19650 [31] standard, LODs were defined, gradually moving toward a LOIN (level of information need) perspective shifting from a prescriptive to a performance approach, based on information granularity depending to predetermined specific BIM uses.

As mentioned, the production of the BIM models followed a number of phases coming from a low degree of definition (LOD 100 [32]), useful in preliminary and outdoor concept stages, up to a LOD 400 (Figure 4, right), according to the BIMForum, "2013 Level of Development Specification" (AIA/AGC, 2013), [32] for indoor energy analysis and simulations purposes as described in Figure 6.

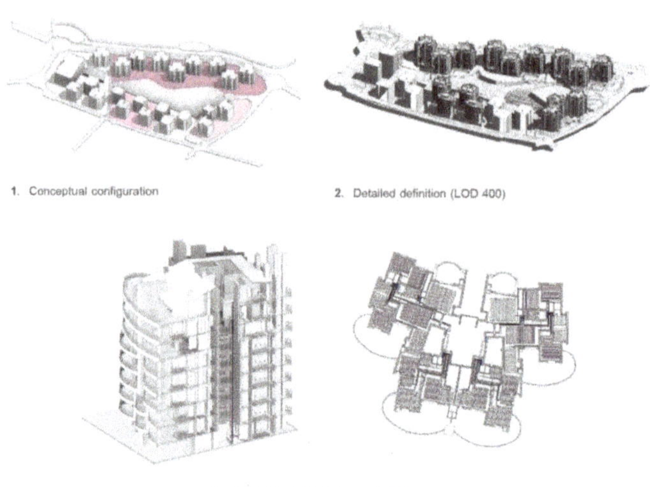

Figure 6. The evolution phases of the BIM model LOD, according to the objective definitions and energy uses.

As configured, the so-called sixth BIM dimension (6D) was achieved since the identified BIM use was connected to energy efficiency and sustainability analyses and simulations [33]. Developing a BIM-oriented methodology allowed to assess the energy performance of the building system, providing relevant support to decision-making processes.

In this section, it is necessary to detail the data and boundary conditions necessary to run the energy analysis through the 6D BIM model [33]. The thermal characteristics of the building envelope technical systems as well as the related data contained in the BIM model are reported in Table 3.

Table 3. Characteristics of the buildings' thermal envelope in the BIM model.

Building Envelope	Thickness (mm)	Thermal Transmittance (W/m^{2}K)	Solar Factor	Threshold Value 2021 (W/m^{2}K) (Italian Regulations)
Facade wall	445	0.29	-	0.32
Roof	480	0.26	-	0.26
Floor structure	300	0.44	-	-
Basement floor	300	0.32	-	0.32
Windows	68	1.37	0.35	1.9

In this case, DT reproduces the energy characteristics of the building envelope and technical plants, which combine a component of renewable energy as described in Section 3.5. In Table 4 the technical components of the main HVAC plants, as well as the controlled mechanical ventilation system are reported.

Table 4. Building's thermal system configuration detailed in the BIM model.

System	Generator	Distribution	Terminal Equipment	Energy
Heating and cooling	GSHP (COP 3.8 winter/5.5 summer)	Water	Radiant floor	Electricity
Ventilation	Centrifugal fans	Filtered air	Air vent	Electricity
Hot sanitary water	Boiler (High efficiency)	Water	-	Gas

3.5. Building Energy Model (BEM)

The main objective of the DT-based developed methodology is using data models across different simulation and monitoring processes [34], combining data from different sources (BIM, GIS, IoT, etc.) in a three-dimensional model, which is aligned almost in real-time with the reproduced system [35,36].

In order to create a building energy model (BEM) [37], each component of the information model was associated with the corresponding products in a BEM software connected to BIM data (MC4 Suite for Revit), defining different thermal zones and boundary conditions.

Once the energy model was generated using a specific and authorized software, [38] it followed the validation phase.

In particular, according to Italian regulation DLgs. 30 May 2008 on "calculation methodologies and requirements for the execution of energy diagnoses and energy certification of buildings" if the deviation between the values estimated by the model and the real consumption does not exceed 5% on average, then the model is validated.

In the pilot project described in the present study, the building complex is supplied by the largest European residential geothermal plant with GSHP (COP of 3.8 in winter configuration and 5.5 in summer configuration), equipped with 200 vertical geoprobes, 150 m deep.

The components of the total energy consumption of Rinascimento district are reported in the following schemes (Figure 7) and divided into four main categories: (1) winter air conditioning; (2) summer air conditioning; (3) hot water; and (4) electric power supply.

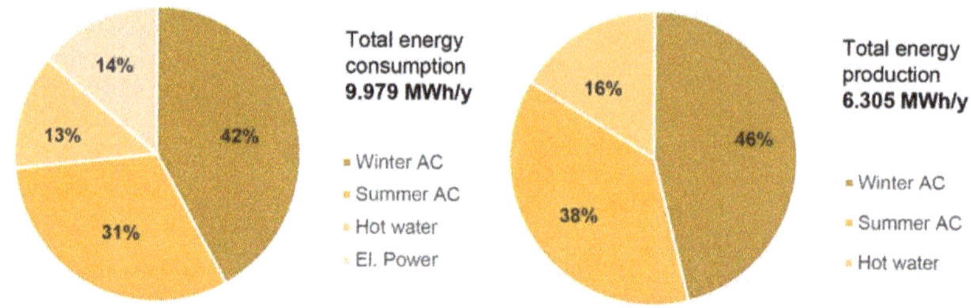

Figure 7. Total primary energy consumption and renewable energy sources (RES) energy production.

In fact, the energy production coming from renewable energy sources (RES) and particularly from the geothermal plant could be estimated in about 6305 MWh/y on 9.979 MWh/y consumed, subdivided as shown in Figure 7. Consequently, 63% of the total energy requirement of primary energy is produced by the geothermal system.

In this case study, the energy diagnosis was conducted on one single building (Figure 8) of about 3648 m^2, using the Revit Suite of Mc4 Software through BIM data, for a dynamic simulation of the building behavior, supplied by a modular portion of the geothermal plant. Since the highlighted building is currently the only one being fully occupied by residents (who permitted the implementation of sensing devices for DT configuration), it was selected for energy modeling and real-time monitoring.

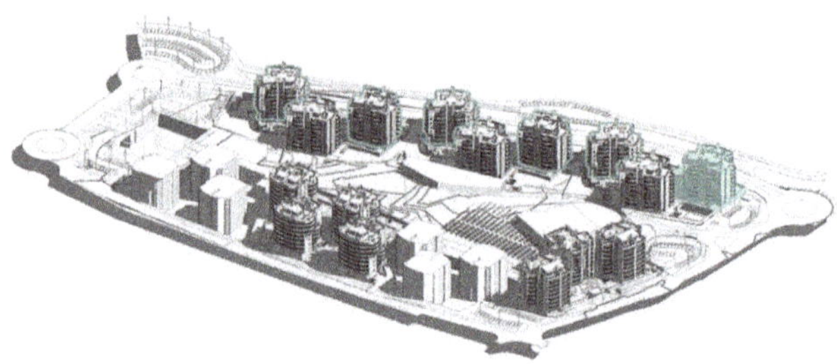

Figure 8. Selected building for the performed energy analysis.

Moreover, since the geometry and spaces subdivision are almost identical for all the buildings, the modeled building is expected to share similar boundary conditions about solar radiation (Figure 9) and ventilation with the other five highlighted in Figure 8, positioned on the outer perimeter of the district without any shading.

Figure 9. Solar radiation analysis.

The performed simulations led to the evaluation (according to the Italian classification of Legislative Decree 48, 10 July 2020) [39] of an A2 class with a specific consumption of 26.8 kWh/m^{2}y; the comparison with the real value building consumptions coming from an average evaluation of 3 year bills (26.6 kWh/m^{2}y) validated the simulation model.

The aim of the DT model was also to simulate the increasing of the RES production percentage, in order to reach the goal for Rinascimento to become a near zero energy district (nZED). The energy simulation in the model were performed considering new installation of photovoltaic panels for the production of electricity and solar collectors for the production of domestic hot water.

In particular, the model was implemented with the integration of 312 kWp of monocrystalline photovoltaic modules in the building façade able to produce 276,000 kWh/y of electricity; and the realization of an area hosting 405 high-efficiency flat-plane solar collectors able to produce 410,000 kWh/y.

The simulations outputs lead to a final result of 6991 MWh/y of energy coming from renewable energy sources (RES) (geothermal+solar), which means about 70% of the district energy consumption directly produced in place by the RES microgrid of the complex.

However, the obtained results so far were focused on the building as a whole, specifying some different thermal zones created according to differences in use, occupation hours, types of HVAC installed, or types of external envelope and sun exposure.

Considering the analysis on a smaller scale, focusing on indoor environmental quality [40] such as thermal-hygrometric conditions, the BIM model was detailed with HVAC systems to develop computational fluid dynamics (CFD) analysis [41].

The standard k-ε model was deployed according to the limited need of calculation power and time for iterations (less than 300) as well as for the absence of high-pressure gradients in the rooms.

The following input conditions have been set:

Average outdoor air temperature equal to 5 °C; radiant floor water temperature equal to 40 °C; underfloor heating surface temperature is between 24 and 29 °C; radiative model discrete ordinates; and 1 s timestep.

Four control probes were temporary fixed and positioned in the center of each room in a typical apartment at 1.50 m from the ground, which is the same height of the DT temperature and humidity monitors fixed in all the apartment rooms (Figure 10).

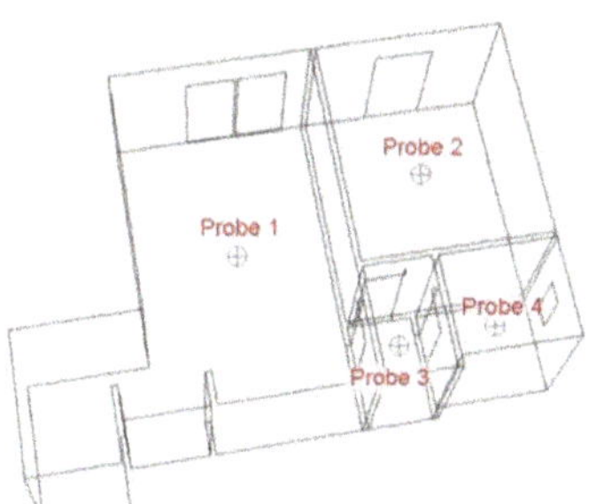

Figure 10. Control probes positioning.

Fluid-dynamics analyses were developed from the BIM model to study the temperature gradient and convective air flows in rooms, triggered by the operation of radiant floors in winter heating mode in order to evaluate comfort parameters in each room, experimenting data interoperability from BIM model to CFD analysis (Figure 11).

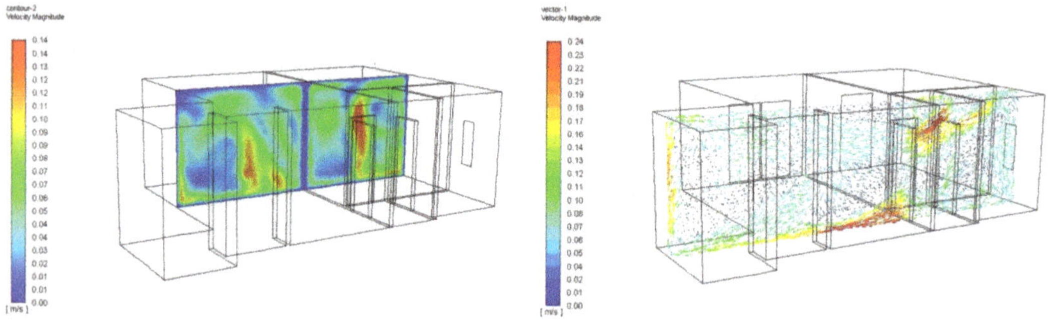

Figure 11. Temperature gradient and air velocity vectors in different rooms.

3.6. Artificial Intelligence

Machine learning is a form of AI providing systems the capability to learn from data without the use of explicit programming. ML produces models where there are some kind of regularity in data [42]. Like human children's learning processes, it is driven by "experience" [43].

As a general rule, training a model requires computer resources which are orders of magnitude bigger than those required to execute the model [44,45].

In this specific case, data are collected and analyzed in order to devise one or more model for energy-efficiency purposes using AI while allowing normal comfort and living habits. The general architecture of the system is shown in Figure 12.

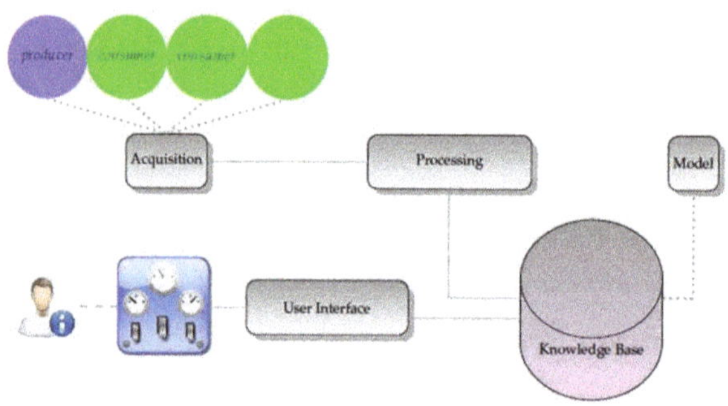

Figure 12. System architecture.

The goal was achieved through two phases: (1) design and implementation of the infrastructure and (2) obtaining data, training, and model testing.

3.6.1. Design and Implementation of the Infrastructure

Energy data are simple time series of power consumption or production, coming from real sensors in a given time lapse, each one transmitting data with its own application programming interface (API); moreover, they are obviously located close to energy loads or near power sources.

This means that data are not all in the same place at the same time, which is a necessary condition to perform the analysis that led to the desired algorithms.

The first problem is therefore to plan and deploy a cost-effective IT (information technology) infrastructure able to provide reliable data to be processed.

Each apartment was implemented with monitoring sensors, so that every device energy consumption could be considered to define the control solution of the overall energy requirement in each apartment.

All the implemented metering sensors produce a huge amount of data requiring significative computational resources to obtain acceptable analysis performances; therefore, the best solution for reducing installation expenses would be to control the system acquiring all the information in a data center or a service in a data center.

This architecture leads to the necessity of setting a local system for interconnecting IoT sensors and actuators over a geographical network (such as the Internet), executing sort of local computation and buffering data in case of connection blackout, using the known "ubiquitous and pervasive computing" [46] techniques to deal with the computational problems of centralized intelligence.

Following this approach, two distinct problems had to be solved designing the infrastructure:

- Have uniform data;
- Have data where they have to be physically processed.

The first element in the infrastructure is a subsystem able to cope with several transmission protocols and time frames, whose output is the synchronized power consumption (or production) of the smart metered devices. This subsystem accepts instruction from the second element to switch on and off some of the controlled devices.

This element needs to be connected with all sensor networks; therefore, it has to be physically placed next to them, minimizing transmission problems and monitoring local environment even in absence of communication with the central control system. This kind of elements is called "elettra" in the following section.

The second architecture element is another subsystem, composed of a different "proxy", and each proxy receives the outputs of the first subsystem as an input. The proxies deliver the data to the central unit and receive back data from the same device, taking care of bandwidth problems and unreliability of the network.

These proxies have to be physically close to the first subsystem while the central unit can be remote; the central control system is a centralized unit able to store and process data, operating building digital simulation models and delivering commands back to the proxies.

The logic model of the designed infrastructure is based on three elements, as shown in Figure 13.

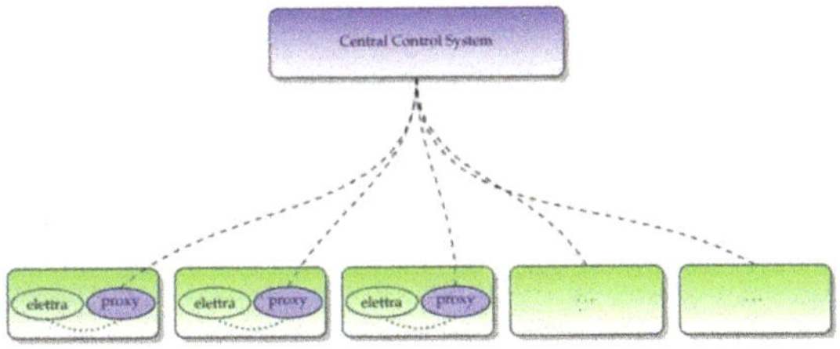

Figure 13. Logic model of the infrastructure.

Following this logic infrastructure, a series of "cheap" small computer or SoC (system on chip) had to be equipped, containing both the "elettra" and "proxy" subsystems; all those computers are connected to a high-performing server in a data centre able to run the software of the central control system. The operative concept of this infrastructure is exemplified in Figure 14, where only a few energy consumer devices are reported as an example.

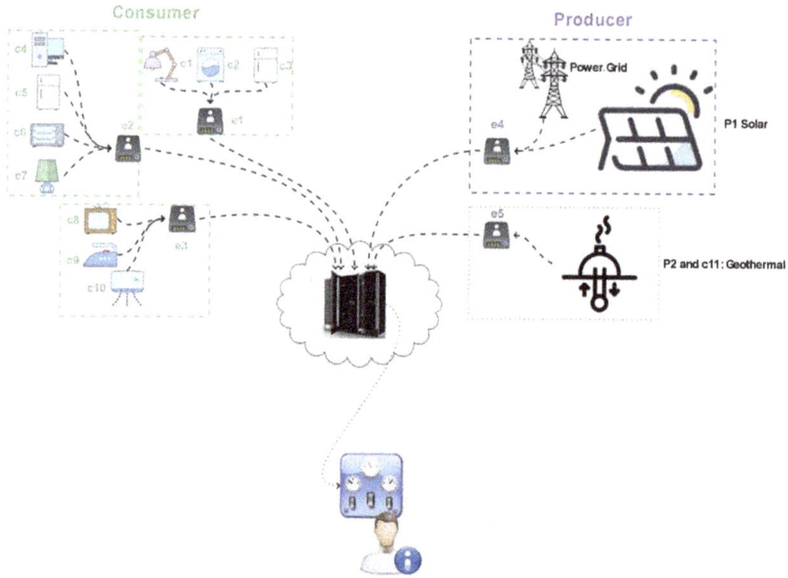

Figure 14. Operative concept.

Elements e_1, e_2, e_3, e_4, and e_5 are the cheap computing containing the elettra subsystem and the proxy, while elements c_1 to c_{10} are energy load examples, and P_1, P_2 are photovoltaic panels for electric power production and geothermal plant.

3.6.2. Obtain Data, Train, and Test Models

Once the data are stored in the central control system, they can be analyzed to build digital numerical models able to simulate and optimize all the main parameters of the smart energy grid. All data have a similar form, so that they can be viewed as a series of {location, date-time, object, value}.

Considering a single location, using ML techniques and rule-based methods such as association rule learning, it is possible to deduce which device is active at a certain time for each selected location [46].

In the present application, it was not possible to consider all the locations as equivalent one to the other, as detailed in Section 4.

A possible general solution is the adoption of best practices, which are hard to define due to the different final uses (home, office, and mixed use) and layouts; if grouped by location and similarities parameters, AI becomes able to automatize processes attributing each location to the most appropriate group or cluster. Therefore, it is necessary to run a ML technique known as "clustering" to automatically create groups of similar apartments used for mathematical representation of each unit: to create the feature vector of each unit, each and every energy consumer and producer was counted and grouped together by type [47].

Given the vector representation of each apartment, we used the well-known unsupervised technique known as K-means, to automatically extract groups of energy-similar apartments.

After a period of observation, a sample for each homogeneous group in a single location was chosen. These local samples were used to extract behavioral rules to be applied to the others belonging to the sample group.

Analyzing the configuration of each location at a given time, it is possible to compare any apartment "Ai" with the sample one "As". As an example, a general association rule can be expressed as follows: "at time tk, make a comparison of device type dj of flat i (dAij) with the correspondent device type of the reference one (d As j). If they are in a similar status, then do nothing; otherwise, switch it on or off, so that it is in the same state of the reference one's.".

An association rule is something in the form X → Y that in a smart grid should assume the simplified form TheSolarPanel IsOn → TheWashingMachineIsOn. We achieved this using the "Apriori Algorithm" which is an influential algorithm for mining frequent item for Boolean association rules. It identifies the frequency of individual items in the dataset, extending them to larger item sets, according to their appearance in the dataset [48].

Nevertheless, every automated system can easily fail if the digital representation of the built environment does not match reality. Assuming that, inevitably during the lifetime of an apartment, some smart plug will be connected to different devices, affecting the digital model reliability and accuracy.

In order to keep the digital model continuously up-to-date, AI techniques transform a power absorption curve of a single device in a sequence of characters named "energy words of the device" [48], using analytical processes similar to those of text analyses; then, a supervised learning method named "Naïve Bayes classifier" automatically identifies the type of each energy load, so that the system can detect a mismatch between the digital representation and what is actually connected to the network.

The dictionary of different energy words exceeded the size of 60,000, with the major number appearing less than three time in the energy footprint; therefore, we set this threshold to avoid dimensionality problems. The resulting predictive model elaborated using the Naïve Bayes classifier was validated using both a 66% train 33% test split and a 10-fold cross validation technique, taking advantage of the tool named "Weka", an open source ML software (using the class weka.classifier.bayes.NaiveBayes).

4. Results

As a consequence of the energy efficiency improvement based on the implementation of renewable energy systems, in winter conditions, the geothermal power plant supplies every building both with heating and domestic hot water; solar collectors integrate the system, while the photovoltaic system powers the external lighting system around the perimeter of the buildings. In summer conditions, domestic hot water is produced through solar collectors covering 100% of the actual needs, while the geothermal power plant only works for the production of chilled water for cooling (through the absorber), while the photovoltaic system powers the entire lighting system of the complex.

The energy diagnosis conducted on a single building using the BIM model through the Revit Suite of Mc4 Software led to the transition from an A2 class (with a specific consumption of 26.8 kWh/m^{2}y) to an A4 class (with a specific consumption of 16.1 kWh/m^{2}y). Moreover, in order to further validate the results and the obtained energy diagnosis, the calculation was also repeated with two other numerical simulation tools: (1) Termus BIM, basing on the BIM model and (2) ArchiEnergy, a semidynamic software developed by Sapienza University of Rome (Table 5).

Table 5. Energy diagnosis results (kWh/m^{2}y): software comparison.

	ArchiEnergy	Termus BIM	MC4 Software	Standard Deviation (SD)	Bills
Ante operam	28.6	24.9	30.2	2.7	26.6 *
Post operam	16.1	15.7	18.7	1.6	16.3 **

* Average of 3 year consumptions of the district; ** 3 month summer bills of the analyzed building.

Once the results and deviation values were obtained, they were evaluated and compared to the following chart in Figure 15, which reports results from other energy diagnosis conducted on similar building systems.

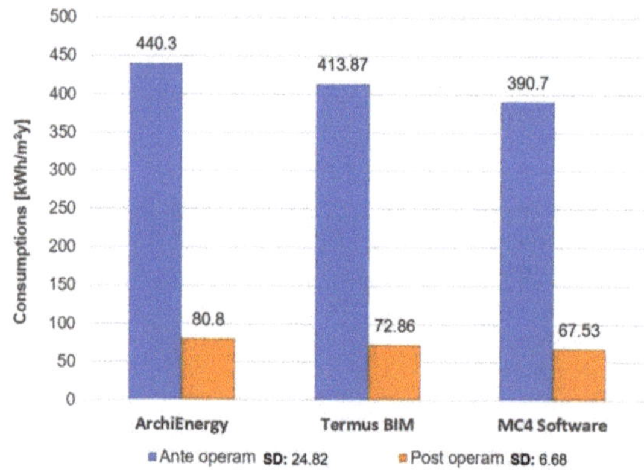

Figure 15. Software comparison through energy diagnosis results on similar buildings.

From the analysis, it is shown that the diagnoses made with the energy software led to similar results with a maximum deviation of 12%, and the difference between the two BIM-based, Mc4 Suite for Revit and Termus BIM, is 5% (Table 5).

Moreover, the fluids-dynamic analysis performed in specific rooms of a single apartment was confirmed by the data coming from sensors, showing that there is no discomfort in any area due to the configuration of the radiant floor equipment.

In fact, large masses of moving air can be observed as previously shown in Figure 11. This is mainly due to the temperature difference between the floor and the environment. Convective motions affecting all the areas are generated; however, the temperature gradient is fully compliant with the regulation requirements, and the air velocities are very low, falling within the range of comfort conditions.

It was also monitored the temperature in each area, where the internal temperature was initially 5 °C (equal to the external temperature), until the achievement of the internal comfort temperature of 20 °C. The temperature transient is shown below (Figure 16).

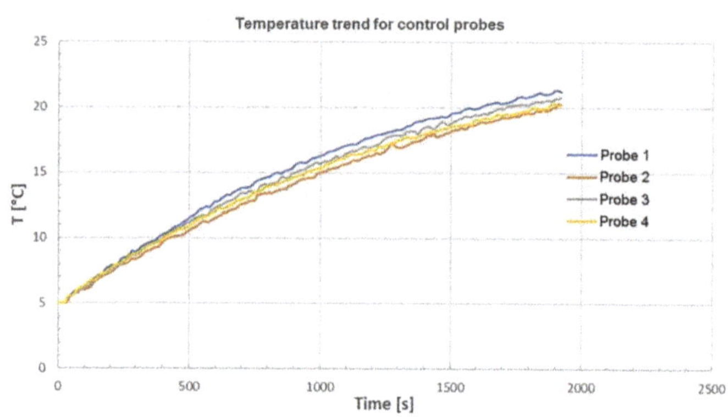

Figure 16. Temperature transient.

It can be noticed that the air heating trend is almost the same for all the rooms, and the comfort temperature is reached in about 1900 s (just over 30 min).

Moreover, another obtained result was the implementation of an intelligent energy management model, i.e., an automatic ML system capable of modulating loads (mainly electrical) according to the expected self-production of energy; for this purpose, information from the European Copernicus [49] earth observation system are acquired in order to have accurate predictive meteorological data.

In this regard, the energy-smart-grid system realized with solar collectors and photovoltaic panels needs a set of rules to establish priorities regarding energy production and consumption loads:

Production: electricity from solar sources, being totally free, must be the first to be fed into the distribution network, followed by the energy coming from the geothermal power plant (which needs electricity to power the circulation pumps); as a last option, it is possible to use energy coming from the public electrical net or use gas.

Consumption: the priority of power supply must be given to the lighting system, followed by the electromotive force circuit, while the air conditioning systems can be regulated and modulated in the event of a lack of energy, by lowering or raising the optimal temperature up to 2 °C.

Therefore, the AI system contributed to reach the goal of increasing the efficiency of the entire energy system by more than 10%, limiting the dependence of the building complex from the electricity and gas distribution networks to a maximum of 20% of the total energy consumed. The system for energy loads forecasting and managing was created in a single apartment (Figure 17) according to the following two logical steps: (a) the creation of a synthetic method to group the plants based on the similarity of results in terms of energy efficiency and (b) metering, evaluation, and analysis of consumption data of the selected plant.

Figure 17. Energy loads forecasting and DT managing through AI.

The inevitable use mutability of the apartments was also considered, as well as the variations in energy loads over time; consequently, an algorithm able to automatically deduce which devices are used in each power outlet was adopted, analyzing the hourly trend of current absorption.

Some energy sensors (as detailed in Material and Methods) were applied, and data were collected in a central system. The different typology of energy loads was considered, and then submetering was performed, as shown in Table 6 and Figure 18.

Table 6. Working day energy metering in a typical apartment (Wh).

	8:30	9:30	10:30	11:30	12:30	13:30	14:30	15:30	16:30	17:30
Boiler	42.99	51.54	15.52	36.61	112.38	71.21	69.27	51.65	12.23	111.32
Lights Room 1	16.00	24.15	24.19	24.26	24.29	24.15	06.12	24.19	24.23	24.75
Mini PC	9.89	14.01	14.34	8.72	13.11	11.63	11.19	12.22	14.22	12.17
Lights Room 2	90.98	106.17	105.46	103.77	104.77	104.73	104.63	104.94	103.93	104.89

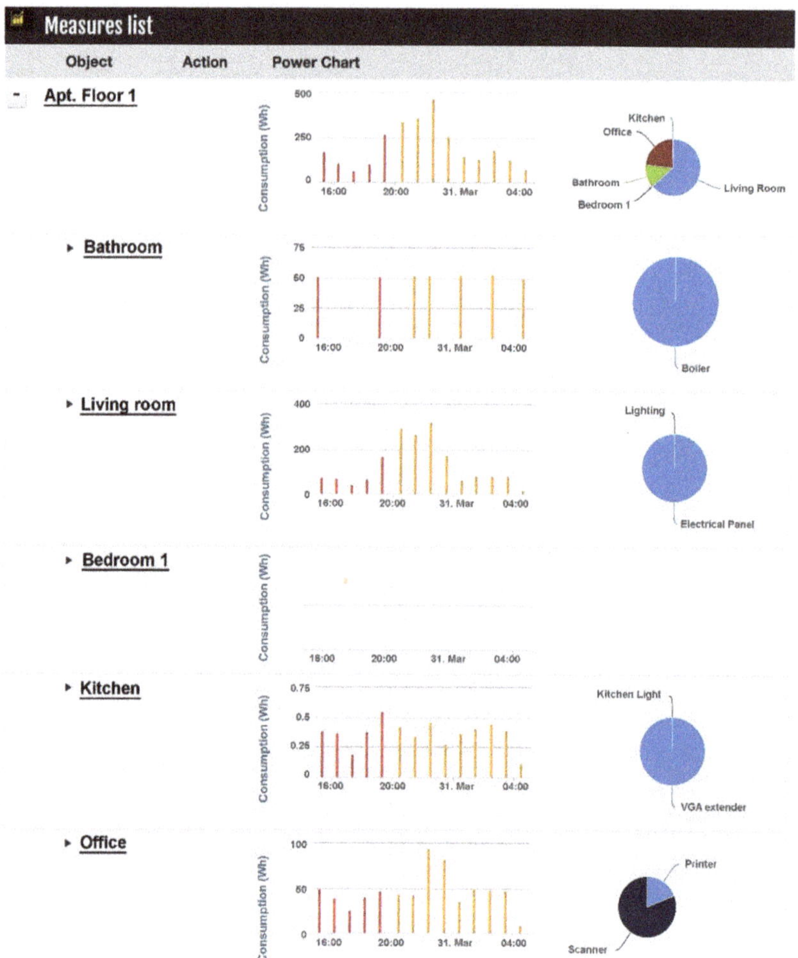

Figure 18. Submetering in a typical apartment.

Energy consumption of each device varies according to its power absorption, as shown in Table 6 and Figures 18 and 19, which report some controlled measures on a typical working day, detailing both the apartment and the single rooms.

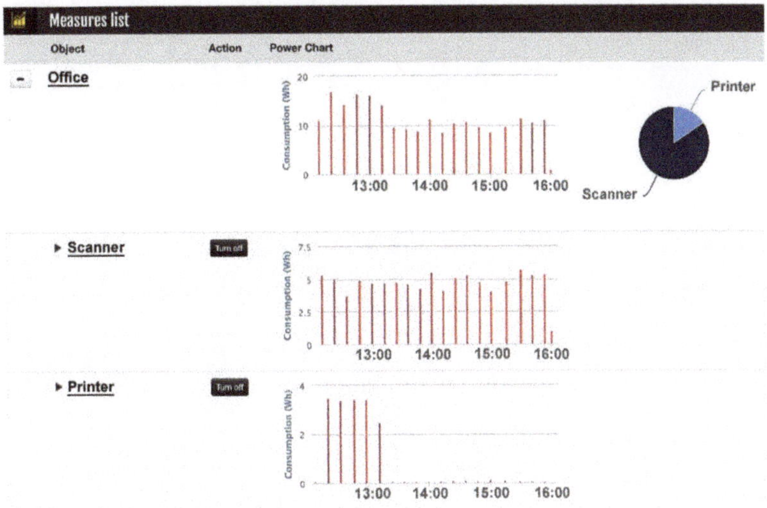

Figure 19. Consumption submetering (Wh) in an office room.

The use of these energy sensors led to another result: the so-called "submetering" It was possible to detect the biggest single load both in the apartment and in a single room. In this way, the analysis and decision of how to save energy becomes simpler, devising strategies affecting the most consuming items, effectively contributing to the overall energy saving.

5. Discussion

The concept of DT is extremely transversal and widely suitable to both microscales such as apartments and macroscales at the district levels. As the new and future buildings will be directed to near-zero-energy building standards (nZEB), or even zero-energy buildings (ZEB), they therefore need tools suitable for the new design requirements, i.e., digital systems able to predict and simulate both global energy consumption and internal behavior [50].

It is quite impossible to define a validation process able to ensure the reliability of the calculation method by 100%.

For a full comprehension of the model and interoperability process accuracy, it was necessary to proceed with a comparison methodology based on the overall final outputs, (kWh/m^{2}y) between three different software (1) Termus BIM by Acca Software, (2) ArchiEnergy from Sapienza University of Rome, and (3) Mc4 Suite for Revit.

The developed analysis was focused on the comparison of results coming from different processes basing on both traditional and BIM approaches. On the one hand, Termus BIM used IFC BIM standards, while in Mc4 Software a plug-in approach was developed directly connecting the Revit BIM model with Mc4 analysis tools. On the other hand, the ArchiEnergy software is a traditional system calculating energy consumption based on inputs by the user about the plant and the building envelope.

Following the validation phase, the DT led to the evaluation of the smart-grid implementation effects. In particular, in Figure 20, the reduced energy consumption and the relative reduced CO_2 consumption coming from the Mc4 Suite for Revit analysis are shown.

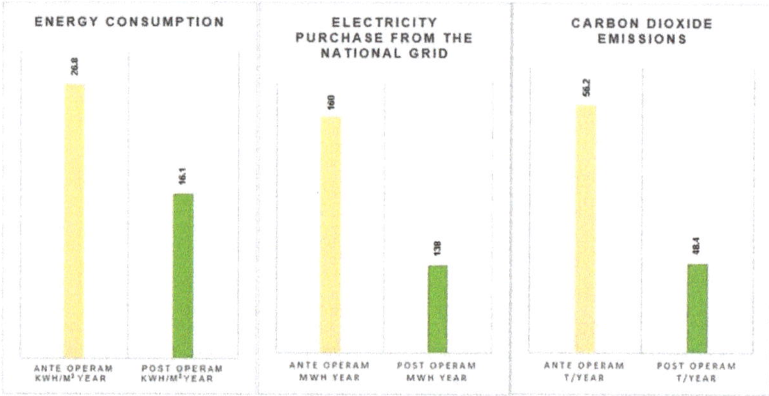

Figure 20. Results of the energy-efficiency interventions.

At the same time, the work carried out highlights how in highly urbanized contexts characterized, it is very difficult to achieve high performances as required by the nZEB Italian Decree [51], even if significant energy requalification interventions are developed, improving both the building envelope and air conditioning systems.

As a consequence, it became necessary to consider building complexes not only as consumers, but also as energy producers in a local, block, district, or neighborhood smart grid: the concept of "prosumer".

By such a logic, the role of AI in smart-grids management and optimization of both energy production and consumption becomes decisive, being able to make reliable forecasts on possible scenarios.

Analyzing similar energy efficiency interventions on buildings and residential complexes, it is shown how efficient technologies are now available, well defined, and widely known. Therefore, the parameters of selection between different interventions are essentially (a) climatic parameters, (b) regulatory restrictions and constraints on interventions, and (c) the availability of government grants for the use of RES, compensating the payback time, which is still too long for certain technologies.

As previously shown, the use of BIM-based systems [16] for building energy efficiency drives no substantial improvements in terms of accuracy of results compared to traditional methodologies [18].

However, the real innovation contribution of DT-enabled systems concerns the definition of digital technologies able to reduce the gap between the expected performance of buildings and their real behavior. These goals are mentioned in the strategies of National and International R&D Programs such as Next Generation EU (Recovery and Resilience Facilities) [52], Strategic Energy Technology (SET) Plan [53], and Italian National Integrated Energy and Climate Plan (Dimension 5 Research, Innovation and Competitiveness) [54].

In this case, DT becomes a key element for research and development on second-generation smart buildings entirely based on electricity consumption and characterized by energy autonomy, high flexibility, block chain, and smart contract dialogue systems with the grid, assisted by digital monitoring methods.

Artificial Intelligence

Although optimizations on energy consumption have been studied in depth [55], when dealing with residential compounds or SOHO (small office home office) buildings, we cannot directly borrow general solutions from research experiences [56,57]. In fact, the overall consumption in these environments is the sum of small contributions by a considerable amount and variety of devices [58], while, mostly in industrial environments, there are generally few big powers draining that can be controlled one by one.

Moreover, these small consumers are operated by people which do not follow any procedure, since they have their personal habits: dealing with both technical and human factors through data analysis techniques becomes a fundamental strategy [59].

DT was coupled with AI to investigate building behaviors as a whole, and supervised learning techniques are used to produce an efficient and intelligent storage system management in the whole complex.

The problem of energy savings in buildings is strictly connected to the need of measuring and controlling energy loads in an efficient way, which can evolve complex scenarios. For instance, if nobody is at home and it is already late morning, both the coffee machine in the kitchen and the air conditioning are wasting energy if they are still switched on, while if someone is still there then both appliances should be still operational. Consequently, several sensors and actuators can be involved and their data should be interconnected so that an ad hoc algorithm derives the correct energy saving policy (e.g., a motion sensor shares data with electrical relays able to switch on/off the correct devices).

Real-time building management system incorporate model-based control through ML [60] to extend the use of mathematical models even to the management of human-related factors. In fact, thermal, humidity, acoustical comfort, and occupants model are combined and connected to ML.

While the first model depend on facts, the latter depend on humans: the behavioral model is a probabilistic one [60]: the probability that an occupant takes specific behavioral decisions or actions is defined as a function of the occupant's characteristic and the current environmental conditions, and "predicting the residents" actions toward a specific situation is not easy".

Considering the apartment microscale, instead of the whole building, our approach was to envision the automatic definition of best practices [61]; if grouped by location and similarities parameters, thanks to unsupervised learning techniques, it was possible to automatize the processes of attributing each location to the most appropriate group or cluster. In our approach, the most efficient and performing apartment for each group or cluster was found considering the energy bill over a few months, confirmed by the energy data collected over a given period [62].

Given these "sample" location, personal actions in apartments can be modeled with behavioral rules [63]: the definition of rules was given using a formal logic that allows exceptions [48] through AI, using Apriori algorithm to automatically learn the rules.

The automatic update of the BIM model to ensure the validity of the DT, based on an up-to-date information model, was dealt with by using web services [64]. Specifically, it was necessary to ensure that information about energy loads coming from smart plugs were up-to-date in the model. A supervised learning technique (named "Naïve Bayes classifier") combined with a novel energy load information coding [65] was used to achieve the goal.

6. Conclusions and Further Developments

The configured DT methodology gives buildings the capability of improving and enriching their knowledge and available data, receiving input and signals from sensors that constantly monitor them, developing self-learning capabilities and predictivity through the integration with AI systems.

Moreover, the paper focuses on how the concept of DT is extremely transversal and applicable both to macroscopic and microscopic scales (from district to apartment), as demonstrated for the use of energy management systems. It can be related, for example, to specific components of technological systems, to the digitalization of infrastructures and real estate assets, to technological systems, or networks of technological systems, etc.

The objective of the research was to exploit ML systems to manage and to simultaneously integrate self-production and supply system in an energy smart grid, in terms of both thermal and electrical loads.

The results of the DT-based real-time monitoring are able to reduce the gap between the energy performance of the buildings (simulated through energy diagnosis) and the real building performance. This is possible thanks to data analysis, which allows one to get more refined energy management strategies, even highlighting inadequate users' behaviors and policies.

As far as load forecasting is concerned, the configured DT is able to calculate thermal loads on a daily basis [60], integrating them with algorithms capable to calculate in advance building consumption based on historical data transmitted by sensors; in this way the system, on the one hand, acquires real-time data from smart metering [61] and environmental quality sensors; on the other hand, it integrates historical data (bills, consumption, etc.) and IoT with a real-time simulation approach [62]. The purpose is aimed at updating and refining the database, tailoring the energy profile of consumption on real users

These intelligent systems implemented also provide an active control on the energy balance; in fact, once the system becomes sufficiently confident, it takes control itself of the energy production systems, as well as of the loads modulation and regulation in order to optimize the energy balance system, limiting nonessential loads in case of production deficit.

Even the optimization of thermo-hygrometric wellbeing parameters in the indoor environment is considered as fundamental. In fact, through the analysis of data from environmental quality sensors and after an appropriate self-learning period, the DT becomes able even to set operations times and levels of the systems to optimize the thermo-hygrometric wellbeing of users.

Moreover, spreading the proposed research to an urban approach, developments in the BIM-GIS synergy, as both large- and small-scale digital information system configuration, would allow for the integration of each urban energy cell with the national power distribution grid, with particular focus on electric mobility and storage systems of smart grids, urban metabolism, etc. Predictions about the impacts on neighboring areas and profiling functional integrations would be performed, providing essential digital tools for the implementation and real-time monitoring of municipal and district energy plans.

In addition, in this regard, further developments of the present research would reach the optimization of the operations using a data model as a process core, replicating reality in real time, limiting or even eliminating system malfunctioning, grid unbalance, or even power breakdowns. With the aim of reducing malfunctions and breakdowns on energy services, the proposed methodology would be applied even to the facility management of HVAC and electrical plants toward configuring predictive maintenance systems.

Author Contributions: Conceptualization, S.A., F.C., G.G., and C.T.; methodology, S.A. and F.C.; software, S.A. and C.T.; validation, S.A., F.C., G.G., and C.T.; formal analysis, S.A.; investigation, G.G.; resources, S.A.; data curation, S.A. and C.T.; writing—original draft preparation, S.A., F.C., and C.T.; writing—review and editing, S.A. and G.G.; visualization, S.A.; supervision, F.C. and G.G.; project administration, G.G. and F.C. All authors have read and agreed to the published version of the manuscript.

Funding: This research received no external funding.

Informed Consent Statement: Not applicable.

Conflicts of Interest: The authors declare no conflict of interest. The funders had no role in the design of the study; in the collection, analyses, or interpretation of data; in the writing of the manuscript, or in the decision to publish the results.

References

1. Tao, F.; Sui, F.; Liu, A.; Qi, Q.; Zhang, M.; Song, B.; Guo, Z.; Lu, S.; Nee, A. Digital twin-driven product design framework. *Int. J. Prod. Res.* **2019**, *57*, 3935–3953. [CrossRef]
2. Cocchia, A. Smart and Digital City: A Systematic Literature Review. In *Smart City: How to Create Public and Economic Value with High Technology in Urban Space*; Dameri, R.P., Rosenthal-Sabroux, C., Eds.; Springer International Publishing: Cham, Switzerland, 2014; pp. 13–43. [CrossRef]

3. Guillemin, P.; Friess, P. *Internet of Things Strategic Research Roadmap. The Cluster of European Research Projects*; Technical Report; European Commission—Information Society and Media DG: Brussels, Belgium, 2009.
4. Chourabi, H.; Nam, T.; Walker, S.; Gil-Garcia, J.R.; Mellouli, S.; Nahon, K.; Pardo, T.A.; Scholl, H.J. Understanding Smart Cities: An Integrative Framework. In Proceedings of the 45th Hawaii International Conference on System Sciences, Maui, HI, USA, 4–7 January 2012; pp. 2289–2297. [CrossRef]
5. White, G.; Clarke, S. Urban Intelligence with Deep Edges. *IEEE Access* **2020**, *8*, 7518–7530. [CrossRef]
6. Perera, C.; Zaslavsky, A.; Christen, P.; Georgakopoulos, D. Sensing as a Service Model for Smart Cities Supported by Internet of Things. *Trans. Emerg. Telecommun. Technol.* **2014**, *25*, 81–93. Available online: https://onlinelibrary.wiley.com/doi/pdf/10.1002/ett.2704 (accessed on 11 November 2020). [CrossRef]
7. *Pathways to Urban Sustainability: Challenges and Opportunities for the United States*; Technical Report; National Academies of Sciences Engineering and Medicine: Washington, DC, USA, 2016. Available online: https://www.nap.edu/catalog/23551/pathways-to-urban-sustainability-challengesand-opportunities-for-the-united (accessed on 14 October 2020).
8. Rosen, R.; Von Wichert, G.; Lo, G.; Bettenhausen, K.D. About the importance of autonomy and digital twins for the future of manufacturing. *IFAC-PapersOnLine* **2015**, *48*, 567–572. [CrossRef]
9. Manfren, M.; Nastasi, B.; Groppi, D.; Astiaso Garcia, D. Open data and energy analytics—An analysis of essential information for energy system planning, design and operation. *Energy* **2020**, *213*, 118803. [CrossRef]
10. Weekes, S. The Rise of Digital Twins in Smart Cities. 2019. Available online: https://www.smartcitiesworld.net/special743 reports/special-reports/the-rise-of-digital-twins-in-smart-cities (accessed on 20 March 2021).
11. Fuldauer, E. Smarter Cities are Born with Digital Twins. 2019. Available online: https://tomorrow.city/a/smarter-citiesare-born-with-digital-twins (accessed on 14 October 2020).
12. Centre for Digital Built Britain. 2020. Available online: https://www.cdbb.cam.ac.uk/what-we-do (accessed on 12 January 2021).
13. Mohammadi, N.; Taylor, J.E. Smart city digital twins. *IEEE Symp. Ser. Comput. Intell.* **2017**, 1–5. [CrossRef]
14. Ruohomäki, T.; Airaksinen, E.; Huuska, P.; Kesäniemi, O.; Martikka, M.; Suomisto, J. Smart City Platform Enabling Digital Twin. *IEEE Access* **2018**, 3–6. [CrossRef]
15. Park, H.-A.; Byeon, G.; Son, W.; Jo, H.-C.; Kim, J.; Kim, S. Digital Twin for Operation of Microgrid: Optimal Scheduling in Virtual Space of Digital Twin. *Energies* **2020**, *13*, 5504. [CrossRef]
16. Kaewunruen, S.; Rungskunroch, P.; Welsh, J. A Digital-Twin Evaluation of Net Zero Energy Building for Existing Buildings. *Sustainability* **2019**, *11*, 159. [CrossRef]
17. Manfren, M.; Sibilla, M.; Tronchin, L. Energy Modelling and Analytics in the Built Environment—A Review of Their Role for Energy Transitions in the Construction Sector. *Energies* **2021**, *14*, 679. [CrossRef]
18. Pauwels, P.; Zhang, S.; Lee, Y.-C. Semantic web technologies in AEC industry: A literature overview. *Autom. Constr.* **2017**, *73*, 145–165. [CrossRef]
19. Istituto di Ricerca per la Protezione Idrogeologica. Consiglio Nazionale delle Ricerche. Available online: http://www.irpi.cnr.it/project/atlante-geotermico-mezzogiorno/ (accessed on 24 March 2021).
20. Smit, J.; Kreutzer, S.; Moeller, C.; Carlberg, M. *Industry 4.0*; PE 570.007; European Union: Brussels, Belgium, 2016; p. 94. Available online: https://www.europarl.europa.eu/RegData/etudes/STUD/2016/570007/IPOL_STU(2016)570007_EN.pdf (accessed on 12 November 2020).
21. Raval, M.; Aravelli, A.; Dofe, J.; Gohel, H. Smart energy optimization for massive IoT using artificial intelligence. *Internet Things* **2021**, *13*, 100354. [CrossRef]
22. Karami, M.; McMorrow, G.V.; Wang, L. Continuous monitoring of indoor environmental quality using an Arduino-based data acquisition system. *J. Build. Eng.* **2018**, *19*, 412–419. [CrossRef]
23. Corry, E.; Pauwels, P.; Hu, S.; Keane, M.; O'Donnell, J. A performance assessment ontology for the environmental and energy management of buildings. *Autom. Constr.* **2015**, *57*, 249–259. [CrossRef]
24. Sacks, R.; Eastman, C.M.; Lee, G.; Teicholz, P. *BIM Handbook: A Guide to Building Information Modeling for Owners, Designers, Engineers, Contractors and Facility Managers*, 3rd ed.; John Wiley and Sons: Hoboken, NJ, USA, 2018.
25. ISO/IEC 2382:2015. *Information Technology—Vocabulary*; Standard, International Organization for Standardization: Geneva, Switzerland, 2015.
26. ISO20944-1:2013. *Information Technology—Metadata Registries Interoperability and Bindings (MDR-IB) -Part 1: Framework, Common Vocabulary, and Common Provisions for Conformance*; Standard, International Organization for Standardization: Geneva, Switzerland, 2013.
27. ISO16678:2014. *Guidelines for Interoperable Object Identification and Related Authentication Systems to Deter Counterfeiting and Illicit Trade*; Standard, International Organization for Standardization: Geneva, Switzerland, 2014.
28. El-Mekawy, M.; Östman, A.; Hijazi, I. An evaluation of IFC-CityGML unidirectional conversion. *Int. J. Adv. Comput. Sci. Appl.* **2012**, *3*, 159–171. [CrossRef]
29. Kreider, R.G.; Messner, J.I. *The Uses of BIM: Classifying and Selecting BIM Uses*; Technical Report; The Pennsylvania State University: University Park, PA, USA, 2013.
30. Copernicus Europes's Eyes on Earth. 2020. Available online: https://www.copernicus.eu/en (accessed on 24 January 2021).

31. ISO 19650-1:2018. *Organization and Digitization of Information About Buildings and Civil Engineering Works, Including Building Information Modeling (BIM)—Information Management Using Building Information Modeling—Part 1: Concepts and Principles*; Standard, International Organization for Standardization: Geneva, Switzerland, 2018.
32. AIA/AGC, Level of Development Specification. 2013. Available online: https://bimforum.org/resources/Documents/BIMForum_LOD_2013_reprint.pdf (accessed on 15 November 2012).
33. Montiel-Santiago, F.J.; Hermoso-orzaez, M.J.; Terrados-Cepeda, J. Sustainability and Energy Efficiency: BIM 6D. Study of the BIM Methodology Applied to Hospital Buildings. Value of Interior Lighting and Daylight in Energy Simulation. *Sustainability* **2020**, *12*, 5731. [CrossRef]
34. Azar, E.; O'Brien, W.; Carlucci, S.; Hong, T.; Sonta, A.; Kim, J.; Andargie, M.S.; Abuimara, T.; El Asmar, M.; Jain, R.K.; et al. Simulation-aided occupant-centric building design: A critical review of tools, methods, and applications. *Energy Build.* **2020**, *224*, 110292. [CrossRef]
35. Guyot, D.; Giraud, F.; Simon, F.; Corgier, D.; Marvillet, C.; Tremeac, B. Building energy model calibration: A detailed case study using sub-hourly measured data. *Energy Build.* **2020**, *223*, 110189. [CrossRef]
36. Mutani, G.; Todeschi, V. Building energy modeling at neighborhood scale. *Energy Effic.* **2020**, *13*, 1353–1386. [CrossRef]
37. Nagrale, S.; Bais, M. Energy Efficiency Analysis and Modelling of a Green Building Using Revit Software. *Int. J. Res. Eng. Sci. Manag.* **2020**, *3*, 365–367.
38. Lu, M.; Li, N.; He, Y.; Zhang, F.; Huang, C.; Fang, X. Use Side-Supply Airflow to Enhance Heat Transfer of Radiant Heating Ceilings for Producing a Better Indoor Thermal Environment in Winter: A Simulation Study. In Proceedings of the 11th International Symposium on Heating, Ventilation and Air Conditioning (ISHVAC 2019), Singapore; Wang, Z., Zhu, Y., Wang, F., Wang, P., Shen, C., Liu, J., Eds.; Springer: Singapore, 2020; pp. 935–942.
39. Italian Legislative Decree 10 Giugno 2020, n. 48, European Union Directive 2018/844, 2020. Available online: https://eur-lex.europa.eu/legal-content/EN/ALL/?uri=CELEX%3A32018L0844 (accessed on 10 February 2021).
40. Benammar, M.; Abdaoui, A.; Ahmad, S.; Touati, F.; Kadri, A. A Modular IoT Platform for Real-Time Indoor Air Quality Monitoring. *Sensors* **2018**, *18*, 581. [CrossRef]
41. Kwok, H.H.; Cheng, J.C.; Li, A.T.; Tong, J.C.; Lau, A.K. Multi-zone indoor CFD under limited information: An approach coupling solar analysis and BIM for improved accuracy. *J. Clean. Prod.* **2020**, *244*, 118912. [CrossRef]
42. Scannapieco, S.; Tomazzoli, C. Ubiquitous and Pervasive Computing for Real-Time Energy Management and Saving. *Adv. Intell. Syst. Comput.* **2017**, *612*, 3–15. [CrossRef]
43. Bock, H.H. Clustering Methods: A History of k-Means Algorithms. In *Selected Contributions in Data Analysis and Classification. Studies in Classification, Data Analysis, and Knowledge Organization*; Brito, P., Cucumel, G., Bertrand, P., de Carvalho, F., Eds.; Springer: Berlin/Heidelberg, Germany, 2007; pp. 161–172. [CrossRef]
44. Marques, G.; Pires, I.M.; Miranda, N.; Pitarma, R. Air quality monitoring using assistive robots for ambient assisted living and enhanced living environments through Internet of Things. *Electronics* **2019**, *8*, 1375. [CrossRef]
45. Marques, G.; Pitarma, R. An Indoor Monitoring system for ambient assisted living based on Internet of Things architecture. *Int. J. Environ. Res. Public Health* **2016**, *13*, 1152. [CrossRef]
46. Tomazzoli, C.; Scannapieco, S.; Cristani, M. Internet of Things and artificial intelligence enable energy efficiency. *J. Ambient Intell. Hum. Comput.* **2020**. [CrossRef]
47. Tomazzoli, C.; Cristani, M.; Olivieri, F. Automatic synthesis of best practices for energy consumptions. In Proceedings of the tenth International Conference on Innovative Mobile and Internet Services in Ubiquitous Computing, IEEE, Fukuoka, Japan, 6–8 July 2016; pp. 1–8.
48. Cristani, M.; Tomazzoli, C.; Olivieri, F.; Erisa, K. Defeasible Reasoning about Electric Consumptions. In Proceedings of the 30th IEEE International Conference on Advanced Information Networking and Applications (AINA-2016), Crans-Montana, Switzerland, 23–25 March 2016; pp. 885–892. [CrossRef]
49. Gidhagen, L.; Olsson, J.; Amorim, J.H.; Asker, C.; Belusic, D.; Carvalho, A.C.; Engardt, M.; Hundecha, Y.; Körnich, H.; Lind, P.; et al. Towards climate services for European cities: Lessons learnt from the Copernicus project Urban SIS. *Urban Clim.* **2020**, *31*, 100549. [CrossRef]
50. Magrini, A.; Lentini, G.; Cuman, S.; Bodrato, A.; Marenco, L. From nearly zero energy buildings (NZEB) to positive energy buildings (PEB): The next challenge—The most recent European trends with some notes on the energy analysis of a forerunner PEB example. *Dev. Built Environ.* **2020**, *3*, 100019. [CrossRef]
51. Italian Legislative Decree Requisiti Minimi 26 Giugno 2015. Available online: https://www.mise.gov.it/index.php/it/normativa/decreti-interministeriali/2032966-decreto-interministeriale-26-giugno-2015-applicazione-delle-metodologie-di-calcolo-delle-prestazioni-energetiche-e-definizione-delle-prescrizioni-e-dei-requisiti-minimi-degli-edifici (accessed on 23 August 2020).
52. Next Generation EU. Available online: https://ec.europa.eu/info/strategy/recovery-plan-europe_en (accessed on 26 March 2021).
53. European Commission, Strategic Energy Technology Plan. Available online: https://ec.europa.eu/energy/topics/technology-and-innovation/strategic-energy-technology-plan_en (accessed on 26 March 2021).
54. Italian National Integrated Energy and Climate Plan (Dimension 5 Research, Innovation and Competitiveness). Available online: https://www.mise.gov.it/index.php/it/energia/energia-e-clima-2030 (accessed on 26 March 2021).

55. Oh, S.; Haberl, J.S.; Baltazar, J.-C. Analysis methods for characterizing energy saving opportunities from home automation devices using smart meter data. *Energy Build.* **2020**, *216*, 109955. [CrossRef]
56. Westermann, P.; Deb, C.; Schlueter, A.; Evins, R. Unsupervised learning of energy signatures to identify the heating system and building type using smart meter data. *Appl. Energy* **2020**, *264*, 114715. [CrossRef]
57. Uhlemann, T.H.J.; Lehmann, C.; Steinhilper, R. The digital twin: Realizing the cyber-physical production system for Industry 4.0. *Procedia Cirp* **2017**, *61*, 335–340. [CrossRef]
58. Saini, J.; Dutta, M.; Marques, G. Indoor Air Quality Monitoring Systems Based on Internet of Things: A Systematic Review. *Int. J. Environ. Res. Public Health* **2020**, *17*, 4942. [CrossRef] [PubMed]
59. Manfren, M.; Nastasi, B.; Tronchin, L. Linking Design and Operation Phase Energy Performance Analysis Through Regression-Based Approaches. *Front. Energy Res.* **2020**. [CrossRef]
60. Eini, R.; Linkous, L.; Zohrabi, N.; Abdelwahed, S. Smart building management system: Performance specifications and design requirements. *J. Build. Eng.* **2021**, *39*. [CrossRef]
61. Mariano-Hernández, D.; Hernández-Callejo, L.; Zorita-Lamadrid, A.; Duque-Pérez, O.; Santos García, F. A review of strategies for building energy management system: Model predictive control, demand side management, optimization, and fault detect & diagnosis. *J. Build. Eng.* **2021**, *33*. [CrossRef]
62. Agostinelli, S.; Cumo, F.; Guidi, G.; Tomazzoli, C. The Potential of Digital Twin Model Integrated with Artificial Intelligence Systems. In Proceedings of the 2020 IEEE International Conference on Environment and Electrical Engineering and 2020 IEEE Industrial and Commercial Power Systems Europe (EEEIC/I&CPS Europe), Madrid, Spain, 9–12 June 2020; pp. 1–6. [CrossRef]
63. Bourgeois, D.; Reinhart, C.; Macdonald, I. Adding advanced behavioural models in whole building energy simulation: A study on the total energy impact of manual and automated lighting control. *Energy Build.* **2006**. [CrossRef]
64. Cheng, J.; Das, M. A Bim-Based Web Service Framework for Green Building Energy Simulation and Code Checking. *J. Inf. Technol. Constr.* **2014**, *19*, 150–168.
65. Tomazzoli, C.; Scannapieco, S. Machine learning for energy efficiency: Automatic detection of electric loads from power consumption. In Proceedings of the 2017 AEIT International Annual Conference, Cagliari, Italy, 20–22 September 2017; pp. 1–6. [CrossRef]

MDPI
St. Alban-Anlage 66
4052 Basel
Switzerland
Tel. +41 61 683 77 34
Fax +41 61 302 89 18
www.mdpi.com

Energies Editorial Office
E-mail: energies@mdpi.com
www.mdpi.com/journal/energies

www.ingramcontent.com/pod-product-compliance
Lightning Source LLC
Chambersburg PA
CBHW040814250926
51850CB00007BA/1165